Technologies for Integrated Energy Systems and Networks

Technologies for Integrated Energy Systems and Networks

Edited by
Giorgio Graditi and Marialaura Di Somma

WILEY-VCH

The Editors

Dr. Giorgio Graditi
Italian National Agency for New Technologies
Energy and Sustainable Economic Development
ENEA, Department of Energy Technologies and Renewable Sources
Rome, Italy

Dr. Marialaura Di Somma
Italian National Agency for New Technologies
Energy and Sustainable Economic Development
ENEA, Department of Energy Technologies and Renewable Sources
Rome, Italy

Cover Image: © zf L/Getty Images

Library of Congress Card No.: applied for

British Library Cataloguing-in-Publication Data
A catalogue record for this book is available from the British Library.

Bibliographic information published by the Deutsche Nationalbibliothek
The Deutsche Nationalbibliothek lists this publication in the Deutsche Nationalbibliografie; detailed bibliographic data are available on the Internet at <http://dnb.d-nb.de>.

Print ISBN: 978-3-527-34899-2
ePDF ISBN: 978-3-527-83361-0
ePub ISBN: 978-3-527-83362-7
oBook ISBN: 978-3-527-83363-4

Typesetting Straive, Chennai, India

Printed and bound by CPI Group (UK) Ltd, Croydon, CR0 4YY

C112838_140322

Contents

1

Challenges and Opportunities of the Energy Transition and the Added Value of Energy Systems Integration

Marialaura Di Somma and Giorgio Graditi

Italian National Agency for New Technologies, Energy and Sustainable Economic Development, ENEA, Department of Energy Technologies and Renewable Sources, Rome, Italy

1.1 Energy Transformation Toward Decarbonization and the Added Value of Energy Systems Integration

The global energy transformation is already in place, and this represents the main reply of humanity to safeguard global climate and maintain sustainable existence on Earth. The first step toward this energy transformation and the international commitment to combating climate change, increasing energy access, and maintaining biodiversity is represented by the Paris Agreement signing at COP 21 with the goal to maintain global warming lower than 2 °C above the pre-industrial levels. Concurrent to the Paris Agreement, countries committed to the United Nations (UN) 17 Sustainable Development Goals (SDGs), representing the plan toward a better world for people and our planet to be achieved by 2030 [1]. Tackling climate change is a transversal goal for almost all SDGs. Although the international commitment is evident, challenges still remain for the successful implementation of the Paris Agreement and climate- and energy-related SDGs, and the gap between aspiration and reality in combating climate change remains significant.

Meeting these ambitious goals requires the commitment beyond the electricity sector, whereas providing decarbonization across different sectors through an integrated approach can represent a valid solution. This is the main idea behind the concept of Integrated Energy Systems that, according to the ETIP SNET Vision 2050 [2], are defined as an integrated infrastructure for all energy carriers, with the electrical system as the backbone. These systems are characterized by a high level of integration among all networks of energy carriers obtained through coupling electrical and gas networks, heating, and cooling, supported by energy storage and conversion processes. Coupling different sectors indicates increasing efforts in a synergic way by coordinating the planning and the operation of energy systems across multiple energy carriers while also achieving a more flexible, reliable, and efficient energy system as a whole.

The main energy trends toward decarbonization are discussed below along with the added value offered by energy systems integration.

Technologies for Integrated Energy Systems and Networks, First Edition.
Edited by Giorgio Graditi and Marialaura Di Somma.

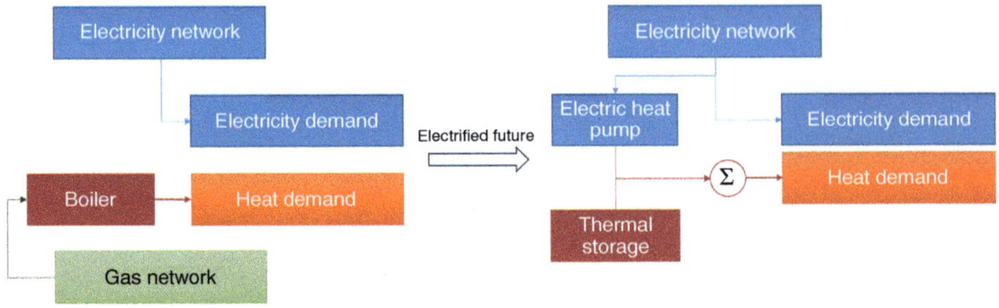

Figure 1.1 Evolution of the current energy system to an electrified energy system.

Electrification is considered a valid cost-effective pathway for decarbonization of final energy consumption. This is mainly due to the fact that several technologies for converting renewable energy into electricity have recently become available at competitive prices such as PV and wind turbines. On the other hand, a large part of CO_2 emissions in industries, transport, and buildings is not related to power sector but to end use of fossil fuels. That is why, a large-scale electrification, characterized by the penetration of an electricity carrier produced by renewable technologies in building, transport, and industry sectors, represents a good pathway for decarbonization. According to the International Renewable Energy Agency (IRENA) Renewable Energy Roadmap (REmap) [3], the share of electricity in final energy consumption amounts to 20% today and will reach the percentages of 29%, 38%, and 49% in 2030, 2040, and 2050, respectively.

Figure 1.1 shows the change from the current energy supply system where the electricity demand is typically satisfied by an electricity network and heat demand by gas-fired boilers supplied by a gas network to an electrified energy system, where the electricity network is used to satisfy all energy demands, including heat demand through Power-to-Heat (PtH) technologies. An electrified future poses important questions such as how much additional power network capacity do we need to satisfy all types of energy demands? Or, what happens if there is a contingency in the power system?

A strong electrification scenario creates a number of challenges for the operation of a power system, which in principle would need additional flexibility, reinforcement, and new investments for the transmission and distribution networks.

In Figure 1.2, the current energy system is compared to an integrated energy system, which is something more than an electrified energy system. In fact, in such system, multiple hybrid energy technologies are managed with high synergy to satisfy the multi-energy demand and services can be provided with the most convenient energy carrier and sector.

If electrification of final consumption is combined with the integration of energy sectors, decarbonization of energy demand would be reached through penetration of renewables in all energy end use sectors while also getting higher flexibility for the whole system by reducing the needs for reinforcing the existing network infrastructures. Moreover, energy systems integration allows increasing efficiency in the energy resources use through exploiting synergies coming from the interplay

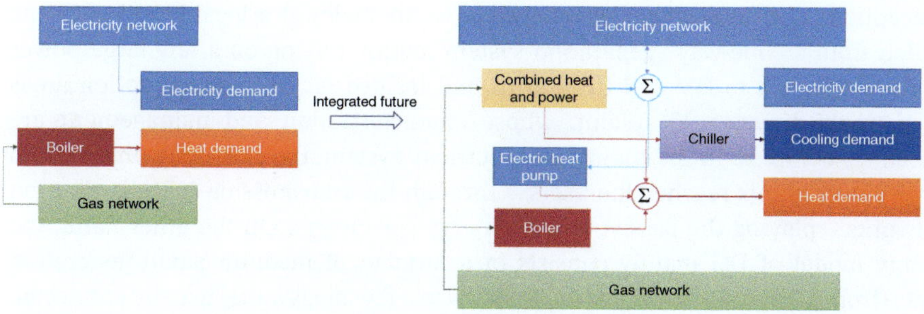

Figure 1.2 Comparison between the current energy system and an integrated energy system.

of different energy carriers and reduction of renewable energy source (RES) curtailment. In practice, for instance, in the case of excess electricity from RES, it can be converted into gas as hydrogen or synthetic methane through Power-to-Gas (PtG) technologies, stored and/or transported by existing gas infrastructures for immediate or later usage, or re-converted again into electricity when renewable electricity supply is insufficient to satisfy the loads. On the other hand, PtH technologies combined with thermal storage can shift production of thermal energy when renewable electricity is in excess, thereby representing another option for reducing RES curtailment [4].

Also, in the transport sector, electrification can be a successful strategy for decarbonization, while making the system as a whole more flexible. In fact, electric vehicles in Power-to-Mobility (PtM) application represent a valid alternative to traditional cars with internal combustion engines and can provide flexibility to the electricity system through smart charging strategies, for instance, by charging batteries during the period of low demands, thereby flattening out the electricity load profile.

Similarly, heat pumps in PtH application represent a cost-effective and more efficient alternative to conventional gas-fired boilers for heating purposes in buildings and also for reducing primary energy consumption thanks to their high conversion efficiency.

According to REmap [3], the number of electric vehicles worldwide will pass from the current 6 millions to 157, 745, and 1166 millions in 2030, 2040, and 2050, respectively, whereas the number of heat pump installations will pass from the current 20 millions to 155, 259, and 334 million in 2030, 2040, and 2050, respectively. The strong expected electrification of transportation and heating sectors could lead to higher peak loads, thereby requiring higher flexibility to match electricity demand and supply. Again, also in these latter cases, the added value of energy systems integration is given by the possibility to store excess electricity from RES and provide back-up supply to cover peak loads, thereby ensuring balance at all times with clean energy in the equation.

Another major trend in energy landscape is represented by the **large-scale deployment of distributed generation (DG).** In the past years, the power system has been affected by a fundamental revolution as compared to its traditional conception. The deployment of renewable technologies at a local level led to the switch from a "one-way" generation system mainly relying on a few large power plants connected to HV and EHV grids and located far from consumption areas to a "multi-directional" system, whose characterization and management are extremely complex. In the traditional electricity system, the electricity produced in large power plants reaches the users – through the transmission and distribution networks – playing the passive role of energy consumers. On the other hand, the energy model of DG mainly consists of a number of medium–small generation units (from a few tens/hundreds of kilowatts to a few megawatts) usually connected to distribution networks. DG units are usually located close to the loads to satisfy and designed to exploit renewable sources spread throughout the territory and otherwise not usable through traditional large-size generation units.

The benefits offered by this new energy model are different:

- increase of the efficiency of the electricity system thanks to the reduction of energy transport loss;
- increase of RES penetration levels and more rational use of energy; and
- optimization of the resources at local level and the local production chain.

According to REmap [3], the renewable energy share in power generation will more than double in 2030, reaching the value of 57% as compared to the current percentage of 25%, to arrive at values of 75% and 86% in 2040 and 2050, respectively. Only in the case of PV systems, the REmap cases foresee that the annual solar PV additions will pass from the current value of 109 GW/yr to 360 GW/yr in 2050, and a similar situation is expected for wind source, for which the annual additions are expected to pass from the current value of 109 GW/yr to 240 GW/yr in 2050.

The increasing intermittent renewables penetration in electricity systems is leading to an increase in the reliability and stability problems. The mitigation of uncertainty, which imperils the balance between generation and demand, urges the search of new sources of ancillary services, traditionally provided by bulky synchronous generators. Energy systems integration and in particular the coupling of the electricity and gas sectors reveals promising flexibility solutions for power systems through energy conversion and hydrogen storage. On the other hand, the operation of PtG technologies in periods of excess electricity supply removes the need for curtailment of renewable electricity generation or the need for additional investments in electricity transmission, distribution, or storage infrastructure.

An important aspect closely related to the changes that are affecting the energy sector is the **evolution of the role of the energy consumer**. Historically, the citizen has been a "passive" user, covering the role of the customer using the energy produced at a centralized level to meet the energy needs. Conversely, the scenario that has been taking shape in recent years sees the emergence of a new type of "active" customer who, thanks to digitization, is more informed about the own consumption and energy prices and is more sensitive to the use of "green" energy resources. Through DG units, the end users have the ability to produce and consume their own energy to store it and sell it back to the grid by exploiting the RES available locally; therefore, from simple consumers, they become "prosumers." The direct consequence is the birth of the "self-consumption" concept, where the consumption of energy produced occurs in the same site where it is consumed, both instantaneously and through storage systems, regardless of the subjects covering the role of a producer and a final customer, provided that they operate in the same suitably defined and confined site and regardless of the source that feeds the generation unit.

Another element through which the end user assumes the active role in the changing energy landscape is represented by Demand Response (DR). The United States Department of Energy (DoE) defines DR programs as changes in electricity consumption by end users in response to changes in the price of energy over time or the payment of incentives designed to lead to lower consumption of electricity in periods when the wholesale market price is high or when system reliability problems occur [5].

According to the aforementioned definition, the DR is an active response from consumers based on the price of energy or on the payment of incentives. In DR programs, the consumers are induced to quickly change their electricity consumption when there is a high energy demand or there are low-reserve margins. The reduction/modulation of energy consumption according to market price trends helps to limit the occurrence of energy price peaks. At the same time, DR services represent an important tool for network operators in maintaining a balance between supply and demand and in ensuring the reliability of the system. The end user can, therefore, temporarily vary the power commitment in response to a price signal (deriving from tariffs or directly from the electricity market) or in compliance with agreements made with subjects such as aggregators and network operators.

It is important to underline that local DG units can also be considered as a DR resource as they also allow for a reduction in the withdrawal of energy from the grid without affecting the absorption and load curves of consumers. The classic actions that DR can adopt can be divided into three main categories:

- reduction of demand in the peak periods of the system;
- shifting of demand from peak periods to off-peak periods, obtaining an effect of leveling the peaks and filling the valleys of the load curve (load shifting);
- self-production or use of energy stored, which does not change the internal absorption profile of the user's system but allows to reduce the energy demand from the network.

Last but not least, the emerging paradigm of energy communities is expected to function as an important tool for engaging end users in renewable generation and low carbon technologies, while also promoting participation in the market of end users that otherwise could not be able to do so.

The added value of energy systems integration in the major trend related to end user engagement and empowerment is mainly given by the possibility to exploit synergies among multiple energy carriers at the local level to increase energy efficiency and RES utilization, as well as to enhance the potential of decarbonization of the energy demand for heating and cooling. For instance, PtH implemented through heat pumps allows to achieve larger flexibility of the energy demand and improve considerably the use of renewables for heating and cooling demands in buildings. Moreover, the high conversion efficiency of this technology can lead to important economic and environmental benefits. The PtH technology coupled with thermal storage could be even more convenient, thanks to the possibility to activate DR services and offer ancillary services to the electric grid. In fact, the excess renewable electrical energy produced could be converted into thermal energy and stored in thermal energy storage, thereby reducing RES curtailment and making the electrical grid more stable to sudden variations of RES. This brings benefits also to network operators through a more efficient use of the existing generation capacity, the reduced need to upgrade the distribution network, the reduction of peak loads, and the more flattened load forms, as well as the reduction of management costs of generation units. In addition to efficient electricity-based (via heat pumps) heating and cooling devices in single houses and small residential buildings, low-carbon district heating and cooling grids can cover the generation and distribution of thermal energy in urban districts. On the other hand, coupling electricity and heating sectors through combined heat and power (CHP) systems would allow to exploit locally the waste heat from power generation processes for thermal purposes in buildings, thereby increasing the efficiency in energy resource use.

1.2 European Union as the Global Leader in Energy Transition

The transition of European Union (EU) to net-zero carbon emissions by 2050 is a big challenge but also a great opportunity to modernize the continent's economy and promote growth, employment, technological advancement, and social inclusion.

An effective demonstration of the EU commitment in combating climate change is represented by the **Clean energy for all Europeans package** [6], which is a fundamental measure to lay the foundations for the realization of "a neutral climate economy" by 2050. It contains a set of measures related to energy efficiency, renewable energy, structure of the electricity market, security of electricity supply, and governance rules for the Energy Union. It consists of eight legislative acts that provide for an update of the European energy policy framework aimed at facilitating the energy transition, defining a modern European energy market, promoting and integrating electricity produced from RES, promoting energy efficiency, and strengthen the regulatory framework in which European and national institutions operate.

In more details, the Clean Energy Package introduces significant changes to the structure of the European electricity market by revising and replacing the provisions contained in Regulation 2009/714/EC [7] and in Directive 2009/72/EC [8], currently at the basis of the regulatory framework relating to the internal electricity market of the Union. These changes actually allow for the creation of an electricity market for the Union characterized by more variable and decentralized production, greater interdependence between individual national markets, and higher opportunities for consumers to participate as active players in the market through demand side management, aggregation, self-generation, and the use of storage systems and digitalization. The new directive 2019/944/EU (Energy Market Directive – EMD II) [9] aims to adapt the current regulatory framework to the new market dynamics taking into account the opportunities and challenges related to the decarbonization objective of the energy system and the possible technological developments, in particular those relating to consumer participation and cross-border cooperation. The main objective of EMD II is the construction of an internal market governed by common rules that can guarantee everyone access to the electricity carrier. In relation to consumers, the EMD II provides an important paradigm shift, aimed at qualifying consumers as "active consumers," who can operate directly or in aggregated manner, sell self-produced electricity, as well as participate in flexibility and energy efficiency mechanisms. The directive states that all consumers should be able to benefit from direct participation in the market, in particular by adjusting consumption according to market signals and, in return, by benefiting from lower electricity prices or other incentives. Another important innovation envisaged by the EMD II directive concerns the introduction of the notion of *Citizen Energy Community* or an energy community which will be guaranteed to operate on the market under equal and non-discriminatory conditions compared to other market players, being able to freely cover the roles of end customer, producer, supplier, or manager of distribution systems.

The innovations introduced by the Clean Energy Package in the field of energy produced from renewable sources are aimed at encouraging the use of these resources for the energy transition up to 2030, setting new objectives at the EU level, simplifying the related authorization procedures, providing stability to the financial supports and strengthening consumer rights. The new Directive 2018/2001/EU (Renewable Energy Directive – RED II) [10] on the promotion of the use of energy from renewable sources applies a substantial revision of the regulatory framework provided for in Directive 2009/28/EC [11]. In detail, RED II pays particular attention to the self-consumption of renewable energy, providing that consumers are allowed to become consumers of renewable energy, capable, also, of producing, storing, and selling the electricity generated in excess, both individually and in aggregated form. Another fundamental innovation envisaged by RED II is the introduction of the notion of *Renewable Energy Community*, that is an energy community with the right to produce, consume, store, and sell renewable energy. Furthermore, these communities will be able to exchange, within the same community, the renewable energy they produce and access the electricity market, directly or through aggregation, in a non-discriminatory way.

The provisions on energy efficiency, introduced by the Clean Energy Package, aim to establish new efficiency targets for both the EU and the Member States, introducing new guidelines and expanding consumer rights in the field of heating and cooling metering, for billing and for domestic hot water production. The new Energy Efficiency Directive [12] amends the previous Directive 2012/27/EU [13], modifying the current provisions directly linked to the achievement of the 2030 targets and introducing new rules aimed at extending consumer rights and improving access to smart metering. A further element of the package, in the field of energy efficiency, is Directive 2018/844/EU (new Energy Performance of Buildings Directive – EPBD) [14], which entered into force on 9 July 2018, amending Directive 2010/31/EU on the energy performance of buildings. The new EPBD contains provisions concerning energy efficiency objectives for buildings, energy certification, methods of verification, monitoring and control of energy consumption, and the definition of obligations related to the installation of charging points for electric vehicles. Furthermore, it introduces the definition of the Smart Readiness Indicator (SRI) and a methodology to calculate this indicator to assess the ability of a building or a property unit to adapt its functioning to the needs of the occupier and the network and to improve its energy efficiency and overall performance. The indicator of readiness of buildings to smartness takes into account the characteristics of higher energy saving, comparative analysis, and flexibility, as well as features and capabilities that are improved through more interconnected and smart devices.

Finally, the EU Regulation 2018/1999 on the governance of the Energy Union and Climate Action [15] aims to encourage cooperation between Member States to achieve the EU energy objectives and targets, in particular by strengthening the programming and reporting obligations of individual Member States in the field of energy, climate, and in relation to the implementation of the measures envisaged by the new structure of the Energy Union. The regulation outlines the five dimensions of the Energy Union, namely, (i) decarbonization, (ii) energy efficiency, (iii) energy security, (iv) internal energy market, and (v) research, innovation, and competitiveness, and defines the obligation for each Member State to send to the European Commission a National Integrated Energy and Climate Plan, covering periods of 10 years. The plan must, among other things, contain

- an overview of the procedure followed for defining the plan itself;
- a description of the national objectives and contributions relating to the five dimensions of the Energy Union;
- a description of the policies and measures adopted to achieve the aforementioned objectives;
- a description of the current state of the five dimensions of the Energy Union; and
- an assessment of the impacts of the policies and measures implemented to achieve the aforementioned objectives.

Promoting secure, reliable, competitive, locally produced and sustainable energy is an increasingly central issue on the agenda of the European Council, which in December 2019 announced the European Green Deal [16], a roadmap whose purpose is to make the EU "a fair and prosperous society, with a competitive and resource-efficient modern economy, in which there are no net greenhouse gas emissions in 2050 and economic growth is decoupled from the resources used."

The Green Deal is divided into a series of macro-actions containing strategies for all sectors of the economy, in particular transport, energy, agriculture, construction, and industrial sectors, including new regulatory provisions and investments, to be implemented in the next years till 2050.

The strategy is divided into eight main objectives:

(1) Making the EU climate goals for 2030 and 2050 more ambitious;
(2) Ensure the supply of clean, economical, and safe energy;
(3) Mobilizing industry for a clean and circular economy;
(4) Building and renovating in an energy- and resource-efficient way;
(5) Accelerate the transition to sustainable and smart mobility;
(6) "From producer to consumer": designing a fair, healthy, and respectful food system of the environment;
(7) Preserve and restore ecosystems and biodiversity; and
(8) "Zero pollution" for an environment free of toxic substances.

The first climate action initiatives under the Green Deal include

- a European climate law to incorporate the goal of climate neutrality into EU law to 2050, which in turn has four objectives: (i) establish the long-term direction for achievement of the 2050 climate neutrality goal; (ii) create a monitoring system of progress and take further action if necessary; (iii) provide conditions of predictability to investors and other economic actors; and (iv) ensure that the transition to climate neutrality is irreversible.
- a European climate agreement, aimed at spreading awareness and promoting action, in a first moment focused on four areas (green areas, green transport, green properties, and green skills), while it may subsequently involve other areas of action, such as consumption and sustainable production, soil quality, healthy food and sustainable nutrition, and so on.
- The Climate Target Plan 2030, with which it is intended to further reduce net emission (setting a new reduction target, for 2030, of at least 55% compared to levels of 1990) but also stimulate the creation of green jobs and encourage international partners to be more ambitious in containing global warming by limiting the global temperature rise to 1.5 ° C.
- A new EU strategy on climate adaptation, with the aim to make adaptation smarter, faster, and more systemic and to step up international action on adapting to climate change so that Europe becomes, by 2050, a climate resilient society fully adapted to the inevitable impacts of climatic change.

The EU "Green Deal" and the related European national requirements set precise targets by 2030 including:

- decarbonization of the building stock, transport, industry, and energy systems;
- involvement of consumers and citizen communities in energy systems;
- digitalization as an enabler of the environmental transition and participative energy markets;
- ambitious reductions in transport emissions; and
- reliability, adaptability, and resilience of the integrated energy systems.

The energy transition taking place in EU is also demonstrated by numbers. The EU energy mix, over the past decade of observation (2009–2019), is changed, with a smaller share of solid fossil fuels (whose share falls from 15% to 11.4%) and oil (which increased from 38.1% to 36.4%), mainly in favor of renewable sources, which in 2019 represented 15.3% of primary energy production (+5.2 p.p. compared to 2009). At the same time, the CO_2 emissions produced in EU have more or less constantly decreased over the course of the past decade, reaching a level of 2400 Mt in 2019, about 12% less than 10 years earlier.

Energy systems integration is the agenda of EU as a possible route to achieve the ambitious targets set to 2030 and 2050. In the near future until 2030, in EU vision, the share of RES, nuclear energy, and carbon-neutral gases and liquids will increase with high contribution to grid stability and uninterrupted energy supply. New energy carriers are being considered in energy, industrial, and transport applications, such as hydrogen and other carbon- neutral liquids and gases. Additionally, the future energy system will also rely on much better balancing capacities including better interconnections, storage capabilities, DR, low-carbon flexible generation units, and effective energy conversion options (Power-to-X). Particular interest is given to the hydrogen as an energy carrier, which will be mainly used for the following applications:

- Energy carrier for industrial applications;
- μ-CHP systems based on fuel cells for buildings;
- Fuel for mobility;
- Power generation; and
- Energy storage.

The concept of coupling electricity to other forms of energy has traditionally been referred exclusively to the electrification of sectors such as heating and transport. With the Clean Energy Package, this concept has been expanded in order to include Power-to-X systems that, starting from the electric vector, involve other energy vectors. These applications can provide flexibility to the energy system by managing to meet the demands for thermal energy, fuels, and mobility through PtH, PtG, and PtM technologies, respectively.

First, the market review concerns the rules relating to electric vehicles. Article 33 of the EMD II [9] states that Member States must provide the regulatory framework necessary to facilitate the connection of public and private charging points to the distribution networks. Also, the new EPBD [14] aims to facilitate the introduction of electric mobility by equipping buildings with infrastructure for electric vehicles. Pursuant to Article 8, Member States must provide for measures to simplify the installation of recharging points in new and existing residential and non-residential buildings and provide for the overcoming of any regulatory obstacles.

Second, the RED II [10] provides a first European target for heating and cooling from renewable sources. According to Article 23 of the RED II [10], Member States have the task to increase their percentage of renewable heat by 1.3% every year until 2030. Waste heat and cold can contribute up to 40% to objective, while district heating and cooling will have to contribute with an average annual increase in renewable energies of at least one percentage point.

The long-term vision to 2050 is well defined by ETIP SNET [2], which considers the electrification of European energy systems as the backbone of its societies and markets. In order to achieve a fully carbon-free energy system, it is needed to exploit in the best possible way integration options between electricity and gas networks as well as count on daily or seasonal storage such as hydro, batteries, hot water seasonal storage, and PtG conversion technologies. A key role in the future energy system will also be played by distributed energy resources that according to this vision will be exploited for their full potential, by helping to maximize the resilience of energy supply for electricity and heating and cooling needs. The future integrated energy system will rely on renewable electricity mainly from hydro, solar, wind, geothermal, and renewable heat and cooling from solar, biomass, biogas, and geothermal, renewable gas as biogas and renewable fuels as biofuels.

1.3 Pillars for the Transition Toward Integrated Decentralized Energy Systems

This book addresses the topic of integrated decentralized energy systems by focusing the attention on the pillars described below that will play a major role in the transition of the traditional energy systems toward this new energy paradigm.

Power conversion plays a key role in future integrated energy systems, where electricity enables for a switch of energy carriers through **Power-to-X technologies**, which provide energy storage and sector coupling by converting electricity into chemical energy and heat, thereby allowing circularity into the energy system. Power-to-X energy can act as a sink for electricity surpluses by using the available energy in a cost-effective way.

By enabling sector coupling while accelerating carbon neutral transition, **hydrogen as a vector** also plays an essential role in integrated energy systems. With high share of variable renewables, the production of carbon-free energy carriers as hydrogen from renewable electricity covers an important role for the decarbonization of the energy system as a whole. The production of hydrogen can provide significant flexibility to the power system, as well as – most importantly – seasonal storage of renewable electricity by blending hydrogen into natural gas grids. Hydrogen as a vector can be seen as an electricity storage method (Power-to-X-to-Power), which can contribute to the increase of stochastic renewable electricity penetration into the grid, but it also represents a versatile cross-vector medium enabling the deep decarbonization of non-electrified hard-to-abate sectors as renewable fuels, sector integration, and mobility.

In an integrated energy system where the locally available energy resources are used for their full economic potential, **storage in all forms and types** plays a crucial role. Energy storage can provide multiple services to the energy system as a whole by storing the energy produced in excess and delivering it on demand. It can smoothen the variability of RES, making the power system more reliable and flexible. Battery energy storage systems are considered among the best suited technologies for short and mid-term flexibility services, such as frequency regulation, spinning reserve, peak shaving, etc. Long-term storage services including seasonal storage are needed in the presence of high penetration levels of solar and wind energy production, and they are generally supplied by bulk energy storage systems, such as pumped-hydro plants or mechanical storage facilities and electrochemical energy storage. Thermal storage solutions can be used in several industrial applications as well as district heating, PtH applications, etc. Besides, they represent a strong support to heating and cooling electrification. These applications are cornerstone to enhance the energy system circularity, thanks to their characteristics of closing energy cycles without energy waste: storing excess electricity that would, in an open cycle, cause the curtailment of renewables, by converting it to other forms, enables new energy streams bending over the cycle toward useful ends, thus increasing circularity.

Digitalization is a key enabler for integrated decentralized energy systems by integrating innovative technologies in the electricity system through interoperable, standardized data architectures and related communication for achieving higher levels of efficiency. Digitalization improves the observability of the power system for stable and secure operation in the presence of high shares of RES, enabling advanced planning, operation, protection, control, and automation of the energy systems, through the availability of real-time information that improves system balancing and resilience at all time scales in the case of any unforeseen and sudden event. Information Technologies including semantic data models, Big Data management, and Artificial Intelligence will enable the optimization and automation of processes and support operators' decisions. Through digitalization, it will be possible to facilitate services and achieve full integration of all types of energy systems. Moreover, digitalization is also key to exploit the full potential of active consumers to contribute to the effective integration of RES in the power system. The massive integration of smart meters and Home Energy Management Systems will allow the implementation of new business models and aggregation schemes (e.g. energy communities) that exploit the flexibility of the active consumers.

Smart mobility plays an important role in accelerating carbon neutral transition. When supported by higher deployment of RES, it contributes with multiple benefits to the sustainability of the transport system. In fact, electric vehicles are expected to play a primary role in the decentralized energy system and represent a driver for increasing RES integration in the buildings to meet their additional power demand. **Smart grids** also support energy transition through reducing CO_2 emissions in a cost-efficient way. By optimizing the asset utilization, they reduce the needs for new investments. Moreover, they enable penetration of renewables and emerging and efficient technologies, thereby allowing minimization of costs and carbon emissions. Another important benefit related to smart grids is the provision of real-time and monitoring control that allows improving stability, resilience, and security of the power system. Last but not least, they enhance the quality of the supplied power through reducing commercial and technical losses.

Efficient energy use in buildings is another constituting factor for integrated energy systems. Moreover, because of the active local energy generation (building-integrated generation) combined with energy efficiency solutions (e.g. insulation and efficient appliances), new buildings in most cases will be **nearly zero-energy and possibly positive-energy buildings**. Employing energy-efficient solutions and renewable energies are critical factors to meet the energy and environmental targets set for the building sector. The first factor can reduce the building's energy consumption, while the second one can reduce the buildings' total energy intensity. Especially, positive energy buildings can be considered advantageous for the decarbonization of the building sector and a promising pathway toward sustainable urban development because of their scalability potential, renewable energy harnessing capacity and high energy efficiency.

Last but not least, **local energy communities** will become increasingly important in the transition toward a low- or even carbon-neutral energy system. Especially in the European context, they represent an emerging paradigm where active consumers and prosumers are engaged and play an active role in aggregated forms through renewable energy communities and citizen energy communities. Moreover, local energy communities can perfectly represent the concept of local integrated energy systems, which, characterized by well-defined boundaries, involve different energy technologies and carriers that can be integrated in order to optimally exploit the synergies coming from this interplay, thereby enhancing energy resources use.

List of Abbreviations

CHP	combined heat and power
DG	distributed generation
DoE	department of energy
DR	demand response
EHV	extra high voltage
EMD	energy market directive
EPBD	energy performance of buildings directive
EU	European Union
HV	high voltage
IRENA	International Renewable Energy Agency
PtG	power-to-gas
PtH	power-to-heat
RED	renewable energy directive
REmap	renewable energy roadmap
RES	renewable energy sources
SDG	sustainable development goals
SRI	smart readiness indicator
UN	United Nations

References

1 McCollum, D.L., Zhou, W., Bertram, C. et al. (2018). Energy investment needs for fulfilling the Paris Agreement and achieving the Sustainable Development Goals. *Nat. Energy* 3 (7): 589–599.

2 ETIP SNET VISION 2050 (2018). Integrating Smart Networks for the Energy Transition: Serving Society and Protecting the Environment.
3 IRENA (2019). *Global Energy Transformation: A Roadmap to 2050* (2019 edition). Abu Dhabi: International Renewable Energy Agency.
4 ETIP SNET (2020). Sector Coupling: Concepts, State-of-the-art and Perspectives. White Paper, January 2020.
5 Eid, C., Koliou, E., Valles, M. et al. (2016). Time-based pricing and electricity demand response: existing barriers and next steps. *Util. Policy* 40: 15–25.
6 Clean energy for all Europeans package. https://ec.europa.eu/energy/topics/energy-strategy/clean-energy-all-europeans_en (accessed 18 October 2021).
7 REGULATION (EC) No 714/2009 OF THE EUROPEAN PARLIAMENT AND OF THE COUNCIL of 13 July 2009 on conditions for access to the network for cross-border exchanges in electricity and repealing Regulation (EC) No 1228/2003.
8 DIRECTIVE 2009/72/EC OF THE EUROPEAN PARLIAMENT AND OF THE COUNCIL of 13 July 2009 oncerning common rules for the internal market in electricity and repealing Directive 2003/54/EC.
9 Directive 2019/944/EU on the internal electricity market. https://eur-lex.europa.eu/legal-content/EN/TXT/PDF/?uri=CELEX:32019L0944&from=EN (accessed 18 October 2021).
10 Directive 2018/2001/EU on the promotion of the use of energy from renewable sources. https://eur-lex.europa.eu/legal-content/EN/TXT/PDF/?uri=CELEX:32018L2001&from=EN (accessed 18 October 2021).
11 DIRECTIVE 2009/28/EC OF THE EUROPEAN PARLIAMENT AND OF THE COUNCIL of 23 April 2009 on the promotion of the use of energy from renewable sources and amending and subsequently repealing Directives 2001/77/EC and 2003/30/EC.
12 Directive 2018/2002 on energy efficiency. https://eur-lex.europa.eu/legal-content/EN/TXT/PDF/?uri=CELEX:32018L2002&from=EN (accessed 18 October 2021).
13 DIRECTIVE 2012/27/EU OF THE EUROPEAN PARLIAMENT AND OF THE COUNCIL of 25 October 2012 on energy efficiency, amending Directives 2009/125/EC and 2010/30/EU and repealing Directives 2004/8/EC and 2006/32/EC.
14 Directive 2018/844 on the energy performance of buildings. https://eur-lex.europa.eu/legal-content/EN/TXT/PDF/?uri=CELEX:32018L0844&from=IT (accessed 18 October 2021).
15 Regulation (EU) 2018/1999 on the governance of the Energy Union and Climate Action. https://eur-lex.europa.eu/legal-content/EN/TXT/HTML/?uri=CELEX:32018R1999&from=EN (accessed 18 October 2021).
16 A European Green Deal Striving to be the first climate-neutral continent. https://ec.europa.eu/info/strategy/priorities-2019-2024/european-green-deal_en (accessed 18 October 2021).

2

Integrated Energy Systems: The Engine for Energy Transition

Marialaura Di Somma and Giorgio Graditi

Italian National Agency for New Technologies, Energy and Sustainable Economic Development, ENEA, Department of Energy Technologies and Renewable Sources, Rome, Italy

2.1 Introduction: the Concept of Integrated Energy System

Energy systems play an essential role in the economic and social development of a country and in the life quality of people [1, 2]. With the increase of energy demand on a worldwide scale, depletion of fossil fuels, and growing environment protection awareness derived by the latest Climate Conference COP21, improving the efficiency of energy resource use has become one of the key challenges [3]. In such a context, the European Union (EU) has set ambitious environmental and energy goals to design a low-carbon energy system by the middle of the twenty-first century. The EU climate and energy framework establishes targets to a 40% reduction in greenhouse gas (GHG) emissions (from 1990 levels), 32% share for renewable electricity, and 32.5% improvement in energy efficiency to be achieved by 2030. These targets become even more ambitious for 2050, with the Energy Roadmap 2050 of the European Commission and the Energy Union strategy supporting the aim of fully decarbonizing the European economy by reducing GHG emissions in developed countries below 80–95% of 1990 levels by 2050. These ambitious targets can be achieved by developing energy systems supporting the implementation of three primary goals: mitigating environmental impacts of energy systems, creating affordable and market-oriented energy services, and ensuring security, reliability, and resilience of energy supply.

Mitigating environmental impacts of energy systems has several dimensions, including:

- reducing GHG emissions for mitigation of climate change;
- monitoring sources of pollution originating from activities directly or indirectly linked to energy systems; and
- promoting a circular economy.

The creation of accessible and market-based energy services can instead be achieved by placing the so-called prosumers, who are at the same time producers

Technologies for Integrated Energy Systems and Networks, First Edition.
Edited by Giorgio Graditi and Marialaura Di Somma.

and consumers of energy, at the center of the energy system, while promoting their active role and making them more responsible for energy consumption.

In order to guarantee the security, reliability, and resilience of energy supply, the creation of integrated energy systems plays a crucial role. Also known as multi-energy systems or hybrid energy systems, the main idea behind them is to move from a single energy carrier to multiple energy carriers in order to exploit the synergies from their interplay, thereby increasing the efficiency in the energy resources used [4].

The concept of integrated energy systems is well defined in the ETIP SNET VISION 2050 [5] that foresees this energy paradigm fully implemented by 2050. According to this vision, these systems are characterized by the integrated management of the electric vector, the mobility, the heat and cooling vector for space heating and cooling, and the various types of storage. Considered as "a system of a systems" and represented in Figure 2.1, it can be seen as an integrated infrastructure for all energy carriers with the electrical system as a backbone, characterized by a high level of integration between all networks of energy carriers, coupling electrical networks with gas networks, heating and cooling, supported by energy storage and conversion processes.

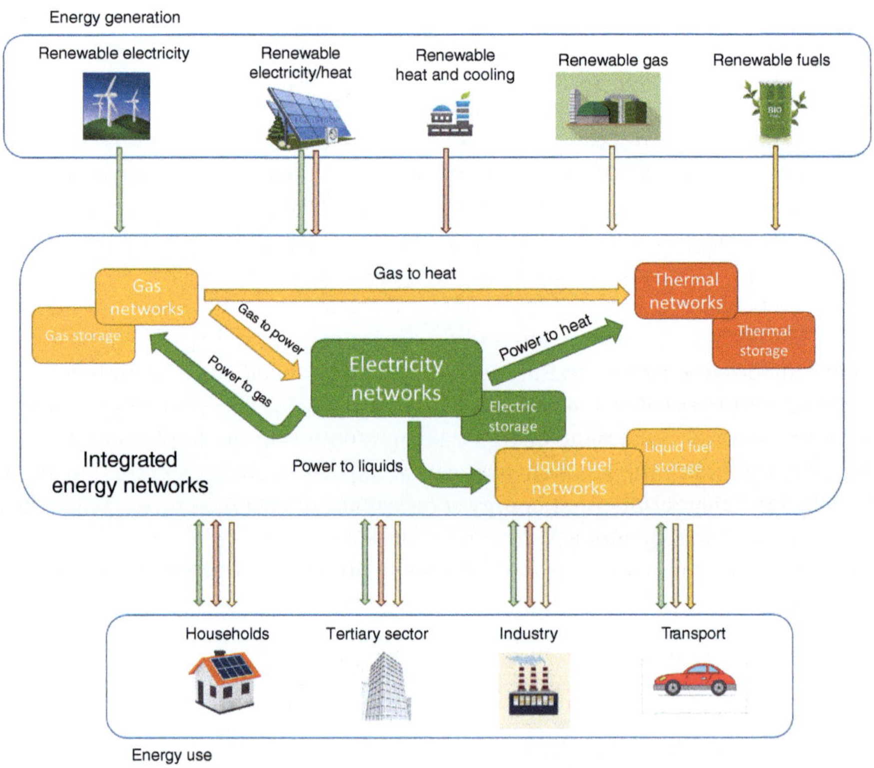

Figure 2.1 Scheme of an integrated energy network according to ETIP SNET Vision 2050. Source: Based on ETIP SNET "VISION 2050" [5].

The key elements characterizing this emerging energy paradigm are:

- the full involvement of the end user in the management of the system itself. Under the concept of integrated energy systems, citizens become active consumers and prosumers, using local and user-friendly energy exchanges, as well as peer-to-peer exchanges, for a wide range of services and optimal energy prices. Not only that, but the active role of end users is fully implemented in the mechanisms of demand response, through which they are made participants in the management of network contingencies, as well as in reducing energy consumption through applications such as zero energy buildings or promoting renewable energy projects through renewable energy communities;
- the integration of all energy carriers and the advent of distributed poly-generation fully supplied by renewable energy sources (RES). Obviously, in this context, storage in all its forms and types plays a crucial role in order to deal with the large penetration of RES. The locally available energy resources are used for their full economic potential, partly reflecting the upgrading needs of the electricity transmission and distribution networks and also contributing to maximizing the resilience of supply channels for heating and cooling needs;
- the integration of digitalization enabling new services for energy consumers as decentralized control techniques, peer-to-peer energy trade, and platforms for data exchange and fast decision-making for all actors operating in integrated energy systems, thereby enabling advanced operation planning, control, and automation of energy systems;
- cross-sector integration through which sectors such as heating, cooling, transportation, and industry are all supplied with low-carbon energy, thus reducing significantly GHG emissions. In such context, the cross-sector integration can contribute to the cost-efficient decarbonization of the energy system, by valuing synergy potentials and interlinkages between different parts of the energy system.

Integrated energy systems rely on sector coupling that indicates linking the various energy carriers – electricity, heat, cold, gas, and liquid fuels – with each other and with the end use sectors, such as households, tertiary sectors, industry, and transport. The plan is to deploy various existing and emerging technologies, processes, and business models, such as information and communication (ICT) and digitalization, smart grids and meters, and flexibility markets.

Although recent legislations and regulations fully support the implementation of integrated energy concept, including the European Green Deal [6] and Clean Energy package for all European [7], from the technical point of view, further studies on the requirements for enabling full operation of integrated systems are needed. Indeed, the core feature of the integrated energy system paradigm is the intrinsic interdependency among energy carriers and sub-systems because of their correlated interactions. Relevant optimization, operation, and planning need to consider this aspect in order to exploit the extended flexibility for enhancing the efficiency of the energy resources used [4].

2.2 Key Enablers for Integrated Energy Systems

2.2.1 Storage and Conversion Technologies

The effective integration of energy systems relying on multiple energy carriers is considered as one of the key enablers for increasing the penetration levels of RES into energy systems, thereby allowing the achievement of the European decarbonization objectives. This high level of integration can only be achieved through the deployment of storage units and power conversion systems enabling on the one hand the coupling among the various energy carriers and on the other hand higher security levels of energy supply.

Therefore, a key role in integrated energy systems is played by storage and power conversion processes. Electrical energy carrier can be converted into several other energy carriers through Power-to-Gas (PtG), Power-to-Heat (PtH), and Power-to-Liquid (PtL) technologies (also defined as Power-to-X), which in turn allow transporting large amounts of energy among distant and interconnected hubs in the energy system [8].

In detail, through energy conversion processes, the converted electrical energy (net of conversion losses) can follow the paths explained as follows:

- stored more easily than in the electrical system, in order to be then re-converted into electricity, thereby allowing the shift in time;
- consumed in another sector, if for example it is more convenient from economic or environmental point of view;
- transported as heat/cooling or gas/liquid, if the new transport performances are better than those related to the transmission and distribution of electrical energy.

The various paths feasible for the electrical energy carrier are shown in Figure 2.2. Depending on the specific objectives, electricity can be directly consumed, converted, stored, or transported. For instance, for decarbonization purposes, electricity can be directly consumed through electrification as an alternative to fossil fuels or can be converted and then consumed, for example, through PtH technologies, as an alternative to using directly fossil fuels.

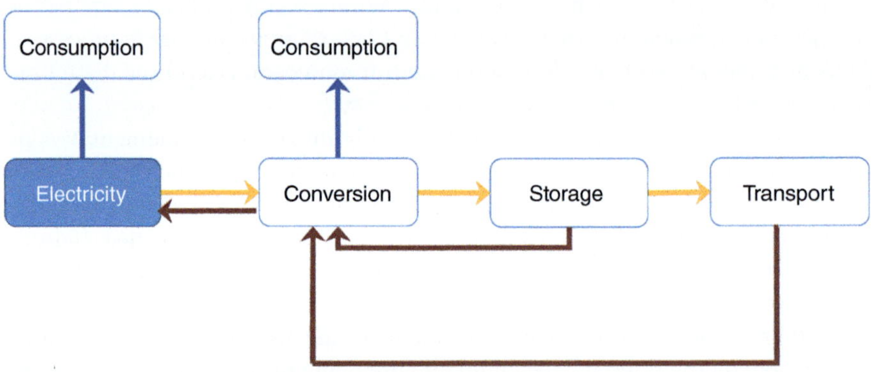

Figure 2.2 Various paths feasible for the electrical energy carrier.

In order to increase the whole flexibility of the system, electricity can also be converted and then stored in other energy systems as an alternative to traditional storage solutions. This process is bi-directional because the "new" energy carrier can be re-converted into electricity.

Finally, in order to optimize the system use or the infrastructure development, electricity can be converted and transported as an alternative to power lines. Also in this case, the process is bi-directional because the "new" energy carrier can be re-converted into electricity, if needed.

Energy storage represents an essential means for providing the needed flexibility to the electrical systems, especially in the presence of high penetration levels of intermittent and variable RES.

One of the critical aspects of electrical systems is certainly the need to ensure a balance between supply and demand at all times; electrical energy storage systems, by allowing energy to be converted into a "storable" form, storing it in this form, and then reconverting it, represent an answer to this problem. Storage can also play an essential role in reducing costly interventions on the transmission and distribution networks that would otherwise be necessary to adapt the system to the increasing levels of renewable production. The benefits that derive from the use of this technology are thus different:

- stable generation system;
- enhanced programming capacity of the generation system;
- provides a reserve for future needs;
- mitigates the impact of demand peaks.

Traditional storage technologies for different energy carriers are shown in Figure 2.3, whereas their mean features are presented in Table 2.1. Their detailed characteristics are out of the scope of this chapter because in the context of integrated energy systems. The additional options of energy storage offered by the possibility to convert electricity into other energy carriers are described in the following sections.

Power-to-Heat/Cooling (PtH/C) represents a very promising option for decarbonization of buildings, considering that in Europe, they are responsible of around 40% of final energy consumption, and most of energy consumed is related to heating and cooling sectors. In such a context, the essential requirement for decarbonization is related to the rollout of low-carbon heating and cooling technologies as alternatives to traditional fossil fuel-based technologies. Among other technologies, electric heat pumps and district heating and cooling networks all fed by RES, as well as RES-based production of hydrogen through fuel cell-based combined heat and power (CHP) systems, are considered the most promising options to achieve a cost-effective and low-carbon energy supply system obtained by coupling electricity and heat/cooling sectors, while also fostering the integration of variable RES.

The interaction between electricity and heat/cooling brings new level of flexibility to the integrated energy system as a whole thanks to the intrinsic flexibility of the heating/cooling sector and the exploitation of the synergies between the two energy

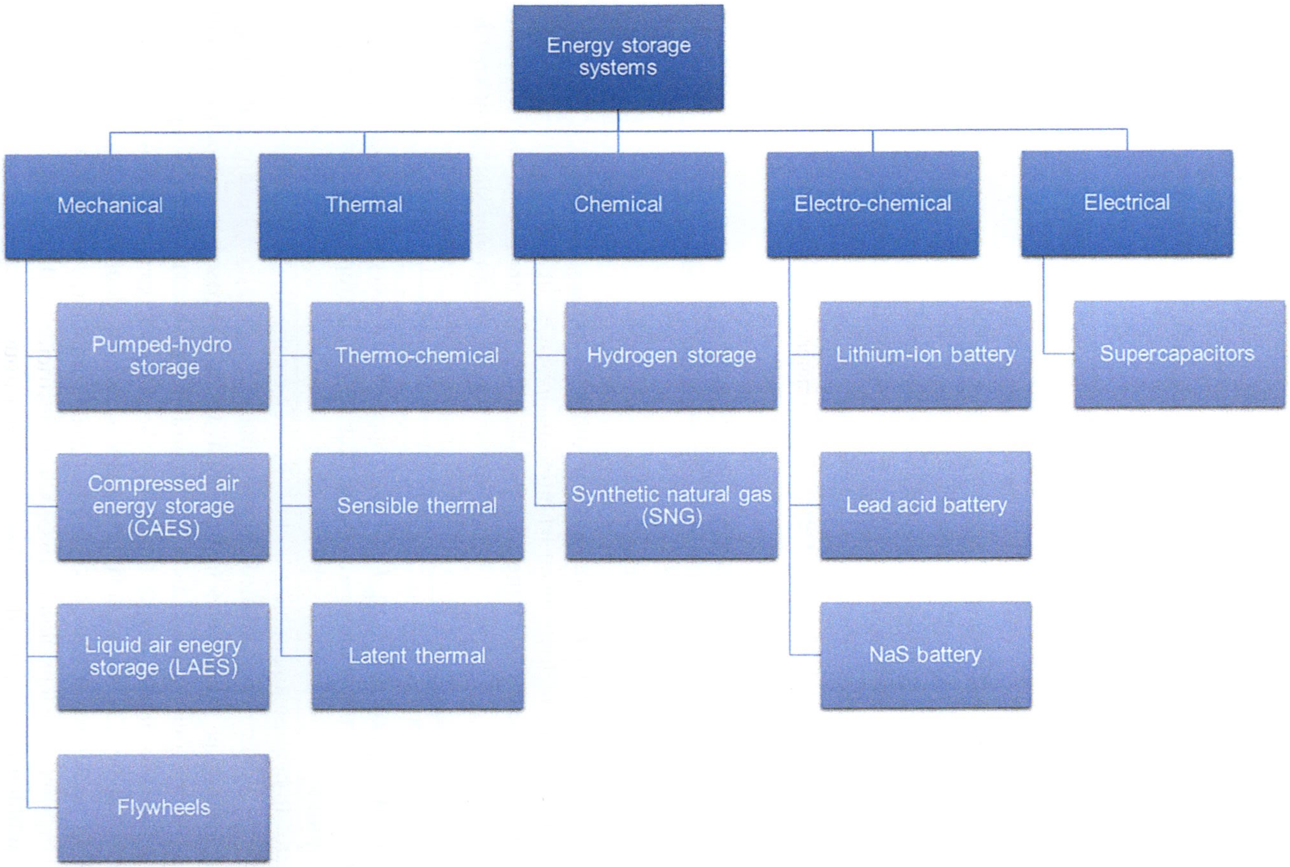

Figure 2.3 Different types of energy storage systems.

Table 2.1 Main features of energy storage technologies.

Storage technology type	Brief description	Main applications in integrated energy systems
Mechanical storage	Consists of several storage principles such as potential energy of water in pumped hydro storage, the volume and pressure work of air in CAES, the stored energy in cryogenic liquids in LAES, and the rotational energy of a mass in flywheels	Ancillary services provision as frequency control, reserve; voltage support; and inertial response Reduction of RES curtailment Network reinforcement deferral Grid stability Load shifting
Thermal energy storage	Mainly includes three types of technologies. Indeed, energy can be stored as sensible heat of materials undergoing a change in temperature. Latent heat storage takes advantage of the energy absorbed or released during the phase change of the storage material and thermochemical energy storage uses the heat evolution of a physical process or a chemical reaction.	District heating; storage in single buildings; concentrated solar power; seasonal storage; and peak shaving
Chemical	Stores energy in chemicals in gaseous, liquid, or solid form and energy is released in chemical reactions. The main characteristic is the high energy storage density	Seasonal energy storage
Electrochemical	Mainly consists of batteries where chemical energy is stored and converted into electrical energy and vice versa thanks to electrical reactions	Electrification of the transport sector Stationary applications
Electrical	Stores electrical energy at an electrode–electrolyte interface. The energy capacity is limited, but the reaction time is fast, whereas the efficiency is very high	Mainly suitable for high power applications; frequency control; and transmission line stability

Source: Based on ETIP SNET [8].

carriers. Suffice it to know that CHPs, heat pumps, and also thermal storage systems can provide ancillary services to the electrical systems. For instance, heat pumps, with the reduction of thermal energy provided without compromising the user comfort thanks to the thermal inertia of buildings, can provide ancillary services.

PtG and PtL technologies are considered a promising option today to facilitate the large-scale integration of carbon-free electricity produced by RES, while also providing flexibility to the power system through the provision of long-term energy storage. With specific reference to PtG, it refers to the process of converting excess production of RES-based energy produced to gaseous energy carriers as hydrogen and methane via water electrolysis. Instead, PtL technologies transform hydrogen produced from electrolysis into liquid fuels such as diesel/gasoil-like fuels, methanol, dimethyl ether, ammonia, or ethanol, which have relatively high energy densities and can be easily stored.

The benefits offered by PtG technologies to the integrated energy system are multiple, but most of others, they allow the consistent reduction of renewable electricity curtailment, thanks to the ability to store for long-time horizons, going from weeks to seasons, renewable electricity during periods of excess wind and solar production, and use it also much later, especially during times where end user demand exceeds renewable generation [9]. At the same time, conversion technologies as electrolyzers and fuel cells could gradually replace conventional power generation units in the provision of ancillary services and capacity reserve, thereby contributing to the reliability of the power networks.

Regarding PtL technologies, beyond the storage possibilities that can offer, they can also play a key role for decarbonizing the transport sector, considering that in the future, it is expected that liquid fuels will remain still necessary especially for heavy-duty vehicles and airplanes.

This type of cross-vector integration provides the energy system with increased flexibility to cope with fluctuations in energy demand and renewable energy supply. Power-to-X energy can act as a sink for electricity surpluses by using the available energy in a cost-effective way. On the other hand, improving the integration of the electricity and gas sectors would also allow an optimized use of the existing gas infrastructure. Gas pipelines could be used to transport renewable energy from supply areas to areas with shortages, reducing the need to expand the electricity transmission capacity. Gas storage could be used to cope with seasonal variations in demand and renewable energy supply. Renewable gas can also be used in gas-fired power plants or fuel cells, providing low-carbon back-up capacity to generate electricity when other renewable energy resources are unavailable.

2.2.2 End User Engagement and Empowerment

As already mentioned, one of the key elements characterizing an integrated energy system is the end user engagement and empowerment. In fact, citizens represent the central actors in the transition phase from a single-carrier energy system based on fossil fuels toward an integrated, low-carbon, accessible, cost-efficient, and market-based energy system.

Demand response represents the main element characterizing the end user engagement, through which the user is called to participate to the management of the energy system itself. Demand response is an articulated program of actions that allows the consumers (industrial, commercial, or residential) to modify their own electrical load (lowering it or translating it horizontally in time) in response to existing problems on the grid, e.g. momentary unavailability of power caused by failures or intermittent production from variable RES, or in response to the dynamics of wholesale electricity prices, or to increase the use of locally or self-produced energy. Demand response can provide several environmental benefits, while making the electric grid more reliable in the presence of high shares of renewables. At the same time, it can contribute to save energy, reduce the use of fossil fuel power plants, and promote integration of RES into the electric grid by also providing increased stability through avoiding peak congestions.

Demand response and consumer engagement are integral parts of the Energy Union and the Clean Energy Package for all Europeans. However, although in Europe consumers are becoming more and more engaged, demand response resources cannot yet completely compete on an equal level with traditional (generation) resources. In addition, for smaller buildings and household consumers, it is not easy to offer demand response services on an individual basis. Therefore, demand response service providers could aggregate demand response of multiple similar end users and sell it to the market as a bundle. In this way, service providers can make a pool of combined loads that they sell as a single resource and as such help individual consumers to value their flexibility potential (acting as a large-scale asset). Europe is now expanding on this model and is in a good track for empowering consumers, giving them the opportunity to make autonomous decisions on how to produce, store, sell, or share their energy [10]. This paradigm of "democratization" of the energy supply is fully expressed in the concept of energy communities, understood as a set of energy users who decide to make common choices for satisfying their energy needs, in order to maximize the benefits resulting from this collegial approach, thanks to the implementation of multi-carrier energy systems for the distributed generation of energy and the smart management of energy flows.

In such a context, the internal Electricity Market Directive (EMD II) [11] and revised Renewable Energy Directive (RED II) [12] proposed the Citizen Energy Community (CEC) and the Renewable Energy Community (REC), respectively, as part of the Clean Energy Package. Beyond the differences between CECs and RECs that will be analyzed in detail in Chapter 11 of this book, the common goals of these energy communities are [13]:

- promotion of the public acceptance of renewable projects;
- development of renewable technologies at the local level;
- promotion of participation in the market of end users; and
- addressing problems related to energy poverty and vulnerability by reducing energy supply costs.

In more details, the Clean Energy Package aims at empowering the end user to foster energy transition by putting this latter at the center of the energy system.

The EMD II aims to construct a true internal market governed by common rules that can guarantee a wide range of electricity accessible to all. In relation to consumers, this directive provides an important paradigm shift, aimed at qualifying the consumers as "active," who can operate directly or in an aggregated manner, sell self-produced electricity, including through agreements for the purchase of electricity and participate in flexibility and energy efficiency mechanisms. In such a context, the directive states that all consumers should be able to take advantage of direct participation in the market, in particular by adjusting consumption based on market signals while benefiting from lower electricity prices or other incentives.

According to this directive, the CEC must be able to operate on the market on equal and non-discriminatory conditions with respect to the other subjects, being able to freely assume the roles of final customer, producer, supplier, or manager of distribution systems.

The RED II Directive also pays particular attention to the role of consumers that are allowed to become consumers of renewable energy and also to be able to produce, store, and sell the electricity produced in surplus, both individually and through aggregators, while guaranteeing the consumer's rights. RED II in fact introduces the concept of REC, which must have the right to produce, consume, store, and sell renewable energy. Consumers will also be able to exchange, within the same community, the renewable energy produced and access all the appropriate electricity markets, directly or through aggregation, in a non-discriminatory way.

The new Energy Efficiency Directive (EED) [14] also extends the consumer rights and improves access to smart metering tools, smart billing, and consumption information, whereas the new Energy Performance of Buildings Directive (EPBD) [15] contains provisions concerning, among other things, energy efficiency targets for buildings, energy certification, verification methods, monitoring, and control of energy use and the establishment of obligations relating to the installation of electricity recharging points.

2.2.3 Digitalization Enabler

Historically, the electrical system has been centrally controlled with a small amount of data available. Since the 2000s, with the use of digital sensors on networks and smart metering systems, the quantity and quality of data have grown considerably. The true potential of digital systems lies in the ability to break down the boundaries among the energy sectors and increase the flexibility allowing integration among systems. Therefore, digitalization represents one of the most important enablers for integrated energy systems by providing extensive services to all kinds of actors for planning, operation, and monitoring issues, while promoting information and connectivity among users. As such, digitalization is a powerful means for increasing efficiency, productivity, and energy savings in a number of sectors as industry, transport, and buildings. Especially with reference to these latter, digitalization offers a great potential to improve energy services and user comfort, while also increasing the efficiency in the energy resources used through the full involvement of end users. Indeed, it contributes to the energy savings, ensuring that energy in buildings is consumed when and where it is needed by enhancing the responsiveness of energy services and predictability of user's behavior.

Digitalization can also enable the active participation of consumers from all demand sectors in the energy system operation, being a key enabler for demand response services to reduce peak loads by shifting the time of use of electric appliances, to shed loads by reducing the set point for temperature to lower energy demand, as well as to store energy in response to specific external signals. Moreover, digitalization allows predicting and monitoring in real time the energy performances of buildings, giving the possibility to the interested stakeholders as consumers and network operators to identify when and where maintenance is needed or where energy savings can be achieved. Active control systems can also create a bridge for linking building energy services with information from the grid, thus improving the management of supply and demand and increasing the energy efficiency [16].

Digitalization can also facilitate the deployment of DER at the local level by improving their management and optimizing their operation strategies in order to maximize economic benefits for prosumers. A greater potential is for sure related to the emerging paradigms of energy communities and peer-to-peer energy trading among users.

Beyond its importance in these specific sectors, the highest potential of digitalization is offered in the context of integrated energy systems thanks to its ability to eliminate the boundaries among energy sectors, increasing flexibility and enabling integration across technologies. Connectivity is the core. It allows the monitoring and active control of large numbers of individual energy systems and the interplay among them.

Among the potentially most impacting digitalization technologies on integrated energy systems, there are Big Data, Machine Learning and Artificial Intelligence (AI), the Internet of Things (IoT), and the block chain, which are briefly described in the following paragraphs.

Big Data can be defined as information resources with a high volume, speed, and variety that require economic and innovative forms of information processing to enhance understanding, decision making, and process automation. These are sets of data with such a large volume that they cannot be managed by conventional tools but by innovative technologies and methods capable of collecting, processing, and analyzing them in order to be able to exploit them for making predictions and thus more efficient decisions [17].

In the energy sector, their use is expected to be particularly suitable for improving the reliability and use of transmission and distribution networks and, in the future, integrated energy systems will become the major field of application of big data [18].

Data on climatic conditions, generation systems, lines, and operating conditions, data on consumption and price signals, and data deriving from electric vehicles represent some of the resources that reside in the use of big data. Their use in the integrated energy systems will become more concrete when it will be possible to analyze all the data coming from different sources in real time so as to promptly respond to faults and power variations.

In its broadest sense, the term **AI** indicates the ability of machines to make decisions based on a set of information available in order to solve very specific problems in limited areas. The capabilities of AI are different but are generally grouped into three categories: evaluation (receiving and recognizing information), deduction (processing and learning from information), and response (acting by making appropriate decisions). **Machine Learning**, on the other hand, can be considered as a subset of AI and refers to those computer systems that can automatically improve the execution of their tasks thanks to the experience [19]. Basically, Machine Learning algorithms use mathematical–computational methods to learn information directly from data, without mathematical models and predetermined equations. Machine Learning algorithms improve their performance in an "adaptive" way as the "examples" from which they "learn" increase. The integrated energy systems represent a valid application for these tools given their growing complexity, especially as regard their modeling and optimization and, not least, the growing use of forecasting systems through neural networks.

The term **IoT** refers to those objects and equipment which, thanks to a built-in sensor, are directly connected to the Internet. Their number in recent years has increased exponentially, by affecting practically every aspect of human life. The main reason behind this widespread diffusion lies in the fact that in the past decade, both processors and, more generally, communication technologies have become much smaller, cheaper, and more efficient so that they can be incorporated into a vast range of equipment. The applications for these systems are different: consumers, for example, can adopt "smart home" technologies thanks to which they can remotely control every aspect of their home from lighting to heating, from security services to the kitchen. The industrial sector uses IoT systems to manage production lines, to control and optimize the processes, and the commercial sector to optimize supply chains, retail, and warehousing. IoT devices are already used in various fields of the energy sector. The most emblematic and certainly the most familiar case is represented by smart meters which provide a real-time reading of the energy consumption of users and periodically report consumption data to suppliers or system operators [20]. An aspect that must be considered regarding the diffusion of digital technologies is the safety of the electricity system. In the past, in fact, power grids have been relatively resilient to cyber attacks mainly because the control systems were not easily accessible remotely and the digitization rate was quite low. Today, with the advent of smart grids and digital control systems, the risk of attacks has increased significantly with potentially catastrophic consequences.

Distributed ledgers technologies (DLT) are systems based on a distributed register or systems in which all the nodes of a network have the same copy of a database that can be read and modified independently by single nodes. Each node is authorized to update the registers independently from the others but always under the consensual control of the other nodes; the updates are no longer managed centrally but, at the same time, every single transaction, even if managed independently, must be verified, voted, and approved by the majority of the network participants. The consensus management methods along with the register setting logic represent two peculiar characteristics of distributed ledgers.

A subfamily of DLTs is the **block chain**, a tool in which the register is structured as a chain of blocks containing transactions and whose validation is entrusted to a consensus mechanism distributed on all nodes of the network in the case of the "permissionless" (or public) block chain, or on all nodes that are authorized to participate in the transaction validation process to be included in the register, in the case of "permissioned" (or private) block chains. Interest in the use of block chain in the energy sector has grown considerably in recent years, especially thanks to the possibility of making more secure and transparent peer-to-peer transactions, which do not need a central supervisor. Although they are relatively new systems and still have many critical issues to face, they will be able to provide an important contribution in the transition to an integrated energy system [21].

DLT systems have also the potential to change completely the energy market with consumers who could acquire energy directly through an exchange platform. The sale of energy would be established between the parties and recorded in the block chain, and its transfer along with the payment would take place automatically at the time of delivery through a smart contract. Transactions recorded in the block chain would be visible to all parties including the system operator. The block chain can also be implemented in smart metering systems so as to record consumption data and in the provision of demand response services so as to allow, for example, a reduced consumption when there is a peak demand on the network.

2.2.4 Emergence of an Integrated Energy Market

With the advent of integrated energy systems, the evolution from a single energy market to an integrated energy market that integrates multiple energy carriers such as electricity, gas, heat, and cooling becomes a compelling need. This market consisting of multiple coupled energy markets improves the efficiency in the energy resources used through the exploitation of the synergies among different carriers, resulting from the complementarity and the coordination of multiple energy resources [22]. Supported by digitalization and advanced information technologies, storage and power conversion, and with different types of energy users as participants, the integrated energy market is based on the concept of multi-energy complementary, indicating that all kinds of energy co-existing in such type of market can complement each other based on their equality and substitutability [23, 24].

The main benefits and challenges of the integrated energy market are shown in Table 2.2.

First of all, this market fosters the participation of distributed market players, by following the needs of an energy system with large penetration of DER as wind power, PV, and distributed storage. Moreover, this type of market strongly supports peer-to-peer interconnections and energy sharing among prosumers within communities. This market also fosters smart energy consumption. In fact, the end users' engagement and empowerment lead the way to their participation in the energy markets to sell (as prosumers) and buy energy and flexibility services to satisfy their

Table 2.2 Main benefits and challenges of integrated energy market.

Benefits	Challenges
Participation of distributed market players	Increase of market complexity
Fosters smart energy consumption	Strong dependence on advanced digitalization services
Comprehensive trade supporting multiple types of energy	—

Source: Based on Dong et al. [22].

needs. In such a context, energy suppliers develop customized energy services to promote smart and flexible energy consumption.

On the other hand, the deployment of multi-energy technologies at large scale makes the market trading subjects and objects much more diversified, and this supports the comprehensive trading of different types of energy carriers.

In the face of these benefits, the integrated energy market also poses several challenges, such as the increase in the market complexity because of its opening to various types of participants and the resulting larger competitiveness. Moreover, it is evident that in order to ensure an effective integration of this market, digitalization and information technologies play a key role. Indeed, collection, transmission, and analysis of a large amount of information are essential to provide decision support to market operators, and security in information sharing represents a prerogative for the well-functioning and reliability of integrated energy trading.

2.3 Integrated Energy Systems at the Local Level

2.3.1 Conceptualizing Local Integrated Energy Systems

Integrated energy systems at the local level present the largest potential for deployment in the short run, by integrating different energy systems through a variety of local generation of electricity, heat and cooling, flexible demand, and all types of storage including electric vehicles.

The integrated local energy system is characterized by well-defined boundaries, being a specific unit within the energy system, which allows the integration of energy resources distributed on different scales. Generally, integrated energy systems can be implemented locally by combining, for example, photovoltaic systems, small wind turbines, and cogeneration plants, with a district heating network and distributed energy storage. The advantage of extending to more buildings, intended as users of the local energy system, lies in the availability of having multiple generation resources and consumption units available, thus increasing the flexibility of the system as a whole and the total extractable value. Integrated local energy systems promote local balancing and strategic exchange with external systems such as the electricity and gas networks. In this way, they have the possibility to interact with other systems, such as the electricity system, and this interaction ensures that the local system receives energy when local generation is not sufficient to meet the needs of the users.

The main features of local integrated energy systems are defined as follows [25]:

- **Modular:** these systems are characterized by a high level of modularity, offering the possibility to include new technologies at the level of individual buildings or system to cope with the increase in users' energy needs.
- **Flexible**: one of the fundamental criteria for an integrated local energy system is flexibility, to be achieved through optimized resource management, demand response services, and local balancing. This flexibility can also be used to provide energy services and functional system services.

- **Smart**: these systems are smart to allow the coordination of energy and information flows, ensuring the balance between energy supply and demand at the local level.
- **Efficient**: these systems are efficient and sustainable by nature from both energy and economic point of view because foster efficiency in the energy resources used by coordinating the different energy carriers and networks.
- **High degree of synergy**: these systems guarantee a high degree of synergy between the different energy carriers and associated technologies.
- **End user involvement**: these systems involve end users through tools such as the sharing and exchange of local energy and the use of economic incentives.

Local energy exchange is another important attribute of local integrated energy systems, giving the possibility to end users to exchange all types of energy locally in order to meet their needs.

2.3.2 Map of Enabling Technologies

The progress that has interested smart grids in recent years in the development of technologies for the management of decentralized energy systems and Information and Communication Technologies (ICT) provide the basis for enabling technological solutions for local integrated energy systems. In detail, the creation of such a type of complex system requires the adoption of distributed generation technologies and technological solutions for the smart management of energy flows and related information. Each enabling technological solutions can be characterized in terms of functionality, degree of centralization, and degree of technological maturity.

With reference to the functionality that the specific technology performs in a local integrated energy system, it is possible to identify three different categories [26]:

- **Production and use of energy**: This category includes technologies that allow producing on-site the energy needed by the users of the local system and to consume this energy in an efficient and smart way.
- **Management, control, and monitoring of energy flows**: This category includes technologies that allow to remotely control the production, distribution, storage, and energy consumption assets within the local energy system and to monitor the energy flows. These technologies are in turn divided into software and hardware systems. The software systems allow, in the forecasting phase, to elaborate the forecasts of energy consumption by end users and production by plants powered by variable RES. Moreover, in this phase, they allow the planning of the optimal operation strategies of the energy production, storage, and consumption assets; during the operation of the plants, they allow the optimization of the operation of the technologies on the basis of the actual operating conditions. The hardware systems for the management, control, and monitoring of energy flows contribute to the governance of the local energy system, imparting the relative operating modes based on the choices made by the management software and the on-site measurement of the main operating parameters of the local system.

Table 2.3 Enabling technologies for local integrated energy systems categorized according to the degree of centralization [25, 26].

Category	Technologies	
	Building level	**System level**
Local generation/conversion/storage	Internal combustion engine/reciprocating engine (CHP) Micro-gas turbine (CHP) Fuel cells Solar collectors Rooftop PV Micro-wind Heat pumps Absorption chillers Thermal energy storage systems Batteries	Internal combustion engine/reciprocating engine (CHP/CCHP) Gas turbine (CHP/CCHP) Fuel cells Solar thermal plant PV plant Wind farm Geothermal Hydro Distributed storage
Management systems	Home/building energy management system	Energy management system

- **Distribution of energy and information flows**: This category includes technologies that allow to distribute energy and information flows between the assets of production, conversion, distribution, storage, and consumption of energy present within the local energy system and the related management systems. In detail, this category includes the physical networks for the distribution of electricity, gas, heat, and cooling (through district heating and cooling network) and the communication infrastructure that enables the exchange of information between the various assets of the local energy system to ensure their proper functioning.

With the degree of centralization, reference is made to the field of application of the technology, or its applicability to a single energy user (building) and/or to system level serving multiple energy users. Table 2.3 shows the categorization of the technological solutions enabling an integrated local energy system according to their degree of centralization.

Regarding management systems, local integrated energy systems are characterized by an active management of information and energy flows across the units of distributed generation, conversion, storage, consumption, and flexible loads, while their correct operation is ensured by energy management technologies, such as home/building energy management systems, which in turn promote the active involvement of end users in the management of the system itself.

The degree of maturity indicates the expected improvement in technical–economic performance compared to the current performance. Figure 2.4 shows the graphical representation of the enabling technologies for local integrated energy systems as a function of the degree of centralization and technological maturity.

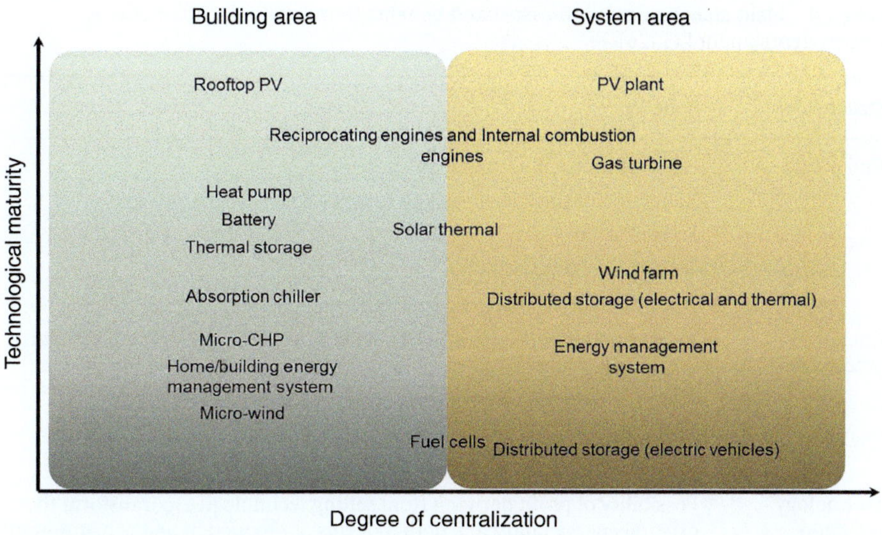

Figure 2.4 Graphical representation of the enabling technologies for local integrated energy systems as a function of degree of centralization and technological maturity. Source: Based on Refs. [26, 27].

It must be said that from technical point of view, one of the main issues to be addressed for achieving deployment of local integrated energy systems is characterized by the different maturity level of the technological solutions involved in such emerging solutions. Indeed, as shown in Figure 2.4, some of them are not fully technically mature and have not yet been widely adopted.

2.3.3 Key Stakeholders and Related Benefits from Local Integrated Energy Systems Deployment

The supply of energy to end users is based on a series of processes ranging from generation, transmission, and distribution of energy to conversion and storage, up to final consumption. These processes in turn involve a multiplicity of interdependent stakeholders in the realization of their objectives. Even within an integrated local energy system, there are different stakeholders with different interests and objectives, which can often be in conflict with each other. For example, end users want to have low-cost energy, aggregators seek to maximize the value of user flexibility in the various forms of market, while policy makers are interested in ensuring a sustainable energy supply chain with low environmental impact, while promoting the energy transition. Table 2.4 provides a detailed summary of the various stakeholders with benefits deriving from the deployment of integrated local energy systems.

It must be said that most of these stakeholders are going to evolve radically their roles and responsibilities in local integrated energy systems as compared to the current energy supply systems, as shown in Figure 2.5. Local integrated energy systems imply new roles for the end users called in first person to contribute to the management of the energy system through energy efficiency measures and energy

Table 2.4 Main stakeholders with associated benefits from local integrated energy systems deployment [25, 26].

Stakeholder	Benefits
End users	On-site, affordable, and clean energy use
	Reduction of the cost of energy carriers and local energy supply
	Self-sufficiency
	Security of energy supply
	High level of resiliency
Energy producers	Investment opportunities in local energy systems (profit maximization)
Energy suppliers	Increase in the share of renewables in the resource portfolio
ESCOs	Possibility of profit deriving from the implementation of energy efficiency measures and optimized management of local generation
Technology providers	Possibility of profit deriving from selling technologies to transform the existing energy landscape both in terms of production and consumption
Aggregators	New business models to generate profits by maximizing the value of flexibility in the various market forms
TSOs	Possibility of larger balance between supply and demand at the lowest cost for consumers
DSOs	Distribution of energy to end users through the use of a safe and reliable network
	Avoiding grid congestion
	Postponing investments in the grid
	Balancing of energy islands
Policy makers and regulators	Ensure affordable energy supply for all users
	Promotion of a sustainable supply system
	Transition to a low carbon energy system
	High energy security

communities, whereas energy suppliers are called to take care of the management of integrated local energy systems. A completely new role could be allocated to ESCOs, which could be called to the management and valorization of flexibility collected from active consumers (in cooperation with aggregators) beyond the typical financing, supply, and installation of energy-efficient measures and building refurbishment. Technology providers will contribute to the effective deployment of these systems through the provision of technologies for local power, heat and cooling generation, conversion, and storage and energy management systems. Also, the roles of system operators are going to be transformed with DSOs that have to adapt the system operation as per system needs by cooperating with TSOs and ensuring a joint operation with gas networks and TSOs that will use the flexibility coming from the interplay of multiple energy carriers for system balance. Finally, energy policy and regulators need to eliminate barriers by fostering investments, policies, and subsidies for the deployment of integrated energy systems.

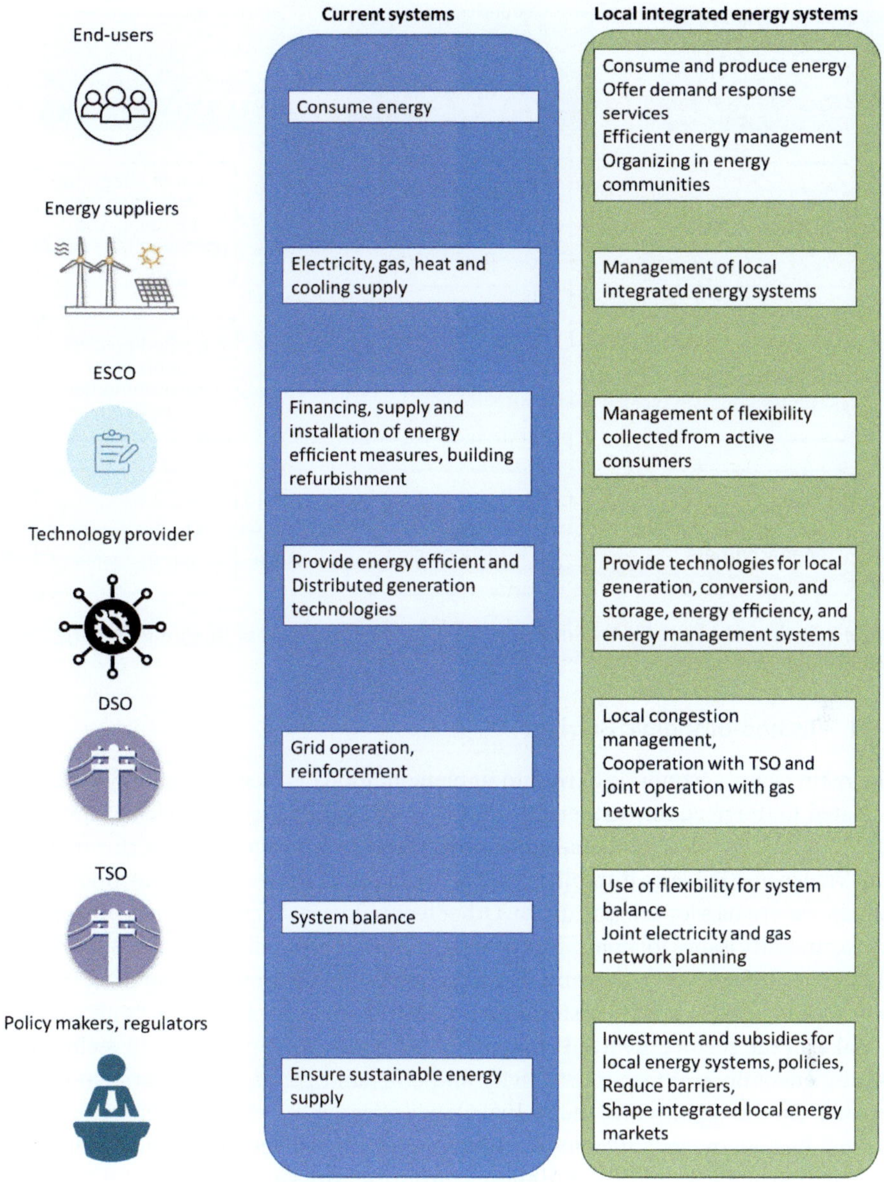

Figure 2.5 New roles and responsibilities of key stakeholders in local integrated energy systems vs. current energy systems [25, 26].

2.4 Main Barriers for Implementation

Integrated energy systems need to face with several techno-economic, socioeconomic, policy, and regulatory barriers to achieve an effective implementation and deployment, as reported in the scheme (see Figure 2.6).

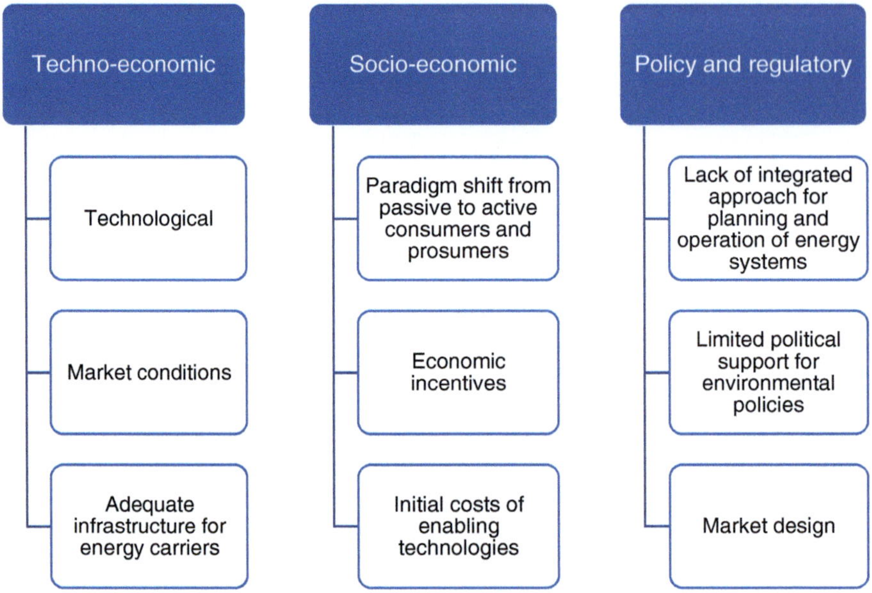

Figure 2.6 Main barriers for the implementation and deployment of integrated energy systems.

2.4.1 Techno-economic Barriers

The main techno-economic barrier to implementation of integrated energy systems is related to the need for further innovation in the various technologies of the sectors involved, which is essential to improve their techno-economic performances [28]. Values of efficiency, durability, and degradation of some enabling technologies still represent a barrier to break down. Other technological barriers are also related to interoperability issues for communication and control aspects, which result from the large variety of system management and control software options operating in the context of integrated energy systems. In order to overcome this issue, it is necessary to make all the sectors involved compatible in terms of components and technologies, by enabling a proper connectivity and efficient operation of all system components. On the other hand, the lack of standardization hinders the deployment of the enabling energy technologies, the interoperability between manufacturers, and the insertion of the technologies in energy infrastructures such as gas grids, while the lack of efficiency labels impedes the comparison and adoption of the most efficient technologies by consumers.

Another example of the techno-economic barrier to break down is represented by market conditions [28]. In fact, although some of the enabling technologies could be considered competitive in certain applications, they can result less competitive in relation to market conditions. For instance, although PtH/C technology can be considered very mature and competitive as compared to conventional gas-fired boilers, its convenience strongly depends on the spread between electricity and gas market prices. Another example is characterized by PtG applications for injection of gas in the distribution network, which is still not convenient considering the low market price of gas.

Last but not least, another barrier belonging to this category is characterized by the adequacy of infrastructures for the multiple energy carriers in integrated energy networks [28]. In fact, although integrated energy systems can lead to delay of investments in the existing infrastructures through adding flexibility potential, other developments related to the increased capacity or the need of adaptation of current infrastructure still need further investments. The related costs represent another crucial factor to consider, especially in the case of creation of new specific infrastructures. The infrastructure issue is also related to the exploitation and adaptation of current infrastructures in order to work with new energy carriers, which strongly depend on the existing technical regulations and standards. A typical example is represented by the injection of hydrogen into the gas distribution network. To break down this barrier, there is the compelling need to define clear regulations and technical rules for admixture and infrastructure connection for local producers.

2.4.2 Socioeconomic Barriers

Citizen engagement is considered to be the best way to obtain public acceptance for integrated energy systems. Indeed, as mentioned earlier in the chapter, involvement of end users in the management of the system itself represents a driving force for the deployment of this new energy paradigm. The shift from passive to active consumers and prosumers can only be achieved through developing an attitude of citizens toward new energy and ICT technologies, inducing changes in consumption patterns and investment in integrated energy systems. In this sense, it is expected that the best opportunities for realizing flexible and integrated energy networks will come from stronger engagement of citizens. It is therefore needed to achieve this goal by increasing consumers' awareness about the tangible and intangible benefits deriving from such types of implementation. Another non-negligible issue, which is strongly related to the public acceptance of integrated energy systems, is the lack of economic incentives to invest in local energy efficiency and renewable projects, especially considering that the initial costs of these solutions can be very high. In such a context, priority should be always given to investments that facilitate the uptake of low-carbon technologies and flexibility solutions.

2.4.3 Policy and Regulatory Barriers

An important policy barrier that needs to be eliminated to foster deployment of integrated energy systems is surely represented by the lack of integrated planning and operation of the different parts of the energy sector that present potentials for synergies, such as electricity and gas sectors [28]. Such an integrated approach also requires a proper enabling regulatory framework. Moreover, it has to consider all energy sectors by taking into account the whole energy supply chain from production to final consumption, all energy carriers and levels, starting from local projects toward the pan-European vision. Therefore, planning strategies should look at the energy system as a whole so that new developments could contribute to facilitate a least-cost transition to a carbon-neutral system. Moreover, through an integrated planning approach, all sectors would receive the same share of investments, leading to the same level of maturity of enabling technologies.

Another issue to address is the limited political support for market-based policies to price externalities through carbon taxes that could increase the awareness on benefits of integrated energy systems, thereby fostering investments in such systems. The pricing of carbon emissions should regard all energy carriers and installations, whereas currently, the heating/cooling and mobility sectors are almost completely neglected. A more coordinated approach, based on a similar CO_2 levy on all energy carriers and installations, would facilitate reaching the climate targets at a lower cost.

Other key barriers are related to market design to enabling flexibility provided by end users and low-carbon technologies. Limited access to the various market options for demand and DER is still a reality in several EU Member States with an almost complete absence of support schemes for fostering penetration of emerging technologies in the markets. Conversely, a revision of the regulatory framework according to the concept of "technology-neutrality" to guarantee the supply of network services from demand side and low-carbon technologies, which is also transparent to the type of energy carrier, would be needed.

2.5 Conclusions

This chapter discusses the concept of integrated energy systems as the engine for the energy transition by analyzing the main key elements characterizing this emerging energy paradigm. The integrated energy system can be seen as "a system of systems," namely, an integrated infrastructure for all energy carriers with the electrical system as a backbone, characterized by a high level of integration between all networks of energy carriers supported by energy storage and conversion processes. The key enabler for the implementation of integrated energy systems are also discussed. In detail, a crucial role is played by power conversion processes because an electrical energy carrier can be easily converted into several other energy carriers through PtG, PtH, and PtL technologies, which in turn allow to transport a large amount of energy among distant and interconnected hubs in the energy system. This type of cross-vector integration also provides the energy system with increased flexibility while increasing energy efficiency, considering that Power-to-X technologies can act as a sink for renewable electricity in excess. Another key enabler of integrated energy systems is represented by end user engagement and empowerment. In fact, citizens represent the central actors in the transition phase from a single-carrier energy system based on fossil fuels toward an integrated, low-carbon, accessible, cost-efficient, and market-based energy system. Digitalization represents one of the most important enablers for integrated energy systems by providing extensive services to all kinds of actors for planning, operation, and monitoring issues, while promoting information and connectivity among users. Moreover, with the advent of integrated energy systems, the evolution from a single energy market to an integrated energy market that integrates multiple energy carriers as electricity, gas, heat, and cooling becomes a compelling need. This market consisting of multiple coupled energy market improves efficiency in energy resources through the exploitation of the synergies among different carriers, resulting from the complementarity and the coordination of multiple energy resources.

Integrated energy systems at a local level present the largest potential for deployment in the short run, by integrating different energy systems through a variety of local generation of electricity, heat and cooling, flexible demand, and all types of storage including electric vehicles. The main features characterizing local integrated energy systems are discussed, along with a map of enabling technologies categorized according to their functionality, degree of centralization, and level of maturity. The key stakeholders with associated benefits from local integrated energy systems deployment are identified as end users, energy producers and suppliers, ESCOs, technology providers, aggregators, system operators, and policy makers. Their new roles and responsibilities in the context of local integrated energy systems are also analyzed as compared to the current system. Finally, the main barriers for implementation of integrated energy systems are discussed. In detail, these innovative systems need to face with several techno-economic, socioeconomic, policy, and regulatory barriers to achieve an effective implementation and deployment. The techno-economic barriers are mainly related to technological barriers such as the need for further innovation in the various technologies of the sectors involved or the need to have an adequate infrastructure for the multiple energy carriers in integrated energy networks. Instead, from social point of view, the citizen engagement is considered to be the best way to obtain public acceptance for integrated energy system, whereas an important policy barrier is surely represented by the lack of integrated planning and operation of the different parts of the energy sector that present potentials for synergies, such as electricity and gas sectors. Such an integrated approach also requires a proper enabling regulatory framework that takes into account the whole energy supply chain from production to final consumption, all energy carriers and levels, starting from local projects toward the pan-European vision. Other key barriers are related to market design to enabling flexibility provided by end users and low-carbon technologies. A revision of the regulatory framework according to the concept of "technology-neutrality" to guarantee the supply of network services from emerging technologies, which is also transparent to the type of energy carrier, would be needed.

Finally, it must be said that although recent legislations and regulations in Europe fully support the implementation of integrated energy concept, from technical point of view, further studies on the requirements for enabling full operation of integrated systems are needed. Indeed, the core feature of the integrated energy system paradigm is the intrinsic interdependency among energy carriers and sub-systems because of their correlated interactions. Relevant optimization, operation, and planning need to consider this aspect in order to exploit the extended flexibility for enhancing the efficiency of the energy resources used.

List of Abbreviations

AI	artificial intelligence
CAES	compressed-air energy storage
CCHP	combined cooling, heating, and power
CEC	citizen energy community
CHP	combined heat and power
DER	distributed energy resources
DLT	distributed ledgers technologies
DSO	distribution system operator
EED	energy efficiency directive
EMD	electricity market directive
EPBD	energy performance of buildings directive
EU	European Union
GHG	greenhouse gas emissions
ICT	information and communication technologies
IoT	Internet of Things
LAES	liquid air energy storage
PtC	power-to-cooling
PtG	power-to-gas
PtH	power-to-heat
PtL	power-to-liquid
PV	photovoltaics
REC	renewable energy community
RED	renewable energy directive
RES	renewable energy sources
SNG	synthetic natural gas
TSO	transmission system operator

References

1 World Commission on Environment and Development (WCED) (1987). *Our Common Future*. Oxford: Oxford University Press.

2 Kari, A. and Arto, S. (2006). Distributed energy generation and sustainable development. *Renew. Sustain. Energy Rev.* 10: 539–558.

3 Yan, B., Di Somma, M., Graditi, G., and Luh, P.B. (2020). Markovian-based stochastic operation optimization of multiple distributed energy systems with renewables in a local energy community. *Electr. Power Syst. Res.* 186: 106364.

4 Neyestani, N. (2021). Modeling of multienergy carriers dependencies in smart local networks with distributed energy resources. In: *Distributed Energy Resources in Local Integrated Energy Systems* (ed. G. Graditi and M. Di Somma), 63–87. Elsevier.

5 ETIP SNET "VISION 2050 (2018). Integrating Smart Networks for the Energy Transition: Serving Society and Protecting the Environment.

6 A European Green Deal Striving to be the first climate-neutral continent. https://ec.europa.eu/info/strategy/priorities-2019-2024/european-green-deal_en (accessed 20 October 2021).

7 Clean energy for all Europeans package. https://ec.europa.eu/energy/topics/energy-strategy/clean-energy-all-europeans_en (accessed 20 October 2021).

8 ETIP SNET (2020). Sector Coupling: Concepts, State-of-the-art and Perspectives, White Paper, January 2020.

9 Berger, M., Radu, D., Fonteneau, R. et al. (2020). The role of power-to-gas and carbon capture technologies in cross-sector decarbonisation strategies. *Electr. Power Syst. Res.* 180: 106039.

10 Deliverable D1.1 (2020). Analysis of directives, policies, measures and regulation relevant for the Active Building EPC concept and business models. Public report, AmBIENCe Project.

11 Directive 2019/944/EU on the internal electricity market. https://eur-lex.europa.eu/legal-content/EN/TXT/PDF/?uri=CELEX:32019L0944&from=EN (accessed 20 October 2021).

12 Directive 2018/2001/EU on the promotion of the use of energy from renewable sources. https://eur-lex.europa.eu/legal-content/EN/TXT/PDF/?uri=CELEX:32018L2001&from=EN (accessed 20 October 2021).

13 Energy Community Definitions (2019). Public report, 2019, COMPILE Project Integrating Community Power in Energy Islands, H2020-824424.

14 Directive 2018/2002 on energy efficiency. https://eur-lex.europa.eu/legal-content/EN/TXT/PDF/?uri=CELEX:32018L2002&from=EN (accessed 20 October 2021).

15 Directive 2018/844 on the energy performance of buildings. https://eur-lex.europa.eu/legal-content/EN/TXT/PDF/?uri=CELEX:32018L0844&from=IT (accessed 20 October 2021).

16 International Energy Agency (2017). Digitalization & Energy.

17 Sagiroglu, S. and Sinanc, D. (2013). Big data: a review. *2013 International Conference on Collaboration Technologies and Systems (CTS)*, pp. 42–47. IEEE.

18 Tu, C., He, X., Shuai, Z., and Jiang, F. (2017). Big data issues in smart grid–a review. *Renew. Sustain. Energy Rev.* 79: 1099–1107.

19 Das, S., Dey, A., Pal, A., and Roy, N. (2015). Applications of artificial intelligence in machine learning: review and prospect. *Int. J. Comput. Appl.* 115 (9): 31–41.

20 Khan, M., Silva, B.N., and Han, K. (2016). Internet of things based energy aware smart home control system. *IEEE Access* 4: 7556–7566.

21 Caruso, M., Gallo, P., Ippolito, M.G. et al. (2021). Challenges and directions for Blockchain technology applied to Demand Response and Vehicle-to-Grid scenarios. In: *Distributed Energy Resources in Local Integrated Energy Systems* (ed. G. Graditi and M. Di Somma), 207–230. Elsevier.

22 Dong, H., Zeng, M., Wang, L. et al. (2020). Integrated energy market mechanism and integrated energy service design. *Curr. Sustain./Renew. Energy Rep.* 7: 1–9.

23 Yu, B., Sun, H., Xiang, T. et al. (2016). Planning design method of integrated energy system. *Electr. Power Constr.* 37 (2): 78–84.

24 Jia, H., Wang, D., Xu, X. et al. (2015). Research on some key problems related to integrated energy systems. *Autom. Electr. Power Syst.* 39 (7): 198–207.

25 Koirala, B.P., Koliou, E., Friege, J. et al. (2016). Energetic communities for community energy: a review of key issues and trends shaping integrated community energy systems. *Renew. Sustain. Energy Rev.* 56: 722–727.

26 Chicco, G., Di Somma, M., and Graditi, G. (2021). Overview of distributed energy resources in the context of local integrated energy systems. In: *Distributed Energy Resources in Local Integrated Energy Systems* (ed. G. Graditi and M. Di Somma), 1–29. Elsevier.

27 Energy & Strategy Group (2014). Smart Grid Report "Le prospettive di sviluppo delle Energy Community in Italia", July 2014 (in Italian). http://www.energystrategy.it/eventi/le-prospettive-di-sviluppo-delle-energy-community-in-italia-3072014.html (accessed 20 October 2021).

28 European Union (2018). Sector coupling: how can it be enhanced in the EU to foster grid stability and decarbonise?

3

Power Conversion Technologies: The Advent of Power-to-Gas, Power-to-Liquid, and Power-to-Heat

Joshua A. Schaidle, R. Gary Grim, Ling Tao, Mark Ruth, Kevin Harrison, Nancy Dowe, Colin McMillan, Shanti Pless, and Douglas J. Arent

National Renewable Energy Laboratory, 15013 Denver West Parkway, Golden, CO 80401, USA

3.1 Introduction

3.1.1 Motivation for Power-to-X

As part of the United Nations Framework Convention on Climate Change and the Paris Agreement, over 190 countries have committed to limiting global warming to well below 2 °C compared to preindustrial levels. Achieving this goal will require significant cross-sector reductions of greenhouse gas emissions around the globe. To make this a reality, many countries are setting aggressive targets for decarbonizing the power sector, as they see this as a critical first step in this transition. For example, the United States has committed to a carbon-free electricity sector by 2035. This strategy leverages the rapidly decreasing levelized costs of renewable electricity generation and the significant surge in installed capacity (i.e. greater than 2500 GW globally as of 2019) over the past 10 years [1]. Recent analyses evaluating high renewable power systems have elucidated the importance of long-duration (e.g. seasonal) storage for reliable power.

One approach to addressing this energy storage need while also providing a pathway to translate the rapid decarbonization of the electricity grid into other energy-intensive sectors such as transportation and manufacturing is Power-to-X (Figure 3.1). Power-to-X technologies provide energy storage and sector coupling by converting electricity (and other requisite feedstocks such as H_2O, CO_2, and N_2) into chemical energy and heat. The direct conversion of electrical energy into chemical energy stored in chemical bonds enables both production of high-volume fuels (e.g. hydrogen, methane, and ethanol) and other chemicals (e.g. ethylene and ammonia), as well as long-term energy storage. Although batteries may store up to 200 Wh/kg, which will work for transient storage on the order of seconds to weeks (with economics indicating optimal sizing configurations of less than 12 hours), species such as CH_4 (15 000 Wh/kg) that can be produced through Power-to-X may store energy on a seasonal basis, are easily transportable, and are compatible with the existing infrastructure [2]. The direct conversion of electrical energy into

Technologies for Integrated Energy Systems and Networks, First Edition.
Edited by Giorgio Graditi and Marialaura Di Somma.

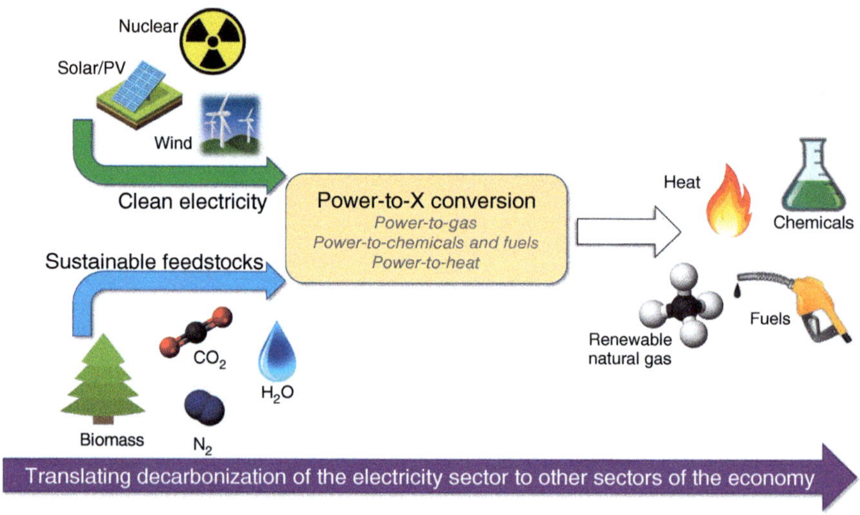

Figure 3.1 Power-to-X brings together clean electricity and abundant, sustainable feedstocks to translate decarbonization of the electricity sector to other sectors of our global economy.

heat (e.g. through electric boilers and heat pumps) enables the decarbonization of industrial and building heating.

Overall, the value proposition of Power-to-X is composed of two main components: (i) deep decarbonization through sector coupling and (ii) power system flexibility. As an example of the potential for deep decarbonization, it has been estimated that the chemical industry could reduce its annual greenhouse gas emissions by up to 3.5 Gt CO_2-eq in 2030 by leveraging Power-to-X technologies [3]. Regarding power system flexibility, Power-to-X enables the decoupling of generation and demand, which is critically important as generation diversifies due to decentralization (e.g. rooftop solar) and demand expands as a result of overall electrification. It should also be noted that Power-to-X supports the broader effort to digitize the power grid as it transitions to bidirectional energy flow, especially in the context of real-time response to changes in generation and demand. Ultimately, Power-to-X is an outcome of four key global trends: electrification, decarbonization, decentralization, and digitalization [4].

Pertaining to the development and deployment of Power-to-X platforms, these technologies exist today, and many of them have been demonstrated at scale [5]. These technologies are defined in Section 3.1.2 and discussed in detail in Section 3.2 of this chapter, but it is important to note that renewable hydrogen from water electrolysis or other renewable sources (i.e. biomass), although not explicitly discussed within this chapter, is a contributor and enabler of this Power-to-X platform. Many of the Power-to-X technologies for the production of fuels and chemicals consume hydrogen, and the overall economics of the process are typically strongly correlated with the cost of hydrogen. In terms of scale, although Power-to-X addresses curtailing renewable electricity by utilizing "excess" and underutilized solar and wind resources, utilizing "excess" renewable electricity alone will be insufficient to reach the scale of greenhouse gas emission reductions required to achieve the goals of the Paris Agreement. For example, achieving the aforementioned 3.5 Gt CO_2-eq reduction in greenhouse gas emissions for the chemical industry in 2030 will require over 18.1 PWh of renewable electricity [3]. Thus, installation of renewable electricity infrastructure alongside Power-to-X facilities needs to be considered concomitantly, with overall capital utilization as a key driver.

3.1.2 Defining Power-to-X Categories

Power-to-X is an all-encompassing term for strategies that convert electricity (and other requisite feedstocks such as CO_2, N_2, and H_2O) into energy carriers, chemicals, fuels, heat, and food. For the purposes of this chapter, we define three specific Power-to-X categories: Power-to-Gas, Power-to-Chemicals-and-Fuels, and Power-to-Heat.

Power-to-Gas technologies convert electricity into gaseous energy carriers such as H_2 and CH_4. Given that Chapter 4 is dedicated to the role of H_2 in a low-carbon energy future, Power-to-Gas in this chapter focuses solely on CH_4 generation, with specific emphasis on biological and thermochemical processes because of the advanced maturity of these technologies.

The Power-to-Chemicals-and-Fuels category focuses on the direct and indirect (i.e. through energy carriers such as H_2) conversion of electricity and small molecules such as CO_2, N_2, and H_2O into fuels and chemicals. These technologies include both direct production of end products and production of intermediates (e.g. formic acid) that can be further upgraded downstream to end products. In addition to carbonaceous products, this category also covers generation of NH_3.

Power-to-Heat technologies utilize electrical energy as a direct power source by converting electricity into heat for utilization in industry and buildings. This section covers both centralized (i.e. supply level, in which heat is generated and distributed to several processes or buildings) and decentralized (i.e. process level, in which heat is generated for a specific industrial process or just an individual residence) Power-to-Heat technologies.

There are three final considerations relevant to these definitions. First, these Power-to-X categories (and technologies) are intertwined. For example, CH_4 generated through a Power-to-Gas technology could be utilized for industrial or residential heat. Further, an industrial Power-to-Heat technology such as an electric boiler could supply heat to a thermochemical process for converting CO_2 and H_2 into chemicals. Thus, these categories should be taken as a means to organize the chapter rather than strict, independent technology groups. Second, there are tangential Power-to-X concepts emerging beyond the categories defined here, including artificial food and data storage. These broader Power-to-X concepts are not covered in this chapter. Lastly, the term Power-to-X can also encompass Power-to-Mobility through charging of battery electric vehicles; this approach is not discussed explicitly within this chapter.

3.1.3 Goal of this Chapter

A key tenet of the Power-to-X value proposition is flexibility. It introduces upstream flexibility for the grid by decoupling generation and demand in response to increasing renewable penetration. It also introduces downstream flexibility in terms of the broad array of technologies available to meet the needs of energy-intensive sectors such as transportation and manufacturing. Taken together, this flexibility indicates that there are options to meet the specific needs of low-carbon or carbon-neutral integrated energy systems. Thus, the goal of this chapter is to help the reader understand these existing and emerging Power-to-X technological options and the associated challenges and opportunities in development and deployment. From our perspective, it is not a matter of if, but when, this Power-to-X strategy will become ubiquitous within our global economy. We need to understand these challenges and opportunities so that we chart the best path forward and retain this flexibility upon implementation.

3.2 Power-to-X Technologies

3.2.1 Power-to-Gas

Many Power-to-Gas technologies start with H_2 production because of its inherent nature to donate electrons to a reaction. Growing penetration of low-carbon electricity production is providing opportunities to capture affordable – and in some instances otherwise curtailed – energy and produce H_2 via water electrolysis. Whenever possible, consuming the pure H_2 product directly in a fuel cell for transportation or power generation, for example, without further transformation would minimize losses. However, where direct H_2 use is limited because of lack of storage, pipeline, or local industrial use, performing an additional conversion may open up new opportunities to move the electricity through H_2 to other molecules. Wherever low-carbon electricity is used as a feedstock to a water electrolysis process, the resulting product would be considered renewable (green) H_2.

One possible solution to move renewable H_2 into industrial sectors is to use carbon as a carrier. In the United States, natural gas infrastructure is ubiquitous and offers terawatt-hour energy storage as pressurized methane stored in millions of miles of pipeline and geological formations. Further decarbonization is possible if waste streams containing biogenic CO_2 are the source of carbon. For example, biogas sources such as landfills, wastewater treatment facilities, and animal farming produce and release a mixture of CH_4, CO_2, and other trace gases.

Power-to-Gas technologies that move beyond H_2 production to CH_4 are possible through biological and thermochemical processes, which can recycle the biogenic carbon from our waste streams. These two methanation technologies are different from gas separation systems that remove CO_2 from waste streams to provide renewable natural gas (RNG) to the existing markets. The RNG produced by separations or methanation is a drop-in direct replacement for anthropogenic natural gas pulled from underground. Unlike the separation process that releases CO_2, the methanation process converts CO_2 to CH_4 using H_2 as a feedstock.

3.2.1.1 Natural Gas Market Demand

3.2.1.1.1 Fossil

According to the International Energy Agency (IEA), global natural gas demand in 2020 was approximately 3910 billion cubic meters[1] or equivalently 148.2 quadrillion British thermal units (BTU; 1 quadrillion BTU = 1 quad = 1×10^{15} BTU). Lawrence Livermore National Laboratory produces energy flow diagrams that quantify the complex interrelationships between resources. Their 2020 diagram for the United States shows that 12.4 quads of electricity flowed through the electrical network and 31.5 quads of natural gas moved through the natural gas network [6].

Although data are provided for an outlier year (2020), these data hold true throughout the years and provide a sobering reminder with respect to decarbonization of our energy system. The fact is that annually, the natural gas network carries an average of roughly 2.5 times the amount of energy as the electrical network. In 2019, for example, the US final energy consumption included 12.7 quads of energy in the form of electricity, and the primary supply included 32.1 quads of natural gas. This reaffirms that decarbonizing fossil natural gas resources in the United States poses a challenge 2.5 times larger than that of electricity.

3.2.1.1.2 Renewable

Anaerobic digestion is the biological process of breaking down organic waste to a mixture of constituents, primarily CO_2 and CH_4. This gas mixture is known as biogas and is itself a source of energy for localized heat and power. Biogas can be upgraded further using (i) separation processes such as membrane separation, pressure swing adsorption, amine scrubbing, and water wash to separate CO_2 from CH_4 or (ii) methanation processes that take CO_2 and CH_4 and produce a product gas – RNG or biomethane – that meets the natural gas specifications for injection into a pipeline. Depending on the source of waste, some RNG is considered a carbon-negative fuel. RNG producers can participate in programs such as California's Low Carbon Fuel Standard and the federal Renewable Fuel Standard, which provide incentives to produce low- to negative-carbon transportation fuel. RNG is viewed as having the potential to deliver low-carbon energy to a wide array of users through the existing infrastructure.

The United States currently has 2200 operating biogas systems, representing less than 20% of the total potential [7]. Of those operating biogas facilities, slightly more than half are located at wastewater treatment facilities and a third are at landfills. Unfortunately, two-thirds of the landfill biogas sites and many more of these facilities simply flare these gases. Argonne National Laboratory's RNG database lists 157 facilities that are producing RNG using gas separation technologies to split the gas mixture – keeping the CH_4 and releasing the CO_2 – for a total of nearly 46 trillion

---

1 As far as possible, these data represent standard cubic meters (measured at 15 °C and 1013 mbar), as they are derived directly from the measures of energy content using an average conversion factor and have been standardized using a gross calorific value of 40 MJ/m^3.

BTU energy production [8]. According to IEA's March 2020 biogas and biomethane report [9], worldwide production of these gases in 2018 was around 35 million tonnes of oil equivalent (Mtoe), or 1.39 billion BTU, only a fraction of the estimated overall potential of 540 Mtoe (21.4 billion BTU) for biogas and 730 Mtoe (29 billion BTU) for RNG. Almost two-thirds of biogas production in 2018 was used to generate electricity and heat (with an approximately equal split between electricity-only facilities and co-generation facilities). Although biomethane represents about 0.1% of natural gas demand today, there is significant potential for growth: IEA estimates a 40% increase in production worldwide by 2040. The growth in RNG production will be heavily influenced by decarbonization policies toward natural gas.

3.2.1.2 Technology Identification and Overview

Methanation processes that produce CH_4 from CO_2 and H_2 can be performed through two conversion approaches: chemical and biological. Both processes are based on the Sabatier reaction (Eq. (3.1)), in which one mole of CO_2 and four moles of H_2 react to form one mole of CH_4, two moles of H_2O, and heat.

$$CO_2 + 4H_2 \rightarrow CH_4 + 2H_2O \ \Delta H = -165\ \text{kJ/mol} \tag{3.1}$$

The chemical processes generally use nickel-based catalysts and operate at temperatures between 200 and 500 °C and elevated pressures. With the biological process, hydrogenotrophic methanogenic microorganisms serve as the catalyst that consumes CO_2 and H_2 to make CH_4, metabolic H_2O, and heat. Several review papers [10–17] summarize the catalysts, reaction kinetics, limitations, techno-economic analysis (TEA), and status of current industrial projects, most of which are in Europe.

Each approach to CO_2 methanation has its positives and negatives. Chemical methanation from CO_2 is a more mature technology with faster reaction rates and better volumetric mass transfer coefficients, and the systems are mostly compatible with industrial plants in operation today. The main technical challenge faced by the chemical approach is the need to reject large amounts of heat, which dictates catalyst, reactor, and heat integration design. Catalyst deactivation is always a concern with these processes, and feed gas cleanup is necessary if using a biogas source.

By comparison, an advantage to the biological approach is the biocatalyst's high tolerance to biogas impurities, particularly H_2S, which can be used as a sulfur source and reducing agent. The biocatalyst can be highly selective, producing only CH_4, water, and cell mass. Reaction temperatures are in the 30–60 °C range, so thermal management is easier. The main technical challenge of the biological process is the H_2 gas mass transfer in the liquid. The typical gas bioreactor is a tall, agitated bubble column designed to maximize H_2 solubility and retention for the biocatalyst to consume. There is an energy loss to the agitator. Next-generation bioreactor designs modeled after catalytic processes that immobilize the biocatalyst could improve the H_2 mass transfer issues by eliminating water, which is the barrier to improved gas mass transfer.

Both systems produce large quantities of water as the side product, which can poison the chemical catalyst and necessitates continual addition of nutrients to the biocatalyst to maintain viability. Very little information is available on a side-by-side comparison of the two processes. One study compared the cost of producing methane from either chemical or biological methanation using sunlight, renewable H_2, and direct air capture CO_2 and found no difference between the two processes [18]. Adoption of either method will likely be made based on specific sites and the availability of low-cost electricity, type of CO_2 source, available infrastructure, and regional markets for the product gas.

3.2.1.3 Unique Integration Challenges and Opportunities

Biogas sources, such as anaerobic digesters, typically operate at low pressures (<1 bar gauge pressure), whereas both biological and thermochemical reactor vessels can operate at elevated pressures (9–30 bar gauge). This difference in pressure will require at least one compression stage to boost the biogas feedstock for use in these pressurized methanation vessels. Distribution-level natural gas systems can operate up to about 15 bar gauge and transmission networks typically range between 70 and 100 bar gauge. Besides the improved gas mass transfer at elevated pressures, it might also make sense to operate the methanation reactor vessel at a level above the pressure of the type of natural gas network.

It is implied that the overall methanation process will utilize low-carbon electricity to power a water electrolyzer to produce the hydrogen feedstock. Only then will the resulting RNG have a lower carbon intensity than the average electricity grid in most regions of the world. Of the two steps, electrolysis and methanation, it is the hydrogen production that consumes the vast majority (>90%) of the energy to produce RNG from carbon dioxide. In order for the resulting RNG to compete economically, the low-carbon electricity must also be low cost. In fact, it is the cost of electricity that has the greatest influence on the resulting cost of hydrogen and therefore the final production cost of the RNG [2].

Because of continued improvements in distributed renewable electricity generation (primarily wind and solar photovoltaics), on-site production may provide the lowest cost alternative to moving electricity over the grid. On-site production system sizing becomes an important consideration to avoid the (relatively) low capacity factors of solar and wind – generally assumed to be in the range of 10–20% for solar and 30–50% for electricity produced by modern wind turbines located in good wind resource locations. However, producing electricity on-site avoids additional costs of moving those electrons across transmission and distribution networks.

To date, most hydrogen-producing, low-temperature water electrolyzers have been designed and installed to provide cooling for thermal power plants and for ultra-high-purity gas sources to reduce fuel inventory by replacing cylinders in laboratories. The primary contaminant of concern for water electrolyzer systems is water vapor. Most systems employ a pressure or temperature swing adsorption system to reduce water vapor content down to single-digit parts per million by volume (ppm_v). In the United States, for example, fuel cell electric vehicles require the hydrogen be dried to less than 5 ppm_v. Methanation systems do not require this level of hydrogen purity, and the drying system could therefore be scaled back or eliminated entirely, thus saving the capital cost and improving the system efficiency.

3.2.2 Power-to-Chemicals-and-Fuels

Most routes for chemical and fuel synthesis involve fossil carbon sources combined with heat and pressure to drive reactions and create products. However, with the recent rise of low-cost renewable energy sources and the growing desire to utilize more sustainable feedstocks, these conventional pathways are being challenged and the way we produce and consume energy and products is beginning to change. As noted in recent reports from the Intergovernmental Panel on Climate Change, limiting the global temperature rise to less than 2 °C relative to preindustrial levels requires aggressive carbon-neutral and carbon-negative actions. Power-to-Chemicals-and-Fuels via CO_2 offers one attractive option to store the otherwise environmentally harmful gas in the form of chemicals and fuels, sourced either directly from the air or other more concentrated point sources.

In addition to CO_2, other sustainable feedstocks such as lignocellulosic biomass and N_2 are also being considered for Power-to-Chemicals-and-Fuels. As a natural carbon sink, utilization of biomass can help achieve deep decarbonization while also providing all required organic elements (e.g. H, C, O, and N) to reach fuels and chemicals. Early studies on Power-to-Chemicals via biomass have shown promise through upgrading of biomass-derived species such as glycerol, furfural, and 5-(hydroxymethyl)-furfural to reach compounds traditionally difficult to access via conventional methods [19]. Studies also show that in some cases, raw biomass may also be treated electrochemically as a means of depolymerization with the possibility of targeting otherwise difficult to access bond linkages [20]. Electrochemical N_2 fixation – another emerging technology – is also a promising application of sustainable Power-to-Chemicals. As noted, conventional ammonia synthesis via the high-pressure Haber–Bosch process is extremely energy intensive and a strong contributor to global emissions. Utilization of renewable energy, either for H_2 production as a co-feedstock in Haber–Bosch or for direct electrochemical ammonia synthesis, offers a potential pathway to decarbonize the agricultural sector. Within this section, we explore the technologies to enable the utilization of these sustainable feedstocks and identify the challenges and opportunities toward future commercialization.

3.2.2.1 Market and Demand

The conversion of Power-to-Chemicals-and-Fuels can occur either directly or indirectly depending on the flow of electricity within the process. In direct conversion strategies, incoming electricity is consumed within the conversion step to produce chemicals or fuels. Alternatively, electricity may be used indirectly to first produce energy-carrying intermediate species such as H_2, CO, or formate, which are then further upgraded as part of a larger integrated process. Each strategy accesses a different suite of products and offers unique benefits and trade-offs. Direct processes involve minimal processing steps, are often highly modular, and are considered promising for distributed-scale and dynamic load-following applications. However, because of the complexities of the electricity-mediated reaction mechanisms and challenges with selectivity building longer carbon chains in a single step, direct processes are currently limited to mostly products containing one or two carbon atoms, with a few exceptions in the higher carbon range. Alternatively, with indirect Power-to-Chemical processes being integrated with downstream centralized processing, easy-to-access intermediate species can be generated and used within conventional chemical synthesis processes, offering the ability to reach a wider range of products while leveraging established commercial pathways and infrastructure. For example, two of the most widely used building blocks among Power-to-Chemical routes are H_2 and CO, which are easily synthesized from H_2O oxidation and CO_2 reduction electrochemical reactions, respectively. From these two feedstocks alone, essentially, all the top fuels and chemicals are accessible through conventional methods such as Fischer–Tropsch, Haber–Bosch, and other industrial routes. However, the higher capital intensity and comparatively larger footprint of indirect processes make these processes potentially less suitable for small-scale application and more challenging to operate intermittently. Overall, between the two Power-to-Chemical strategies, there are clear pathways to reach essentially all of the chemicals and products we consume today, providing options across all scales and technical maturity levels.

3.2.2.2 Technology Identification and Overview

A variety of both emerging and established pathways exist for Power-to-Chemicals-and-Fuels production, spanning multiple disciplines, technologies, and scales. Among the many routes for converting power to chemicals, six of the most commonly cited pathways include low-temperature electrolysis (LTE), high-temperature electrolysis (HTE), microbial electrosynthesis (MES), non-thermal plasma (NTP) reduction, biochemical conversion (BC), and thermochemical conversion (TC). The characteristics and unique advantages and disadvantages of each pathway are discussed in more detail in the following sections.

3.2.2.2.1 Low-Temperature Electrolysis (LTE)

Low-temperature electrolyzers have long been studied for the production of H_2 via H_2O oxidation reactions. However, interest in alternative electrolysis chemistries has been recently reignited for use with sustainable feedstocks such as CO_2 and N_2. Electrolyzers come in many configurations, from the long-commercialized aqueous alkaline electrolyzers used in H_2 synthesis to emerging polymer–electrolyte membrane electrode assembly (MEA) stacks. Yet despite the numerous shapes and configurations, almost all electrolyzers contain the same four basic elements: an anode, cathode, electrolyte, and ion-conducting membrane separating the anode and cathode chambers, as shown in Figure 3.2. When an external voltage is applied across the two anodic and cathodic electrocatalysts, electrochemical oxidation and reduction reactions occur. Unlike conventional chemical synthesis, which often requires elevated temperatures and/or pressures to drive the reactions, voltage in these systems is typically the sole driving force that allows for operation, often at near-ambient

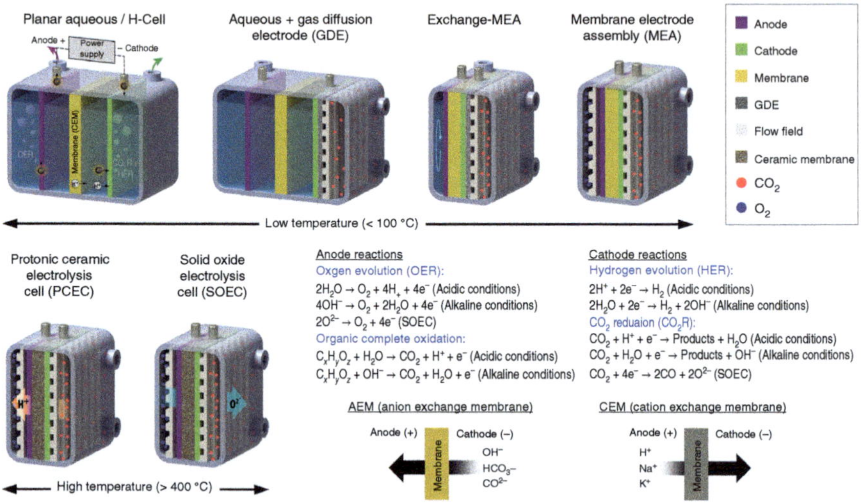

Figure 3.2 The many configurations of electrolyzers and their associated reactions. Source: Reproduced from Grim et al. [21] with permission from the Royal Society of Chemistry.

temperature and pressure. Electrochemical systems are typically evaluated using at least three key metrics: current density, overpotential, and faradaic efficiency. Current density measures the flux of current per unit surface area of the electrode, typically measured in mA/cm^2, and is a measure of the process's productivity. Overpotential, measured in volts, is a measurement of how much the applied cell potential exceeds the thermodynamic minimum voltage required to drive the reaction. Faradaic efficiency reflects the amount of supplied current consumed for the formation of a certain species, similar to the selectivity in conventional catalytic chemical reactors.

LTE demonstrates several promising features such as high theoretical conversion efficiencies, high theoretical energy efficiencies, and the potential for a "tunable" distribution of over 20 products by simply changing applied voltage, providing the possibility for a market-responsive process concept [22]. For some products such as H_2 and CO, electrolysis systems have already been scaled to the commercial or near-commercial stage, representing immediately deployable options for Power-to-Chemicals-and-Fuels. However, for most other products, current LTE systems have generally shown low selectivity to C_{2+} products and a limited product distribution to carbon numbers ≤ 4. Further, most MEA systems currently suffer from rapid deactivation and have not yet shown suitable stabilities required for commercial application.

3.2.2.2.2 High-Temperature Electrolysis (HTE)

High-temperature electrolyzers contain the same four major components as in LTE shown in Figure 3.2, but with a few key distinctions. The polymer-based membranes used in LTE are unstable at the elevated operating temperatures (600–850 °C) used during HTE, which instead require the use of a ceramic material. In most

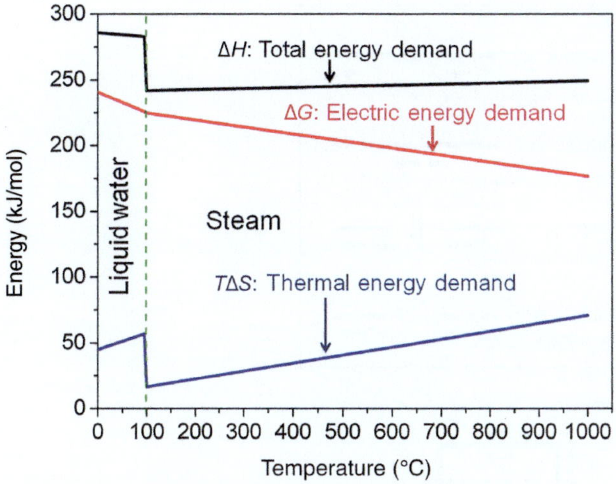

Figure 3.3 Thermodynamics of high-temperature H_2O oxidation. Source: Reproduced from Bi et al. [23] with permission from the Royal Society of Chemistry.

configurations, the ceramic membrane conducts oxygen ions to the anode, which recombine to form O_2. One of the benefits of operating at high temperatures can be seen in Figure 3.3, displaying the thermodynamics of a hypothetical H_2O oxidation reaction. The enthalpy of the reaction, or total energy demand, comprises the Gibbs free energy term ΔG, satisfied by the input electricity, as well as the entropic term $T\Delta S$. As the reaction temperature increases, there is a corresponding increase in the entropic contribution, thereby reducing the required electric demand. In practice, this relationship allows the electrolyzer to operate at lower voltages than the LTE counterparts, often resulting in higher overall energy efficiencies, especially if co-located with a waste heat source such as a nuclear power plant. A trade-off, however, of high-temperature systems is that due to the more extreme temperatures and propensity for coking, the observed products are currently generally limited to H_2 or simple single-carbon products such as CO and CH_4.

3.2.2.2.3 Microbial Electrosynthesis (MES)

Although the LTE and HTE pathways utilize metal electrocatalysts to facilitate the oxidation and reduction reactions, in the MES pathways, microorganisms drive the underlying chemistry and assume the roles of electron donors and/or acceptors. Again comprising the same four elements as the other electrolysis systems, the major differentiator of MES systems are the microorganisms that typically cover either the cathode and/or the anode surface(s), depending on the type of desired reaction, as shown in Figure 3.4. To date, the most productive species for MES has been *Clostridium ljungdahlii*, which facilities CO_2 reduction through the Wood–Ljungdahl pathway and provided access to products such as acetate, ethanol, methane, butanol, and propanol. MES pathways have several advantages among the Power-to-Chemical pathways in that they are capable of forming C—C bonds with a high degree of selectivity while also offering a potential for nearly 99% conversion efficiency. With

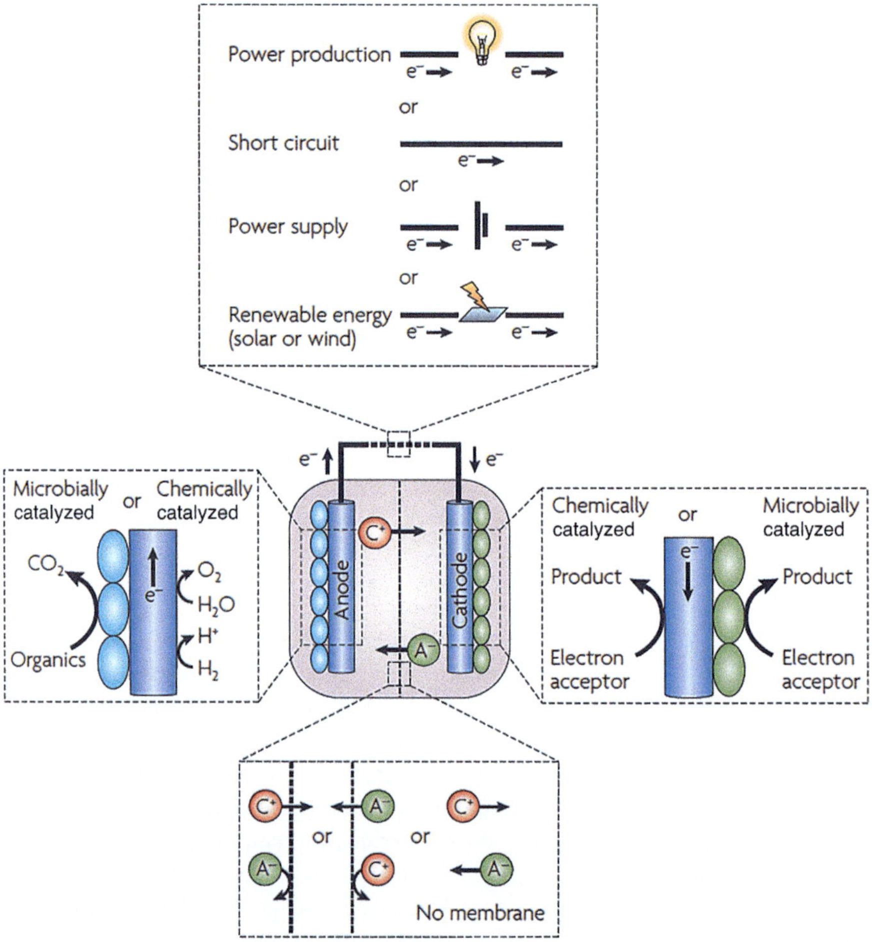

Figure 3.4 Conceptual process diagram for a microbial electrosynthesis reaction. Source: Reproduced from Rabaey and Rozendal [24] by permission from Springer Nature: Nature Reviews Microbiology (Microbial electrosynthesis—revisiting the electrical route for microbial production, Rabaey and Rozendal), Copyright 2010.

the possibility of future genetic engineering, there is also the potential to access otherwise difficult chemistries through specialized and tailorable pathways. However, MES is currently one of the most technically immature Power-to-Chemical processes, limited exclusively to the bench scale, facing challenges surrounding low productivity and often poorly understood reaction mechanisms.

3.2.2.2.4 Non-thermal Plasma (NTP)

Plasma is often described as the fourth state of matter, comprising ionized gas. Although most often observed at extremely high temperatures (e.g. stars and lightning), NTPs have recently been considered as a possible pathway for Power-to-Chemicals-and-Fuels. Using an electronic potential to selectivity activate electrons to energies in the range of 1–10 eV, the NTP system can be operated

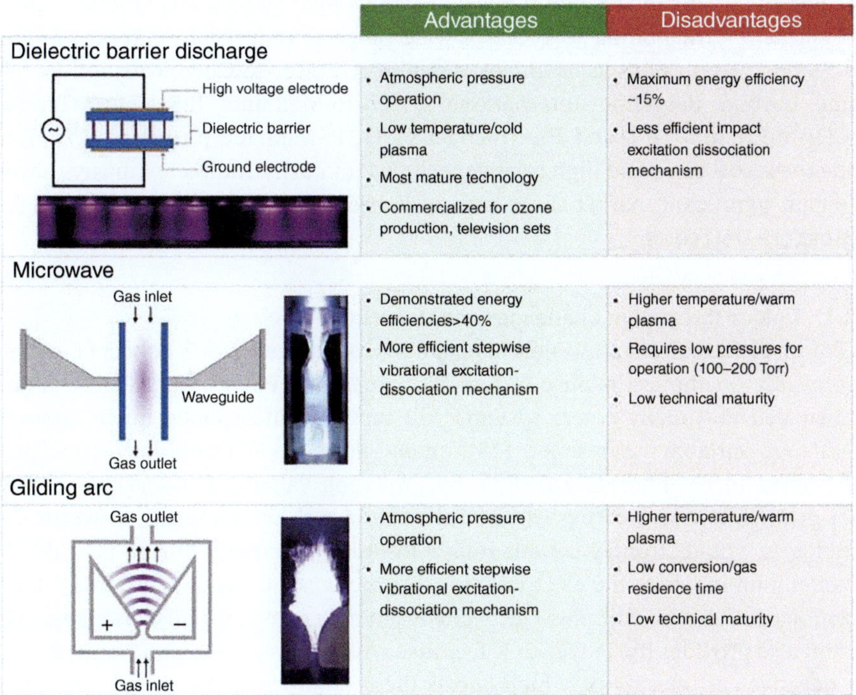

Figure 3.5 Different forms of NTP and their associated advantages and disadvantages. Source: Grim et al. [21]/with permission of Royal Society of Chemistry.

in a state of non-local thermodynamic equilibrium, whereby the bulk of the process is at near-ambient conditions yet still achieves a high driving force for generating free radial chemistry. NTP has found some application outside of Power-to-Chemicals-and-Fuels in wastewater treatment but has not yet been applied commercially for chemical and fuel production. The NTP systems typically utilize one of the three reactor configurations – microwave, dielectric barrier, or gliding arc – as shown in Figure 3.5. Across the three configurations, the NTP processes show several promising attributes in that they typically do not require any rare earth materials, can be operated at near-ambient conditions, are feedstock flexible, and can be ramped up or down quickly for load-following applications [21]. However, NTP technologies are among the least technically mature power-to-chemical processes and critical questions remain around their scalability, high power demand, and costly supporting equipment.

3.2.2.2.5 Biochemical and Thermochemical Conversion

Although the direct Power-to-Chemical pathways are generally considered as having a low technology readiness level (TRL), the bio- and thermochemical indirect conversion pathways are comparatively more mature and in many cases have been demonstrated commercially. By leveraging the use of an intermediate energy carrier such as electrolytic H_2 vs. directly using electricity, indirect pathways take advantage of decades of research and investments on catalytic hydrogenation synthetic routes and biological fermentation to access a wide range of industrially relevant products. Some current at-scale examples of indirect Power-to-Chemicals-and-Fuels include Carbon Recycling International's CO_2-to-Methanol [25], LanzaTech's CO/CO_2-to-Ethanol [26], and Electrochaea's CO_2-to-Methane processes [27]. Yet despite these successes and high relative technical maturity, indirect pathways have yet to gain significant market share, primarily because of challenges around high feedstock (i.e. H_2) costs.

3.2.2.3 Unique Integration Challenges and Opportunities

The Power-to-Chemicals-and-Fuels concept, like other Power-to-X concepts, relies economically on the availability of low-cost renewable energy. Specifically, as it is anticipated that many power systems will evolve to incorporate much greater amounts of variable renewables [28], many Power-to-X technology concepts are designed to be inherently flexible by providing value-added services to the power grid while also creating value-added chemicals. Economically, Power-to-X technologies would strongly benefit if able to utilize electricity during periods of overgeneration in which the electricity would otherwise be lost or "curtailed." The utilization of otherwise curtailed electricity is not only a boon for utilities increased sales but also provides the Power-to-X facilities with access to cheap and sometimes even negative-cost electricity, which boosts the economics of Power-to-Chemicals processes. However, because of the inherent intermittent nature of renewable energy technologies, technologies must be designed for transient operation and should be able to quickly ramp up or ramp down to take advantage of the available electricity. Further, processes should strive to minimize the capital intensity of the process to maximize the economic potential and lessen the burden from idling equipment. Because electricity is only one component of the required feedstock, this balance of energy supply with other chemical feedstocks such as CO_2 or N_2 must also be balanced to match the incoming supplied rates.

3.2.2.4 Implications on Power Generation

Because CO_2 is in a fully oxidized state and contains no inherent energy content, CO_2 conversion – and more broadly Power-to-Chemicals-and-Fuels – performed at scale would currently rank among the most energy-demanding industrial processes, rivaling other known energy-intensive processes such as Haber–Bosch for ammonia production or steel production. Yet, despite the surge in Power-to-Chemicals research over the past decade, there remain significant questions and challenges surrounding the total at-scale power demands, upfront capital cost, and land use required to perform Power-to-Chemicals-and-Fuels, even at smaller scales.

The power demand of electrochemical Power-to-Chemicals conversion (e.g. LTE, HTE, and MES) can be calculated starting from Eq. (3.2) to first determine the total charge requirement (Q) in coulomb, where z is the number of electrons transferred per molecule of product, n is the moles of product formed per unit time, F is Faraday's constant of 96 485 C/mol, and FE is the Faradaic efficiency [29]. In the form of coulomb per second (amperes), multiplying by the operating voltage of each conversion technology yields the total required power in watts.

$$Q = \frac{z \cdot n \cdot F}{\mathrm{FE}} \tag{3.2}$$

As an illustration, Table 3.1 estimates the total energy demand to meet the current annual global production across a variety of products, including CO from HTE; methanol from catalytic TC; CH_4, acetic acid, and ethanol from biochemical fermentation; and C_2H_4 from LTE. These product–pathway combinations represent some of the most widely studied options starting from CO_2 and span a range of electricity demand (i.e. 2–12 electrons/molecule). Note that the pathways that do not utilize electricity directly in the conversion step (e.g. BC/TC) assume that two electrons are consumed per mole of electrolytic H_2 used in the reaction.

For the listed products, the total energy demand to replace conventional processes with Power-to-Chemical alternatives ranges from approximately 410 000 to 68 000 000 GWh/yr. If it is assumed that the energy demand is met by solar photovoltaic power, which as of 2020 has demonstrated an average global capacity factor of 18% [30], an installed generation capacity of ~70 to 43 000 GW would be required depending on the product–pathway combination. For perspective, at the end of 2020, the total estimated installed capacity across all renewables worldwide was approximately 2800 GW [31]. At a current installed cost of \$995/MW and land use rate of 0.032 km^2/MW for solar photovoltaic energy [30, 32], Table 3.1 shows that the estimated upfront capital requirements and land areas required to meet the global production range from \$70 billion to \$43 trillion US\$ and 2300 to 1 400 000 km^2, respectively. It should also be noted that these upfront capital costs only include the cost to build the capacity for renewable energy generation and do not include other capital costs for the Power-to-Chemical process itself, which can also be substantial.

Although these figures represent high-end estimates, these data nevertheless highlight a major challenge in power generation for the widespread adoption of Power-to-X. Namely, scaling Power-to-X processes will be extremely energy-intensive, and as a result, it is likely that curtailed electricity alone would be insufficient to significantly impact the global markets of many products. Instead, a dedicated energy generation infrastructure is needed to achieve decarbonization in a meaningful way.

Although it is anticipated that these metrics will improve over time, future projections in Table 3.1 – which account for falling electricity costs and more efficient conversion processes – still point to a challenging environment for Power-to-Chemicals-and-Fuels with respect to high energy intensity, capital costs, and land usages. Consequently, Power-to-Chemicals processes may find the greatest near-term opportunities as an integrated component of power/energy systems while economically utilizing the existing renewable energy during periods of overgeneration, thereby removing the need for dedicated infrastructure and additional capital investments. Efforts toward electrolyzer capital cost reduction and efficiency gains in Power-to-Chemicals conversion will also be critical so that using surplus electricity with low capitalization becomes more viable. However,

Table 3.1 Calculated energy demand, installed cost, and land usage for promising Power-to-chemical processes under current and future assumptions.

	Product (technology)	No. of electrons required[a)]	Global production (MMT/yr)	Total energy demand (GWh/yr)	Required solar capacity (GW)[b)]	Installed cost solar ($M)[c) d)]	Solar land use (km^2)[e)]
Current[f)]	CO (HTE)	2	150	4.07E+05	258	260 000	8 300
	Methanol (TC)	6	92	1.00E+06	637	630 000	20 000
	CH_4 (BC)	8	2 366	6.81E+07	43 172	43 000 000	1 400 000
	Acetic acid (BC)	8	14	1.11E+05	71	70 000	2 300
	Ethanol (BC)	12	90	1.36E+06	865	860 000	28 000
	C_2H_4 (LTE)	12	156	9.63E+06	6 107	6 000 000	200 000
Future[g)]	CO (HTE)	2	150	3.75E+05	238	140 000	7 600
	Methanol (TC)	6	92	7.59E+05	481	280 000	15 000
	CH_4 (BC)	8	2 366	5.14E+07	32 602	20 000 000	1 000 000
	Acetic acid (BC)	8	14	8.41E+04	53	31 000	1 700
	Ethanol (BC)	12	90	1.03E+06	653	380 000	21 000
	C_2H_4 (LTE)	12	156	3.43E+06	2 173	1 300 000	70 000

a) TC and BC pathways utilize H_2, which is assumed to consume two electrons per mole.
b) Assumes average solar capacity factor of 18%.
c) $995/kW (current).
d) $587/kW (future).
e) $0.032 km^2/MW.
f) Technical assumptions from Ref. [2]. Electrolytic H_2 electrical demand assumed as 58 kWh/kg.
g) Technical assumptions from Ref. [2]. Electrolytic H_2 electrical demand assumed as 43.8 kWh/kg.
Source: Based on Huang et al. [2], Grim et al. [21].

considering the magnitude of the power demand required to produce even a minor fraction of many of the widely consumed chemicals and fuels used today, the volume of products accessible from curtailed electricity scenarios is not likely to meaningfully contribute to a decarbonized economy. Collectively, the implications of greatly expanding the power system should be considered, as well as if there are better, more effective alternatives for the production of green chemicals.

3.2.3 Power-to-Heat

3.2.3.1 Market and Demand

Seventy million American homes and businesses combust natural gas, oil, or propane on-site to heat their buildings and to create domestic hot water. Domestic water heating is the second largest energy use in US residential buildings, and when space heating is added, this represents over 50% of the residential sector's total energy use. In total, this residential sector's on-site combustion represents both a present and emerging target for electrification and decarbonization efforts [33]. Industry electrification advocates [34, 35] suggest that reaching decarbonization goals of 75% reduction in greenhouse gas emissions will require eliminating most of the CO_2 produced by furnaces and water heaters across the residential single family and multifamily industry.

The current approaches to electrifying this large on-site residential hot water and heating combustion load are with a series of heat pump technologies. The growing market for heat pumps for providing hot water and space heating includes a range of air source heat pumps and water source heat pumps with coefficient of performance (COP) efficiencies from 2 to 5. Air source heat pumps for domestic hot water heating are available as package units (see Figure 3.6 as an example [36]) that can extract heat from their surroundings or split between the location of the hot water tank and the outdoors. Air source heat pumps for space heating can be easily added to existing air-conditioning systems with a refrigerant reversing value and utilize the same air handler and coils to provide both space heating and space cooling. Low-cost, mini-split zone level air source heat pumps are the most common used industry wide because of their flexibility in outdoor condenser location, minimal ductwork costs, and mass production of shared components. Air source heat pumps have proven to be a viable space heating solution for decades. The popularity of air source heat pumps has typically been limited to temperate climates because of poor low-temperature performance, the need for air-conditioning in summer months, the availability of natural gas in colder climates, and the prevalence of more cost-effective alternatives. However, with growing interest in renewable energy generation and electrification, variable-capacity, inverter-driven air source heat pumps are a viable option for cold climates in North America.

Similarly, water source heat pumps have often been used for space heating in cold climates, where a waste heat or secondary heat source can provide heat to a shared source water loop to each water source heat pump. Extended temperature range water source heat pumps are often used for when the water loop is heated and cooled through ground-coupled heat exchangers. Historically, high heat exchanger

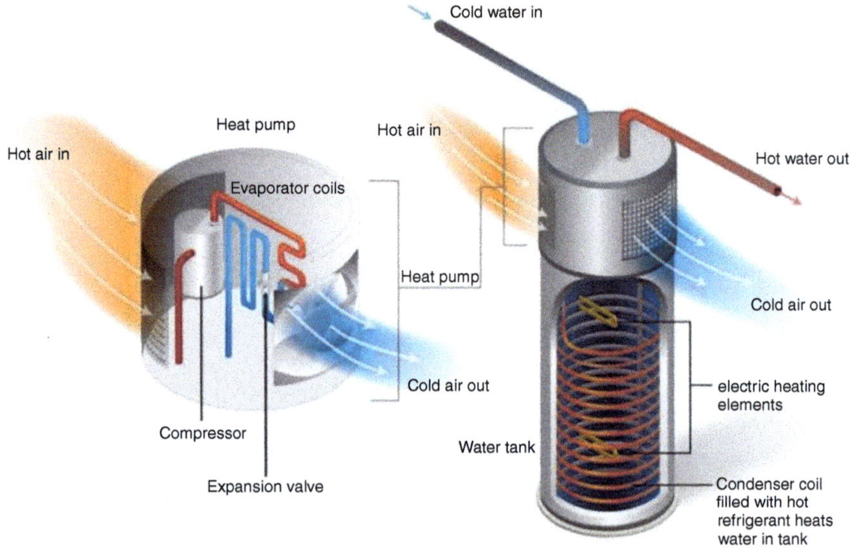

Figure 3.6 Package air source heat pump water heater schematic [36].

and well drilling costs have limited ground source heat pump markets from becoming mainstream.

As the residential market for heat pump water heaters and space heaters grows and becomes more cost-effective and mature, future emerging technologies will be expected to address some of the key industry-wide constraints around the performance in all climate zones, concerns with the global warming potential used within heat pump refrigerant systems, cost-effectiveness of solutions, and growing concerns of a suitable and flexible electrical utility system capable of powering all this new electrified heating and hot water load. Emerging heat pump solutions currently being developed are seeking to overcome these mature market constraints, with active R&D activities [32] and emerging cold-climate heat pump solutions that can provide hot water and space heating, even during the coldest months of the year. Further R&D efforts in heat pumps are focusing on low global warming potential (GWP) refrigerants with a GWP of 750 or less. R-32, at 675 GWP, has approximately 30% lower GWP than typical refrigerants used today [37] and transcritical CO_2 with a GWP of 1 is a possible long-term solution to virtually eliminate the global warming concerns with leaking heat pump refrigerant systems. To address the high first cost issues with ground source heat pump systems, new heat exchangers and sources of waste heat are being considered. Wastewater heat recovery heat pumps are coming onto the market now, as well as low-grade waste heat recovery systems from data centers and other industry cooling loads.

As shown in Figure 3.7, space heating, water heating, and cooking account for most of the direct fossil fuel use in residential buildings.

The industrial sector is a significant energy user and source of combustion and non-combustion greenhouse gas emissions. During the period between the Great

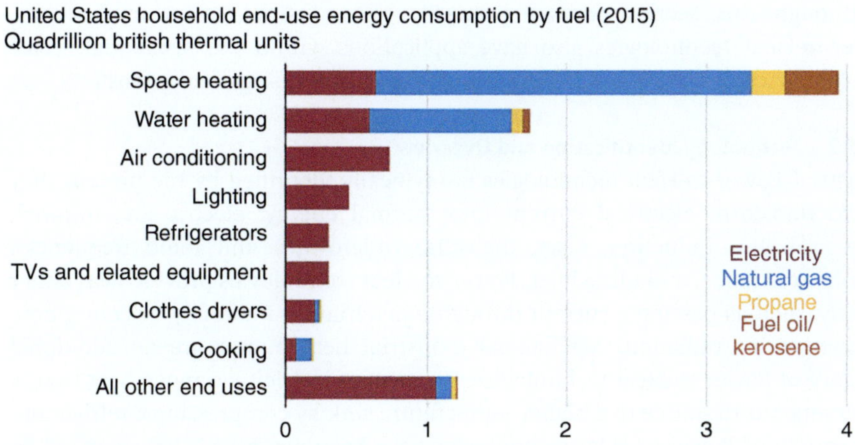

Figure 3.7 Final residential building energy use. Source: US Energy Information Administration [38].

Recession and the COVID-19 pandemic (2010–2019), energy use by the US industrial sector grew by 7%, reaching about 34.58 EJ [39]. Concurrently, the use of electricity by industry shrunk by 5% (including electricity system losses) and natural gas use expanded by nearly 29% [39]. Although the increase in total industrial energy use mirrored the global trend, the decline in electricity use, in absolute terms and as a share of total energy, is a notable departure for the United States. The share of electricity in the US industrial sector is now the lowest since 1979. The concurrent success of US natural gas production is likely behind these trends. Over the same period, gross withdrawals of natural gas increased by 52% to 1.157 trillion cubic meters [40]. This has increased not only the demand in end-use sectors but also the use of natural gas in extraction, processing, and transportation operations.

Of course, these aggregate statistics hide important detail. Between 2010 and 2018, the share of electricity for the US manufacturing sector – responsible for the largest portion of industrial energy – remained largely unchanged at about 17% [41, 42]. However, the share of natural gas did increase at the expense of coal, fuel oils, and by-product fuels, and is now responsible for 6.7 EJ, or 43% of manufacturing fuel energy use [42].

Most fuel energy in manufacturing is used to produce heat – in boilers, combined heat and power, or process heating equipment (collectively, industrial process heat). Disaggregating industrial process heat by process temperature reveals that approximately two-thirds of demand is for temperatures at or below 300 °C [43]. As opposed to very high temperatures (i.e. greater than 1000 °C), which are limited to a small number of energy-intensive industries such as iron, steel, and cement, these low-temperature demands are widespread across many industries. Process temperature is just one characteristic of industrial process heat demand and has implications for the applicability and integration of Power-to-Heat technologies, which are explored in the following sections.

Although this section focuses primarily on industrial sector applications, Power-to-Heat technologies also have applications within the residential sector [44–46]. Space heating is a major contributor to greenhouse gas emissions [47].

3.2.3.2 Technology Identification and Overview

Industrial Power-to-Heat technologies are typically identified by the process they use to transform electrical current into thermal energy: electric arc, infrared, electron beam, induction, laser, dielectric (microwave and radio frequency), plasma, resistance, and ultraviolet. Power-to-Heat technologies provide heat either directly, such as passing a current through a material, or indirectly by convection, conduction, or radiation. Mechanical industrial heat pumps are an additional category of Power-to-Heat technologies. These use electricity to move heat from a low-temperature source to a higher-temperature sink by compressing a refrigerant. The variety of Power-to-Heat technologies have been extensively summarized by others [48–50].

3.2.3.3 Unique Integration Challenges and Opportunities

The emergence of electrification as a decarbonization pathway for industry (assuming a decarbonized electric grid) has renewed interest in analysis that matches Power-to-Heat technologies with industrial and residential sectors. Before delving into this work, however, it is helpful to first introduce a general framework of how alternative sources are integrated with industrial processes. At the most basic level, integration occurs at the supply level (e.g. heat generated by a central boiler and distributed to individual processes via steam) or process level (e.g. heat delivered to a process via an external or an internal heat exchanger or direct steam injection) [51].

Supply-level substitution with Power-to-Heat technologies can be accomplished with electric boilers (electric resistance and electrode boilers) and heat pumps. As mentioned previously, steam and hot water demands from boilers constitute much of the industrial process heat demand in the United States. As a result, there may be significant technical opportunity across many different industries, constrained in part by the capacity limits of commercial electric boilers and heat pumps [43]. Supply-level integration of Power-to-Heat technologies may also face smaller barriers to adoption than process-level integration. This distinction is captured in the concept of "distance to core production processes," which describes a higher perception of risk for the adoption of new technologies that are critical to production processes than technologies at the periphery of production, such as lighting [52].

There are many more relevant Power-to-Heat technologies for process-level integration, as shown in Table 3.2. It is notable that Power-to-Heat technologies may replace the need for supply-level heating, including the heat distribution infrastructure and its associated inefficiencies, by providing heat directly at the process level (e.g. substituting microwave dryers for steam dryers). However, because integration occurs closer to or within core production processes themselves, these Power-to-Heat technologies may face higher barriers to adoption. The high costs of Power-to-Heat technologies, in terms of upfront costs and the

Table 3.2 Industries, processes, and relevant Power-to-Heat technologies for process-level integration [48, 49, 53].

Industry	Processes	Power-to-Heat technology
Food and beverages	Pasteurization, drying, sterilizing, boiling, washing, and baking	Heat pumps, electric resistance air dryers, infrared processing, microwave, and radio frequency heating
Pulp and paper	Drying	Infrared dryers, radio frequency dryers, and heat pumps
Container glass	Melting and refining	Electric resistance furnaces
Chemicals	Separations, reactions, and drying, heating	Heat pumps, electric resistance heaters, microwave heaters, and radio frequency heaters
Plastic and rubber	Melting and drying	Radio frequency processing, resistance heating, and induction heating
Steel	Steelmaking and remelting	Electric arc furnaces
Aluminum	Melting and casting	Induction melting and vacuum melting
Fabricated metal products	Preheating and heat treatment	Induction furnace, electric resistance, furnace, and infrared heating

relative cost of electricity to combustion fuels, are also significant barriers to adoption [53].

Depending on the industry in question, adoption barriers for Power-to-Heat technologies may be much more significant and systematic, as described by carbon lock-in [54]. Additionally, these technologies may be at odds with other decarbonization pathways. Through interviews with chemical industry representatives, Janipour et al. [55] describe the possible incompatibility of electrification and carbon capture and sequestration (CCS) technologies. If CCS is the preferred decarbonization option, its widespread adoption could create a new path dependency that further inhibits electrification.

Madeddu et al. [56] capture the varying integration complexity and technological maturity for Power-to-Heat opportunities in the European Union and the United Kingdom (EU27/UK). The authors identify three stages of electrification with established Power-to-Heat technologies, culminating in electrifying 99% of the fuel energy demand[2] of 11 industries that constitute 92% of EU27/UK industrial CO_2 emissions. The first stage, responsible for the largest increase in electricity use, relies on heat pumps and chillers, electric boilers, infrared heaters, dielectric heaters for space heating, hot water and steam, drying, and curing across all 11 industries. The second stage focuses on induction, resistance, and electric arc furnaces in metals industries. The third stage involves low-maturity technologies of electric steam crackers and

2 Chemical feedstock energy remains, which comprise 40% of EU27/UK industrial energy demand.

reformers for chemicals, thermal plasma generators for cement, and complete substitution of primary steel with scrap-based electric arc furnace.

3.2.3.4 Implications on Power Generation

Power-to-Heat technologies would not require the same magnitude of new electricity generation as Power-to-Gas and Power-to-Chemicals-and-Fuels, although the change would still be significant. For instance, full direct electrification of EU27/UK industry would require two to three times the current industrial electricity use [56]. When Power-to-Gas and Power-to-Chemicals-and-Fuels have been included in the analysis of industrial electrification, they account for 70% of the additional electricity demand [57].

3.3 Overarching Challenges, Opportunities, and Considerations

3.3.1 Feedstock and Energy Sourcing

3.3.1.1 Feedstocks (CO_2, N_2, H_2O, and Biomass)

To date, most Power-to-X research has been conducted at the lab level and under idealized conditions taking advantage of pure feedstocks (e.g. ultra-high-purity CO_2, N_2, and H_2O). However, in practice, many would be feedstock streams such as industrial gaseous point sources (e.g. power plants and petrochemical facilities), lignocellulosic biomass, and even the air contain impurities, particulates, and other diluents that can negatively impact and potentially foul the underlying conversion process, which are often designed to accept feedstocks with a narrow and specific concentration profile.

Upstream purification of feedstocks through processes such as air separation units (i.e. to produce pure N_2), amine-based CO_2 scrubbing, and sulfur removal adsorbents are common techniques available to address these issues. However, the addition of feedstock purification and subsequent transportation alone can be costly and energy intensive, which in some cases can dramatically influence the economic viability of Power-to-X. As an example, the only operating carbon capture coal plant in the United States, the Petranova CCS Plant, recently announced it had closed. It was revealed that in order to effectively capture and purify (scrub) a portion of CO_2 from the coal-fired power plant waste stream, an entirely separate natural gas power plant had to be constructed on-site simply to power the scrubber, and its emissions were not offset by the original CCS technology [58].

Moving forward, the design of more resilient conversion systems capable of handling variable feedstock streams while showing tolerance to contaminants, such as the robust LanzaTech biocatalyst [26], would enable broader applicability and improve the underlying economics of Power-to-X and lessen the burden on upstream purification processes. One emerging option currently under investigation is reactive capture and conversion technologies that combine the feedstock capture and conversion steps, which in some cases could remove the need for supporting feedstock purification processes.

3.3.1.2 Operational Flexibility for Grid Integration and Revenue

The cost of electricity is the largest fraction of the total levelized cost for many Power-to-X technologies. Therefore, the focus of much of the research in this area is on increasing efficiency (i.e. increasing the amount of product generated per unit of electrical energy used). However, unlike traditional chemical processing technologies, which do not have many options to reduce variable operating costs, Power-to-X technologies have an alternative. They can act as dispatchable electricity loads operating when electricity prices are low and minimizing the electricity load when prices are high. Doing so can reduce the total electricity costs.

Traditional chemical operations are operated at high-capacity factors because variable costs are not time sensitive and capital costs are high. High-capacity factors enable the capital costs to be spread over the maximum product volume. Over the last couple of decades, electricity markets have evolved and are often being settled at hourly or subhourly frequencies. Thus, electricity loads can be exposed to real-time prices. In addition, those real-time prices are becoming more volatile because of increases in variable renewable generation. If operated as dispatchable loads, Power-to-X technologies can take advantage of an alternative operational paradigm – reducing the process's electricity cost by operating when electricity prices are low and sitting in standby when electricity prices are high. However, that operational strategy reduces the annual capacity factor, resulting in a higher cost for capital recovery per unit of product. Lower capital costs could mitigate that increase but that may not be possible for the full plant. An alternative is storing intermediates to decouple the electrochemical portion of the process from the remainder. That option can reduce the fraction of the plant that operates with lower capacity factors and overcomes some of the economic challenges, but it adds a cost for storage.

Another opportunity to reduce electricity cost is direct connection to wind, photovoltaic, or hybrid variable renewable system generation. Direct connection would require load to be controlled to match electricity availability (unlike the aforementioned case where load responds to cost). This option could be beneficial in situations where electricity transmission is limited, expensive, or where product offtake can be located at the site of electricity generation (e.g. an ethanol plant that has a dedicated rail connection and surrounding land with good wind resources might be able to transport a product more economically than building new electricity transmission). This opportunity could also reduce costs by reducing required power electronics. Instead of needing power electronics for the renewable electricity generator to provide clean AC power to the grid, fewer power electronics may be needed to directly connect DC power to the electrochemical unit.

Both options share the requirement that the electrochemical reactor must be able to ramp up and down quickly. Because both the number and speed of ramping events can negatively impact the durability, impacts on performance and increased maintenance costs should be considered in TEA. Impacts on durability can be mitigated by increasing catalyst loading. However, increased catalyst loading will also increase the capital cost, thus reducing the economic benefits of being a dispatchable load. Use of electrochemical storage (e.g. batteries) and additional technology development (e.g. limiting ramp rates) could minimize those negative impacts.

3.3.2 Key Considerations from Life Cycle Analysis and Techno-economic Analysis

3.3.2.1 Life Cycle Analysis

As CO_2 utilization technologies for converting thermodynamically stable CO_2 are typically energy-intensive, it is necessary to evaluate the related life cycle greenhouse gas emissions to see whether there are actual emission reduction benefits. Life cycle assessment is used to analyze environmental potentials for different technologies and for alternative process design strategies. One of the primary metrics from life cycle assessment is to quantify the reduction of the life cycle greenhouse gas (GHG) emissions (carbon intensities) when compared with fossil baseline. The learnings from quantitively evaluation of GHG emission reduction can provide guidance to both technology development and deployment of a Power-to-X technology from environmental perspectives.

3.3.2.2 Techno-Economic Analysis

TEA is a methodology utilized by the government, industry, and academia to quantify the impact of research discoveries and engineering advances on the economic viability of an integrated process. When effectively coupled with research and development (R&D) efforts, TEA is an important complementary tool for understanding and identifying the key process attributes that can be improved to minimize overall production costs. The key outcomes of strategic TEA are not just cost numbers but quantitative assessments and evaluations of a technology's cost reduction strategies, environmental impacts, and scale-up challenges for the integrated advanced biorefinery to fuels and chemicals. Moreover, strategic TEA in this modern age has been expanded to complex sensitivity analysis, scenario analysis, comparative analysis, advanced bioenergy concepts with multiple coproducts, biorefinery integration with existing supply chains and infrastructures, and biorefinery optimization.

3.3.2.2.1 How to Apply TEA to Low-, Median-, and High-TRL Technologies?

When performing TEA across a range of TRLs, different analysis strategies apply. For instance, early-stage TEA is used to validate an initial idea, whereas the development of more rigorous process designs and economic evaluations are performed with high-TRL technologies. This analysis is necessary to support R&D directions, identify production and cost drivers, and inform key decisions on cost-efficient strategies for bioderived fuels and chemicals. Cost numbers are often used to map or track individual program accomplishments to evaluate the potential of technical success and allow inclusion of multiple products and processes in the aggregate net value estimate.

3.3.2.2.2 Key Cost Drivers and Key Cost Uncertainties, Economic Viability of Near-Term Products, and Opportunities for Transformational R&D

Near-term viability of Power-to-X technologies are typically high-value specialty chemicals with small market share. This implies significant opportunities for impactful R&D in this space. As technology progresses, many other products, such as chemicals, fuel precursors, and fuels, can be cost-competitive with those made from fossil-based conversion strategies in the long term. Maximum theoretical mass yields as oxygen atoms are removed from the final product, theoretical carbon yield, number of required electrons, or moles of H_2 needed are critical to advance Power-to-X technology and can vary widely depending on the product from the technology.

3.3.2.2.3 Scalability and Tech-to-Market Potentials of Long-Term Options

To move these low-TRL direct pathways toward commercial readiness, transformational R&D is needed to advance the core conversion technology. Previous studies have projected that a significant amount of renewable electricity will be needed to decouple chemical production from fossil resources via Power-to-X technologies [3]. The second challenge for scalability is how to systematically integrate massive, curtailed electricity, dedicated energy infrastructure with regionally deployed Power-to-X strategies to maximize product output per capital invested. More analysis is recommended to move forward to understand total energy demand to drive Power-to-X at industrially relevant scales.

3.3.3 Business Model and Business Innovation

As noted, flexibility is a key tenet of Power-to-X technologies. Flexibility also applies to approaches to business model innovations. Approaching Power-to-X business solutions must recognize the critical trade-off of capital and operating expenses, operational flexibility, and revenue hedging. That is, power-to-X facilities with traditional high capital cost structures will strive for high-capacity utilization and low operating expenses to minimize the cost per unit output. However, these facilities may face increasingly challenging business circumstances in retail power markets that move to subhourly prices and demand charges and incent demand reduction flexibility.

Alternatively, Power-to-X producers may follow recent trends of other large customers, including multiple power purchase structures, or even dedicated facilities as previously mentioned. Multiple financial hedging instruments have also evolved over the past decade that offer risk mitigation options, including weather derivatives, cost for differences contracts, and differential virtual power purchase agreements.

On the other hand, business models built off new technology innovations for low-capital and highly efficient operations may prove to be financially more attractive, particularly if they can hedge expenses through operational flexibility, be hybridized with cost-effective storage, and potentially have multiple product streams. For example, consider a low capital expense facility that includes hydrogen generation (and storage) and CO_2 utilization to produce renewable CH_4 (RNG). Such a facility has the flexibility to provide demand response (and possibly other) remunerated services to the power system; sell hydrogen, potentially cell power (if paired with an H_2 compatible or H_2/CH_4 turbine, or fuel cell), and RNG; as well as be paid for CO_2 utilization. Such financial and technical flexibility has both positive and negative attributes, and specific locational prices and infrastructure will determine the financial viability.

3.4 Concluding Remarks

Power-to-X technologies are recognized as a critical component of a low-carbon energy economy. As a means to accelerate a transition to a low-carbon energy economy, particularly offering solutions for "hard to decarbonize" sectors such as chemicals and fuels, as well as industrial energy demand, Power-to-X technologies are in development for the production of methane (i.e. Power-to-Gas), chemicals and fuels (e.g. ammonia, carbon monoxide, ethylene, methanol, ethanol, and formic acid), and heat (e.g. for industrial and building applications). Many of these technical solutions exist at a fundamental chemical level, but many options are being pursued to increase the efficiency and lower the capital costs of production. We emphasize that Power-to-X technologies do exist today at various stages of development and commercialization and that a transition to lower carbon intensity for the transportation and manufacturing sectors is possible but requires continued technical innovation, and realizing this at scale will require immense growth of the electrical grid, appropriate systems design and operations, and creative technical and business solutions.

Disclaimer

This work was authored in part by the National Renewable Energy Laboratory, operated by Alliance for Sustainable Energy, LLC, for the US Department of Energy (DOE) under Contract No. DE-AC36-08GO28308. The views expressed in the article do not necessarily represent the views of the DOE or the US Government. The US Government retains and the publisher, by accepting the article for publication, acknowledges that the US Government retains a non-exclusive, paid-up, irrevocable, worldwide license to publish or reproduce the published form of this work, or allow others to do so, for US Government purposes.

On behalf of all authors, the corresponding author states that there is no conflict of interest.

List of Abbreviations

BC	biochemical conversion
BTU	British thermal units
CCS	carbon capture and sequestration
EU27/UK	European Union and the United Kingdom
HTE	high-temperature electrolysis
IEA	International Energy Agency

LTE	low-temperature electrolysis
MEA	membrane electrode assembly
MES	microbial electrosynthesis
Mtoe	million tonnes of oil equivalent
NTP	non-thermal plasma
ppm_v	parts per million by volume
R&D	research and development
RNG	renewable natural gas
TC	thermochemical conversion
TEA	techno-economic analysis
TRL	technology readiness level

References

1 International Renewable Energy Agency (2019). *Renewable Capacity Statistics 2019.*

2 Huang, Z., Grim, R.G., Tao, L., and Schaidle, J. (2021). The economic outlook for converting CO_2 and electrons to molecules. *Energy Environ. Sci.* https://doi.org/10.1039/d0ee03525d.

3 Kätelhön, A., Meys, R., Deutz, S. et al. (2019). Climate change mitigation potential of carbon capture and utilization in the chemical industry. *Proc. Natl. Acad. Sci. U.S.A.* 116 (23): 11187–11194.

4 International Renewable Energy Agency (2019). Innovation landscape for a renewable-powered future: solutions to integrate variable renewables.

5 Chehade, Z., Mansilla, C., Lucchese, P. et al. (2019). Review and analysis of demonstration projects on power-to-X pathways in the world. *Int. J. Hydrogen Energy* 44 (51): 27637–27655.

6 Lawrence Livermore National Laboratory (2021). Estimated U.S. Energy Consumption in 2020. https://flowcharts.llnl.gov/content/assets/images/charts/Energy/Energy_2020_United-States.png (accessed 18 October 2021).

7 American Biogas Council (2021). Biogas Market Snapshot. https://americanbiogascouncil.org/biogas-market-snapshot/ (accessed 18 October 2021).

8 Argonne National Laboratory (2021). Renewable Natural Gas Database. https://www.anl.gov/es/reference/renewable-natural-gas-database (accessed 18 October 2021).

9 International Energy Agency (2021). Report Extract: The Outlook for Biogas and Biomethane to 2040. https://www.iea.org/reports/outlook-for-biogas-and-biomethane-prospects-for-organic-growth/the-outlook-for-biogas-and-biomethane-to-2040 (accessed 18 October 2021).

10 Roy, S., Cherevotan, A., and Peter, S.C. (2018). Thermochemical CO_2 hydrogenation to single carbon products: scientific and technological challenges. *ACS Energy Lett.* 3 (8): 1938–1966.

11 Rönsch, S., Schneider, J., Matthischke, S. et al. (2016). Review on methanation – from fundamentals to current projects. *Fuel* 166: 276–296.

12 Lecker, B., Illi, L., Lemmer, A., and Oechsner, H. (2017). Biological hydrogen methanation – a review. *Bioresour. Technol.* 245: 1220–1228.

13 Ghaib, K., Nitz, K., and Ben-Fares, F.Z. (2016). Chemical methanation of CO_2: a review. *ChemBioEng Rev.* 3 (6): 266–275.

14 Lee, W.J., Li, C., Prajitno, H. et al. (2021). Recent trend in thermal catalytic low temperature CO_2 methanation: a critical review. *Catal. Today* 368: 2–19.

15 Thema, M., Bauer, F., and Sterner, M. (2019). Power-to-gas: electrolysis and methanation status review. *Renewable Sustainable Energy Rev.* 112: 775–787.

16 Peters, R., Baltruweit, M., Grube, T. et al. (2019). A techno economic analysis of the power to gas route. *J. CO2 Util.* 34: 616–634.

17 Schildhauer, T.J. and Calbry-Muzyka, A.S. (2020). Direct methanation of biogas—technical challenges and recent progress. *Front. Energy Res.* 8: 356.

18 Welch, A.J., Digdaya, I.A., Kent, R. et al. (2021). Comparative technoeconomic analysis of renewable generation of methane using sunlight, water, and carbon dioxide. *ACS Energy Lett.* 6 (4): 1540–1549.

19 Lucas, F.W.S., Grim, R.G., Tacey, S.A. et al. (2021). Electrochemical routes for the valorization of biomass-derived feedstocks: from chemistry to application. *ACS Energy Lett.* 6 (4): 1205–1270.

20 Dier, T.K.F., Rauber, D., Durneata, D. et al. (2017). Sustainable electrochemical depolymerization of lignin in reusable ionic liquids. *Sci. Rep.* 7 (1): 5041.

21 Grim, R.G., Huang, Z., Guarnieri, M.T. et al. (2020). Transforming the carbon economy: challenges and opportunities in the convergence of low-cost electricity and reductive CO_2 utilization. *Energy Environ. Sci.* 13 (2): 472–494.

22 Kuhl, K.P., Cave, E.R., Abram, D.N., and Jaramillo, T.F. (2012). New insights into the electrochemical reduction of carbon dioxide on metallic copper surfaces. *Energy Environ. Sci.* 5 (5): 7050–7059.

23 Bi, L., Boulfrad, S., and Traversa, E. (2014). Steam electrolysis by solid oxide electrolysis cells (SOECs) with proton-conducting oxides. *Chem. Soc. Rev.* 43 (24): 8255–8270.

24 Rabaey, K. and Rozendal, R.A. (2010). Microbial electrosynthesis — revisiting the electrical route for microbial production. *Nat. Rev. Microbiol.* 8 (10): 706–716.

25 Carbon Recycling International (2019). George Olah Renewable Methanol Plant. https://www.carbonrecycling.is/projects (accessed 18 October 2021).

26 LanzaTech (2018). World's First Commercial Waste Gas to Ethanol Plant Starts Up. https://www.lanzatech.com/2018/06/08/worlds-first-commercial-waste-gas-ethanol-plant-starts/ (accessed 18 October 2021).

27 Electrochaea (2020). http://www.electrochaea.com/technology/ (accessed 18 October 2021).

28 International Energy Agency (2020). *World Energy Outlook 2020.*

29 Martin, A.J., Larrazábal, G.O., and Pérez-Ramírez, J. (2015). Towards sustainable fuels and chemicals through the electrochemical reduction of CO_2: lessons from water electrolysis. *Green Chem.* 17 (12): 5114–5130.

30 International Renewable Energy Agency (2020). *Renewable Power Generation Costs in 2019.*

31 International Renewable Energy Agency (2021). *Renewable Capacity Statistics 2021*.

32 Ong, S., Campbell, C., Denholm, P. et al. (2013). Land-Use Requirements for Solar Power Plants in the United States. *NREL/TP-6A20-56290*. Golden, CO, USA: National Renewable Energy Laboratory.

33 U.S. Energy Information Administration (2015). Energy use in homes. https://www.eia.gov/energyexplained/use-of-energy/homes.php (accessed 18 October 2021).

34 Rocky Mountain Institute (2018). The Economics of Electrifying Buildings. https://rmi.org/insight/the-economics-of-electrifying-buildings/ (accessed 18 October 2021).

35 New Buildings Institute (2021). Advanced Water Heating Initiative Playbook and 2020 Progress Report. https://newbuildings.org/wp-content/uploads/2021/02/AWHI_PlaybookAndProgressReport202102.pdf (accessed 18 October 2021).

36 Macguire, J., Burch, J., Merrigan, T., and Ong, S. (2013). Energy Savings and Breakeven Cost for Residential Heat Pump Water Heaters in the United States. *NREL/TP-5500-58594*. https://www.nrel.gov/docs/fy13osti/58594.pdf (accessed 18 October 2021).

37 Daikin (2021). R-32, Next-Generation Refrigerant. https://www.daikin.com/corporate/why_daikin/benefits/r-32/ (accessed 18 October 2021).

38 U.S. Energy Information Administration (2015). Residential Energy Consumption Survey. https://www.eia.gov/consumption/residential/ (accessed 18 October 2021).

39 U.S. Energy Information Administration (2021). Monthly Energy Review.

40 U.S. Energy Information Administration (2021). U.S. Natural Gas Gross Withdrawals (Million Cubic Feet).

41 U.S. Energy Information Administration (2013). Manufacturing Energy Consumption Survey.

42 U.S. Energy Information Administration (2021). Manufacturing Energy Consumption Survey.

43 McMillan, C., Schoeneberger, C., Zhang, J. et al. (2021). Opportunities for Solar Industrial Process Heat in the United States. *NREL/TP-6A20-77760*. Golden, CO USA: National Renewable Energy Laboratory.

44 Bloess, A., Schill, W.-P., and Zerrahn, A. (2018). Power-to-heat for renewable energy integration: A review of technologies, modeling approaches, and flexibility potentials. *Appl. Energy* 212: 1611–1626.

45 Golmohamadi, H., Larsen, K.G., Jensen, P.G., and Hasrat, I.R. (2021). Optimization of power-to-heat flexibility for residential buildings in response to day-ahead electricity price. *Energy Build.* 232: 110665.

46 Bianco, V., Marchitto, A., Scarpa, F., and Tagliafico, L.A. (2020). Heat pumps for buildings heating: energy, environmental, and economic issues. *Energy Environ.* 31 (1): 116–129.

47 Lucon, O., Ürge-Vorsatz, D., Ahmed, A.Z. et al. (2014). Buildings. In: *Climate Change 2014: Mitigation of Climate Change. Contribution of Working Group III to the Fifth Assessment Report of the Intergovernmental Panel on Climate Change* (ed. O. Edenhofer, R. Pichs-Madruga, Y. Sokona, et al.). Cambridge, UK and New York, NY: Cambridge University Press.

48 Beyond Zero Emissions (2018). Zero Carbon Industry Plan: Electrifying Industry.

49 Chindris, M. and Sumper, A. (2012). Industrial heating processes. In: *Electrical Energy Efficiency: Technologies and Applications* (ed. A. Sumper and A. Baggini), 295–334. West Sussex, UK: Wiley.

50 Cheremisinoff, N.P. (1996). *Electrotechnology: Industrial and Environmental Applications*. Elsevier.

51 Schmitt, B. (2016). Classification of industrial heat consumers for integration of solar heat. *Energy Procedia* 91: 650–660.

52 Fleiter, T., Hirzel, S., and Worrell, E. (2012). The characteristics of energy-efficiency measures – a neglected dimension. *Energy Policy* 51: 502–513.

53 Hasanbeigi, A. and Khutal, H. (2021). *Industrial Supply Chains Decarbonization in Southeast Asia*. Tampa Bay Area, FL: Global Efficiency Intelligence, LLC.

54 Unruh, G.C. (2000). Understanding carbon lock-in. *Energy Policy* 28 (12): 817–830.

55 Janipour, Z., de Nooij, R., Scholten, P. et al. (2020). What are sources of carbon lock-in in energy-intensive industry? A case study into Dutch chemicals production. *Energy Res. Social Sci.* 60: 101320.

56 Madeddu, S., Ueckerdt, F., Pehl, M. et al. (2020). The CO_2 reduction potential for the European industry via direct electrification of heat supply (power-to-heat). *Environ. Res. Lett.* 15 (12): 124004.

57 Lechtenböhmer, S., Nilsson, L.J., Åhman, M., and Schneider, C. (2016). Decarbonising the energy intensive basic materials industry through electrification – implications for future EU electricity demand. *Energy* 115: 1623–1631.

58 Taft, M. (2021). *The Only Carbon Capture Coal Plant in the U.S. Just Closed.* Gozmodo. https://earther.gizmodo.com/the-only-carbon-capture-plant-in-the-u-s-just-closed-1846177778 (accessed 18 October 2021).

4

Role of Hydrogen in Low-Carbon Energy Future

Andrea Monforti Ferrario[1], Viviana Cigolotti[1], Ana Marìa Ruz[2], Felipe Gallardo[3], Jose García[3], and Giulia Monteleone[1]

[1] *ENEA, Italian National Agency for New Technologies, Energy and Sustainable Economic Development, Department of Energy Technologies and Renewable Sources (TERIN-PSU-ABI), Via Anguillarese 301, Rome 00123, Italy*
[2] *Chilean Economic Development Agency (Corfo), Moneda 921, Región Metropolitana, Santiago 8320250, Chile*
[3] *KTH Royal Institute of Technology, Unit of Heat and Power Technology, Brinellvägen 68, Stockholm 10044, Sweden*

4.1 Introduction

To achieve climate neutrality by 2050, significant efforts in terms of decarbonization at a global level must be undertaken. On July 2020, the EU Commission has announced its industrial strategy paving the way for a new, more efficient, safer, and more sustainable energy system oriented to the dual objective of a cleaner planet and toward the complete decarbonization of the economy. The strategy is articulated as follows:

- Promoting a climate-neutral economy: the EU strategy for the integration of energy systems [1]
- A hydrogen strategy for a climate-neutral Europe [2]

Many EU member states have shortly followed, transposing the EU directives into national roadmaps and industrial strategies, as well as key strategic extra EU countries, with exceptional production or consumption potential.

The former sets out a model to accelerate the energy transition, identifying three main pillars: (i) efficiency and more "circular" energy system, (ii) greater direct electrification of end uses, and (iii) utilization of the renewable and low-carbon fuels (including hydrogen) in all those sectors where direct electrification is not feasible (the so-called hard-to-abate sectors). The latter is triggered on this last pillar and will have the role of accelerating, in an integrated energy system, the decarbonization of industry, mobility, electricity generation, and residential energy use, realizing this potential through investments, regulation, value creation, research, and innovation.

So why hydrogen? Hydrogen is the simplest and most abundant element in the Universe. It combines readily with other chemical elements, and for this reason, it

Technologies for Integrated Energy Systems and Networks, First Edition.
Edited by Giorgio Graditi and Marialaura Di Somma.

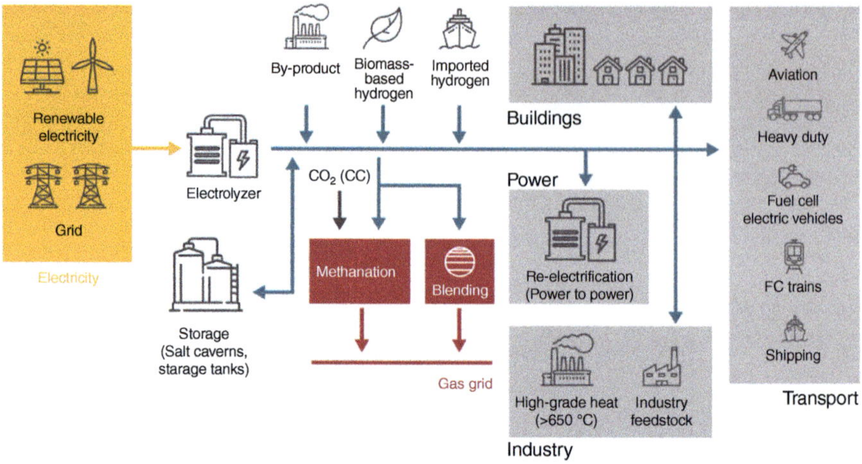

Figure 4.1 Overview of the hydrogen value chain in an integrated energy system. Source: IRENA 2018 [5].

is always found as part of other substances. It is overabundant, volatile, reactive, and with high mass-based energy density. Hydrogen can be used as a raw material, fuel, energy carrier, or energy storage medium and shows possible applications in all sectors of the economy: industry, mobility, electricity, and residential energy use [3]. Its most important feature is the fact that when used, it does not emit CO_2 as it does not contain carbon in its molecule; as a consequence, it does not cause air pollution and therefore allows the achievement of climate neutrality commitments and global acts for the containment of global temperature within 1.5 °C [4].

The complexity and broadness of the value chain explains by itself why hydrogen will play a leading role in the decarbonization of the next energy future, where different levels of implementation (production, storage, transport and distribution, and end uses) are intertwined with different sectors of interest (energy, industry, mobility, and residential) [5]. A synoptic overview of the hydrogen value chain in an integrated energy system is shown in Figure 4.1.

4.2 Main Drivers for Hydrogen Implementation

In a continuously mutating energy network environment, hydrogen technologies have experienced periodic waves of uprising interest and crisp downfalls [3]. In fact, despite representing a non-negligible additional cost to the network (hydrogen cost between 1–3 €/kg for non-renewable hydrogen and above 5–10 €/kg for fully renewable hydrogen [6, 7]), hydrogen provides the opportunity for bulk electrical energy storage and flexibility [8] (especially in high renewable penetration scenarios [9]) where direct electricity storage and management is challenging [10] and represents a gateway for low-carbon energy integration in non-electrified hard-to-abate sectors [11, 12], leading toward the deep decarbonization of the whole primary energy network and not solely of the electricity energy network (emitting around 30–40% of CO_2 in 2018 with respect to a total of 31.86 Gt_{CO2} [13]).

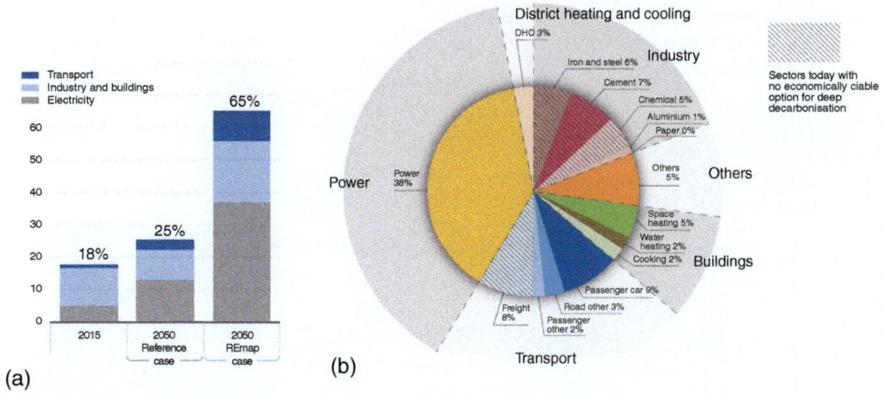

Figure 4.2 Main drivers for hydrogen implementation: (a) increasing stochastic renewable energy and (b) opportunity for sector coupling. Source: IRENA 2018 [5].

The demand for hydrogen (2019) is around 115 Mt_{H2}/yr to 70 Mt_{H2}/yr in the form of pure hydrogen and 45 Mt_{H2}/yr mixed with other gases. The demand is dominated by the industry sector accounting for >85% of the demand. The carbon footprint of the hydrogen value chain today is very high (around 830 Mt_{CO2}/yr) being produced from fossil fuels and the energy-related end uses represent only a marginal percentage of the total demand [7]. Furthermore, the main user sectors are quite well established and unlikely to increase the demand organically.

The current announced policies foresee 2030 global scale-up targets of around 100 GW in terms of installed production capacity (mainly Europe 40 GW by 2030, followed by Chile 25 GW by 2030, Korea 15 GW by 2040) and around 20–50 Mt_{H2}/yr in terms of green hydrogen consumption (mainly Europe 10 Mt_{H2}/yr by 2030), most of which supported by national and international subsidies [14–16]. Although ambitious, the gap between the announced uptake of hydrogen technologies and the required efforts to achieve global hydrogen cost targets and emission reduction targets is still substantial.

Figure 4.2 illustrates the main hydrogen drivers.

4.2.1 Increasing Penetration of Stochastic Renewable Energy

The dramatic increasing trend in stochastic renewable energy sources (RES) in the past two decades is well known [17]. However, because of the variable and non-dispatchable nature, these technologies face technical challenges in order to supply a rigid power demand [10], leading to critical issues for grid balance [18] with excessive amount of RES energy in the electricity mixes.

Hydrogen as a vector enables to decouple the generation from the demand, thanks to the chemical storage properties, becoming a key enabler for the integration of RES energy in the power system [11, 19, 20]. RES-based electrolytic hydrogen can be competitively produced by the RES excess/curtailment with a load leveling approach [21–23], or even with dedicated RES plants in particularly favorable locations in terms of resource availability and electricity cost [24–26].

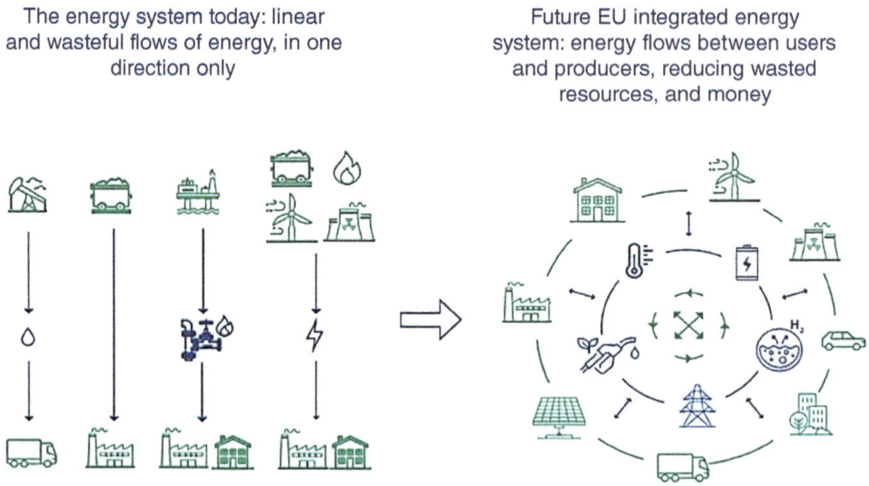

Figure 4.3 Hydrogen as a sector coupling enabler. Source: Adapted from European Commision [1].

4.2.2 Opportunity of Hydrogen as a Sector Coupling Enabler

Hydrogen, differently from other chemical vectors, can be deployed in a multitude of processes both as an energy vector and as a commodity, providing the opportunity to link and flexibly interoperate between different sectors and networks. This enables the so-called *Sector Coupling* (Figure 4.3), where a unit of electricity can be dispatched in other sectors (the so-called hard-to-abate sectors), which heavily rely on fossil fuels and account for 20% of global CO_2 emissions and very little progress in decarbonization to date [27, 28].

In addition, hydrogen as a molecule has an intrinsically high added value as a material input: it can be used as a feedstock or a fuel in chemical, oil and gas, steel manufacturing sectors, among others. The use of hydrogen as a feedstock currently holds the largest market share (around 70 Mt_{H2}/yr, i.e. >85% of current hydrogen demand for fertilizer production and fuel refinement [7]) because of the non-competitivity of alternative technologies.

4.3 Hydrogen Economy and Policy in Europe and Worldwide

Despite the interest in hydrogen technology since before 2000s [29–31], technological efforts have never been supported by an enabling policy framework. From 2020s, several favorable conditions (such as electricity prices below 30 €/MWh, extensive RES penetration, improving electrolyzer technology, growing electrification requirements, foreseen emission regulation and carbon tax, shift to decentralized paradigm, etc.) represent a viable opportunity to support mass hydrogen implementation from an economic and policy perspective [7, 14].

Green hydrogen and RES electricity present a strong synergy – both from a technical and a policy perspective – each one enhancing the development of the other [5, 14, 32], because of the high energy intensity of the electrolysis process. In fact, the conversion of the current market to green hydrogen would require 3600 TWh/yr of RES electricity [7] today and up to 8–30 PWh/yr with the scale-up of hydrogen demand in 2050 [33].

Cost is the prime limiting factor to effectively compete with fossil fuel-derived hydrogen. In contrast to fossil fuel business cases, the largest cost contribution is operational mainly composed of electricity price followed only secondly by the CAPEX of the electrolyzer [24]. According to IRENA [34] (Figure 4.4), for green hydrogen cost to decrease from 5 US$/kg to 1 US$/kg (reference cost for gray hydrogen), the electricity price should drop from 53 US$/MWh to 20 US$/MWh, electrolyzer CAPEX should decrease by 80% (from over 650–1000 US$/kW to 130–307 US$/kW) for an increased installed capacity of up to 1 TW in 2050 with an efficiency increase from 65% to 76% (based on the lower heating value (LHV), equivalent to 51.2 and 43.8 kWh/kg, respectively).

Taking into consideration the hydrogen cost composition and supply chain structure, countries with favorable RES resources and with access to extremely low electricity prices (<35 US$/MWh) can become in the short-term potential strategic exporters, thanks to low hydrogen production costs, as in the case of Australia and Chile [24, 26, 35–39], investing mainly in electrolysis technology. Other countries, such as Japan, have announced import-driven approaches (300 kt_{H2}/yr by 2030) to the hydrogen supply chain, investing in network infrastructure end use applications in local markets, making leverage on a complex and highly integrated energy

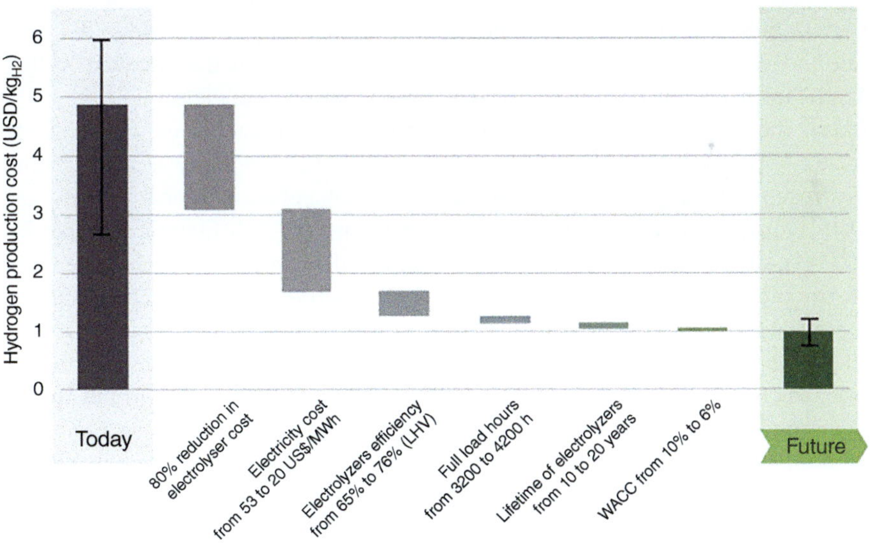

Figure 4.4 Green hydrogen cost reduction scaling up electrolyzers to meet the 1.5 °C climate goal. Source: IRENA 2020 [34].

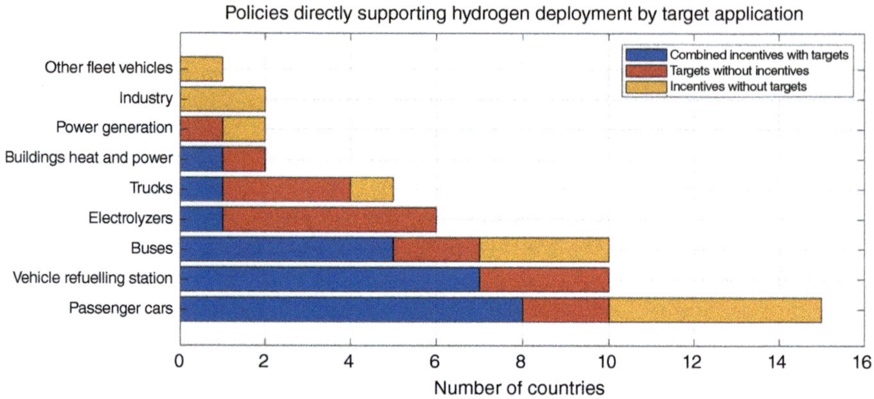

Figure 4.5 Policies directly supporting hydrogen deployment by target application. Source: Adapted from IEA International Energy Agency [7].

system where hydrogen and fuel cell systems are already deeply ingrained [40, 41]. Both approaches (export/import) pave the way for investment attraction and contribute into positioning such countries as key nodes in the international market network.

However, in order to foster a low-carbon, effective, and long-term development of a full "hydrogen economy," it is crucial to also promote investment in a broad spectrum of end use applications. Such aspect is currently often overlooked from a global point of view, with the risk of creating stimulus in the hydrogen production sector without equally enhancing an accessible and profitable market for the hydrogen off-takers, inhibiting a spontaneous and organic market growth. In this sense, demonstration projects (Annex 1) and captive markets are key to bridge the economic and risk gap between research and industry, overcoming the so-called "valley of death" from a technological and system integration point of view.

Today, the most subsidized end use sector for hydrogen is by far mobility (Figure 4.5) [42, 43]. Most countries have set goals on the mobility sector in terms of hydrogen refueling infrastructure and fuel cell vehicles (global targets are 10 000 Hydrogen Refueling Stations and 10 million Fuel Cell Electric Vehicles by 2030 [14]), supported by the EU Directive on the deployment of alternative fuels infrastructure (DAFI) [44]. However, only few countries have set a clear goal on other scalable hydrogen end use applications, such as stationary production and residential combined heat and power, which should be addressed in the near future [7, 42]. In contrast, hydrogen in industry environments is a quick win but might offer a limited market in the future, with limited added value.

Additionally, the proprietary and know-how aspect related to hydrogen technologies in the value chain will be a crucial aspect, with high-value generation potential locally and abroad. In this sense, Europe is the global leader in hydrogen and fuel cell technology, filing about twice as many patents and publications as its nearest competitors – the United States, China, and Japan – over the past 10–15 years [45]. It is foreseen that Europe can maintain such advantage, thanks to the substantial financial effort to support Research and Development (R&D) along the hydrogen value chain, in spite of the potential strong competition from China and other countries that are favored by much larger deployment volumes [42].

Aspects such as (i) *access to low electricity prices*, (ii) *enabling capacity ramp-up* to trigger scale effect, (iii) *R&D efforts* for the improvement of the technology, and (iv) *facilitating off-takers to uptake the use of green hydrogen* all need to be simultaneously addressed as a priority by public policy to stimulate electrolytic green hydrogen production. As a positive externality, (v) *a carbon tax* could increase the competitiveness of both green and blue hydrogen, with respect to competing fossil fuel-based hydrogen production [46]. Also, (vi) *harmonization and standardization of the regulatory framework* would strongly enhance the uptake of hydrogen and fuel cell technologies in the short to medium term, especially if well targeted (natural gas blending limits, simplification of permitting for key infrastructure and strategic hubs, international trade regulation, etc.) [7, 14].

4.4 Main Renewable Hydrogen Production, Storage, and Transmission/Distribution Schemes

4.4.1 Hydrogen Production Pathways

Despite being widely available in nature, hydrogen as an isolated molecule does not exist in nature in a stable form. Instead, it is found combined with other molecules – mainly water and hydrocarbons – which must be decomposed with energy-intensive processes to obtain hydrogen. Overall, three main hydrogen categories are recognized in the community (Figure 4.6): **gray hydrogen** (fossil fuel based, without any carbon capture [CC] solutions), **blue hydrogen** (fossil fuel based but with CC), and **green hydrogen** (based on zero emission RES) [6]. The environmental impact of each "color" of hydrogen is mainly related to the footprint of the feedstock and energy used in the production process [47, 48].

Gray hydrogen mainly derives from the classical thermochemical conversion methods of solid, liquid, or gaseous fossil fuels. The most widespread technology is steam methane reforming (SMR). The SMR process has the advantage to be extremely cheap – obtaining levelized costs of 1–3 US$/$kg_{H2}$ – because of the low natural gas cost; although from an energy perspective, it is quite demanding [7, 34]. Furthermore, SMR plants are typically large scale, with typical capacities of over 1000 kg/h for industrial environments, with limited possibility for downsizing and modulation [49]. Other gray hydrogen production methods are autothermal reforming and partial oxidation of natural gas and crude oil or gasification of coal, which all share the commonality of low cost, low hydrogen purity, and high CO_2 and other emissions [6].

Blue hydrogen complements gray hydrogen production methods with CC technology, resulting in a mitigated CO_2 balance [6]. Blue hydrogen is considered a

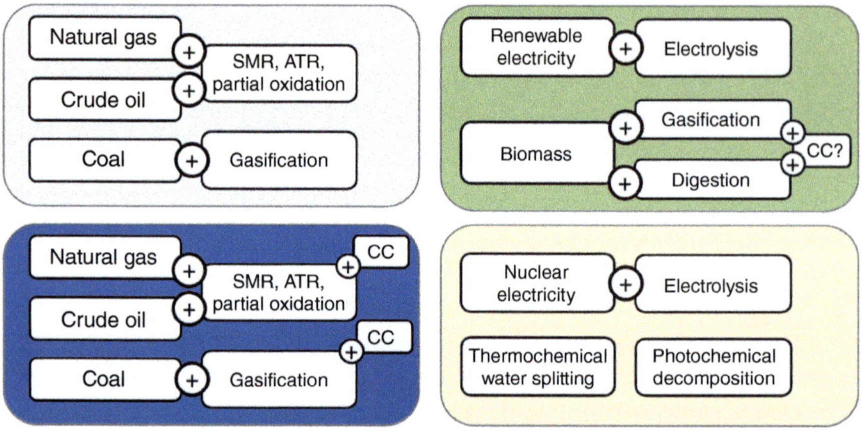

Figure 4.6 Overview of hydrogen production pathways. Source: Noussan et al. [6].

promising alternative for the medium-term transition thanks to the possibility to continue using the current infrastructure, maintaining fossil fuel-based conversion approaches [16, 46]. However, the production of blue hydrogen shares most disadvantages related to gray hydrogen and suppresses the shift to a fully zero emission paradigm.

Green hydrogen is produced from renewable electricity, mainly via the electrolysis process, which is an electrochemical device that decomposes water in its main constituents under a certain applied DC current [50–53] (Table 4.1). The most widespread and commercially mature electrolysis technologies are alkaline electrolyzers (AEL) and polymeric electrolyte membrane electrolyzers (PEMEL) that operate at low temperatures between 60 and 100 °C [7, 11, 24, 56]. The constituent materials of the electrolyzer determine operating envelopes in terms of temperature, pressure, and current density affecting both stack and balance of plant design. High-temperature electrolyzers (HTE), operating at temperatures >650 °C, are a promising upcoming technology that can offer great advantages in terms of energy consumption, power density, and catalyst requirements, thanks to favorable thermodynamic conditions [57].

In general, AELs are more robust and have a slightly higher efficiency than PEMEL, however the present strong limitations in dynamic operation and power modulation (no lower than 20% of nominal power and a current density of around 0.1–0.4 A/cm^2). PEMEL thanks to their enhanced dynamic capability are particularly suitable for coupling to stochastic renewable energy, allowing variable current loading with a more compact design (current density above 0.6–1 A/cm^2) [51, 54, 56, 58]. HTE presents very high conversion efficiencies (below 40 kWh/kg_{H2} with respect to over 50–60 kWh/kg_{H2} for low-temperature electrolysis) in compact devices (sustaining up to 1–2 A/cm^2 with low cell resistance) [59–61].

Green hydrogen can also be produced from biological, biochemical, and thermochemical conversion of biomass, although similar limitations previously described for gray-blue hydrogen are present [62–64]. Also, the utilization of biomass as a

Table 4.1 Summary of water electrolysis characteristics.

	AEL	PEMEL	HTE
Temperature (°C)	60–90	50–80	600–1000
Pressure (bar)	1–10	Up to 30	Atmospheric
Specific consumption [stack - system] (kWh/Nm^3)	[4.2–5.5]	[4.5–5.8]	[3–3.8]
Efficiency [stack-system] (%)	[65%–55%]	[60%[a)]–50%]	[>90%–80%][b)]
Maximum size	4 MW module	2 MW module	kW-scale
Flexibility	20–100% load	Up to 160% var.	100% (reversible)
Start-up/shut-down	Minutes	Seconds	Hours
Maturity level	Commercial	Small scale	Pre-commercial
Main challenges	Dynamic response	CAPEX cost, efficiency	Degradation and cost

a) Peaks 65–70%.
b) Including thermal.
Source: Based on IEA International Energy Agency [7], Buttler and Spliethoff [54], RSE [55].

renewable resource is controversial, except for waste biomass, for which scale-up is a limiting factor [65, 66]. Other renewable hydrogen production pathways such as nuclear-powered electrolysis, thermochemical water splitting via solar or nuclear energy, and direct photochemical decomposition are encompassed within the green hydrogen category, although sometimes can be referred to as yellow hydrogen [6].

4.4.2 Hydrogen Transmission and Distribution

4.4.2.1 Main Hydrogen Storage Technologies

Hydrogen storage technologies can be grouped in two categories: physical storage and material-based storage [67]. Physical hydrogen storage refers to hydrogen in its gaseous or liquid form, whereas in material-based storage systems, it is bound to other elements. While in physical storage methods, hydrogen is in its pure form, the three primary material-based hydrogen storage media are adsorbents, reversible metal hydrides, and chemical hydrogen storage materials [67]. The most appropriate type of storage technology depends on the application (energy, power, and duration requirements), which determines the suitable trade-off between storage characteristics.

Most small- to medium-scale applications almost exclusively use compressed gaseous hydrogen (CGH_2) in cylinders; the low mass density in gaseous phase requires nominal working pressures of 200, 350, and up to 700 bar. The main advantage of CGH_2 storage is the engineering simplicity in terms of system integration and charge/discharge dynamics (similar to current fossil fuel alternatives).

Table 4.2 Hydrogen vessel types.

Vessel type	Features/purposes	Maximum pressures
Type I	• All-metal construction, typically steel, widely available, relatively low cost. • Relatively high mass per unit storage volume. • Mostly used in fixed applications.	175–200 bar
Type II	• Mostly steel or aluminum with a glass fiber composite overwrap. • Structural loads shared between the metal vessel and composite materials. • Higher cost than type I but lighter weight.	260–300 bar
Type III	• Tanks made from a metal liner with a full composite overwrap (e.g, aluminum with a carbon fiber composite). • The composite materials carry the structural loads.	300–700 bar
Type IV	• All-composite construction using a polymer liner with carbon fiber or hybrid carbon/glass fiber composite. • Relatively expensive but lower tank mass per unit volume. • Mostly used in mobile applications.	700 bar

Source Based on LeGault [68], EIHP [69].

However, robust pressure vessels are required to withstand the high-pressure levels (type III or type IV vessels), which can incur significant weight and cost penalties, other than safety and permitting issues, as shown in Table 4.2.

Different hydrogen compression technologies are available in relation to different applications that present different pressure, mass flow, and purity requirements. Dry running piston compressors are well established and are the preferred choices in industry because of their larger flow rates (10–115,000 m^3/h) at discharge pressures up to 1000 bar. Other compression technologies are diaphragm compressors, screw compressors, and ionic compressors, which can present significant technical advantages and modularity for smaller scales but are not yet affirmed in the industry. Finally, turbo compressors and electrochemical compressors are emerging technologies being developed for specific applications [70].

The liquefaction of pure hydrogen is the way to increase most the hydrogen density. The primary concern for the storage of liquid hydrogen (LH_2) is the energy intensity of the process (around 10 kWh/kg; >30% of energy content) because of the extremely low boiling point of hydrogen (−253 °C at 1 bar) [24]. On top of that, the capital cost of a liquefaction plant is high and present limited operational versatility. Thus, this storage option is mainly prospected for large H_2 volumes, where density is a substantial advantage [71].

Conversion to a carrier (chemical hydrogen storage) is an effective way of storing hydrogen at large scale. The most common chemical carriers for hydrogen storage are synthetic natural gas and methanol (reaction with CO_2), ammonia (reaction with N_2), and liquid organic hydrogen carriers (LOHC). The fact that methanol and ammonia are already widely produced bulk chemicals is advantageous as most of the necessary infrastructure required for supply chain is already in place. Dehydrogenation of the chemical carrier at the end user is a challenging technical aspect that must be kept in consideration.

Metal hydrides provide lower containment pressures with low energy consumption in the charging/discharging process, representing a promising technology for storing hydrogen. The stability of the bonds with the metal compound offers high-density storage at near-ambient operating conditions. Although a vast array of metal hydrides has been developed and investigated for thermolysis-based storage, relatively few have been applied for hydrolysis with any significant success at commercial scale.

4.4.2.2 Methods for Hydrogen Transmission and Distribution

Different and complementary means of transportation and distribution of hydrogen are envisioned to connect hydrogen production with consumption. As shown in Figure 4.7, three main methods for moving hydrogen exist: pipes, trucks, and ships. The most suitable method depends on the volume (m^3) that needs to be moved and the distance (km) that needs to be covered [73].

CGH_2 can be moved in pipelines in a similar way to natural gas. While blended pipelines are seen as a short-term solution to move large quantities of hydrogen in concentrations up to 5–20 vol% in the existing natural gas network, new hydrogen pipelines (made of polyethylene) will be needed for large-scale transmission of hydrogen as metal can be embrittled with larger shares of hydrogen [15, 74–76] and the gas mixture properties are worsened with increasing hydrogen content (vol% basis). Large-capacity pipelines are the cheapest option for hydrogen transport if the volume justifies the investment cost, which is very high [77].

Trucks can be used to carry CGH_2, LH_2, or other chemical carriers. Trucks carrying CGH_2 and LH_2 are already in commercial use but are expensive. For low volumes and short distances (less than 300 km), trucks with CGH_2 are the most implemented option today, while for longer distances (300–400 km), carrying LH_2 or other carriers (NH_3 or LOHC) can be more advantageous as the increased transported quantity

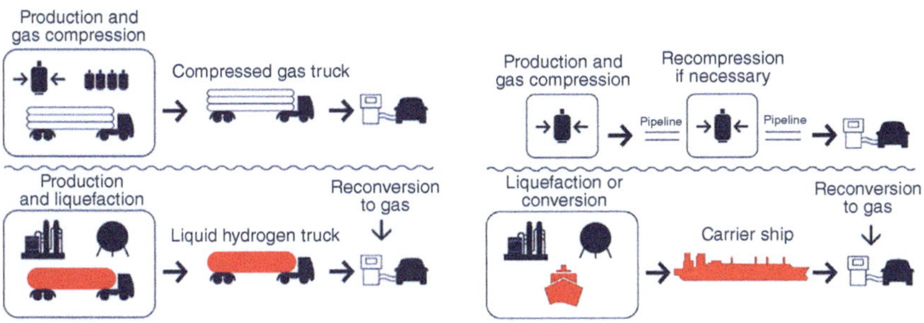

Figure 4.7 Hydrogen transmission and distribution options. Source: Adapted from Hydroville [72].

Table 4.3 Hydrogen carriers properties.

Carrier	Type	Energy for storage (kWh_e/kg_{H2})	Energy for release (kWh_{th}/kg_{H2})	H_2 content (H_2 $\%_{wt}$)	Density (kg/m^3)	Storage for transportation
LH_2	Pure H_2	6 at −253 °C			70	Cryogenic vessels at −253 °C
CGH_2	Pure H_2	1.6 at 700 bar			42	CGH_2 vessel at 700 bar
Ammonia	Chemical carrier	3 at 400 bar, 400 °C	4.2 at 425 °C	17.70%	123	Mineral oil tank, at 10 bar, 25 °C
Methanol	Chemical carrier	1.5 at 50 bar, 250 °C	6.7 at 250 °C	12.50%	99	Mineral oil tank at1 bar, 25 °C
Perhydro-dibenzyltoluene	LOHC	0.7 at 30 bar, 150 C	9 at 300 °C	6.20%	64	Mineral oil tank at 1 bar, 25 °C
Toluene	LOHC	0.7 at 30 bar, 150 °C	11.2 at 350 °C	6.10%	47	Mineral oil tank at 1 bar, 25 °C
Aluminum hydride	Metal Hydride	10 at 70 °C	1 at 100 °C	10.10%	86	Hydrogen storage metal hydride tank
Magnesium hydride	Metal Hydride	0.7 at 30 bar, 300 °C	10.3 at 350 °C	7.60%	110	Hydrogen storage metal hydride tank
Metal organic framework	Adsorbent	6.7 at 40 bar, −176 °C	—	7%	40	Hydrogen storage MOF tank

Source: Based on [70, 71, 74, 77].

counterbalances the increase in energy and cost related to the conversion processes (e.g. hydrogenation and dehydrogenation of the carrier).

Shipping is a transport mode dedicated to large-scale volumes because of the large capacities involved and the need for conversion to either LH_2 or other chemical forms. Hydrogen can also be used for the ship propulsion and auxiliary systems (hydrogen powered ships) and for different energy-intensive demand units of the port infrastructure (hydrogen-based port mobility, CHP supply for port buildings and services, power supply to shore connections, etc.), presenting many possible synergies with the port environment. Table 4.3 summarizes the main options for hydrogen storage and transportation.

4.5 Technological Applications in Integrated Energy Systems and Networks

4.5.1 Hydrogen as an Energy Storage System for Flexibility at Different Scales

H_2-to-Power devices [11, 78] can be integrated in energy systems and networks as generators, combined heat and power (CHP) units, and energy storage systems at different time scales (short-, medium-, long-term charge–discharge times) and size ranges (national level: >MW scale; district level: kW–MW scale; local level micro, off-grid: kW scale) [9, 79–81].

In particular, the highly modular and dispatchable nature of the electrochemical conversion process in fuel cells provides an intrinsic opportunity to provide storage services. High-temperature fuel cell (solid oxide fuel cells [SOFC] and molten carbonate fuel cells [MCFC]) systems are more suitable for stationary energy generation, while PEM systems can be operated with variable power profiles, thanks to the fast dynamic response [9, 82]. The intermittency of storage services can be a criticality for electrolyzer/fuel cell technology cycling degradation; for this reason, it is preferable that such services are provided on top of normal operation to stabilize the utilization factor [22]. The main storage services are described as follows:

Bulk energy storage (weekly to seasonal): conversion of large volumes of electricity (in the order of several MWh or more) to hydrogen in relation to unbalances of resource availability (e.g. solar and wind) or demand patterns (e.g. large users with seasonal consumption). Bulk energy storage can avoid congestion events and provide balancing services in critical nodes of the transmission grid [8, 21, 82, 83].

Flexibility services (intra-day to daily): production/consumption of hydrogen or electricity (in the order of several kWh) in Power-to-Power or Power-to-X schemes in district/microgrid/off-grid environments with RES generation in order to balance the generation (increasing RES integration quotas) and demand side (meeting demand during no-generation hours) and providing optimization and flexibility services in smart hubs with flexible or dispatchable loads [9, 19, 84, 85].

Ancillary grid services (intra-hour to hourly): dispatchable electricity production/consumption as a grid service (frequency/voltage control, local grid balancing, peak shaving, etc.) for time scales from the order of seconds to the order of hours. Such short-term grid services not only constitute an additional technical capability but can potentially result in additional revenue streams for hydrogen plant operators [9, 86, 87].

A critical aspect for the integration of hydrogen systems in the power network is related to the automatic control system. In this sense, energy management systems (EMS) must be implemented in microgrids at small scales while automatic generation control (AGC) managed by system operators should be present at large scale [88–90]. The automatic control system should be tailored to maximize the performances of the integrated system, considering the specific demand requirements and system configuration [84].

4.5.2 Industrial Use as a Renewable Feedstock in Hard-to-Abate Sectors and for the Production of Derivates

Hydrogen is already widely used in the industry as a feedstock rather than as an energy vector. Furthermore, hydrogen can be combined with CO_2-producing synthetic fuels, as well as being a fuel for high-grade heat and CHP generation. In industrial-scale applications, the main challenges are related to the size mismatch (to meet the hydrogen demand of a small refinery, electrolyzer capacities in the multi-MW scale are required, which are currently not yet fully developed) and the overall process and plant flexibility of the industrial processes (which are usually operated in steady state). The main industrial applications for hydrogen are discussed below.

Chemical Plants: The feedstock stream of H_2-based chemical products (e.g. ammonia, methanol, and other petrochemicals [91]) – currently fed by SMR-derived hydrogen – could be switched to green hydrogen with limited technical adaptations, although challenges related to process and plant flexibility should be addressed [24, 92].

Oil Refineries: Several refining processes (e.g. hydrocracking and treatment of heavy products), with the aim to increase the fuel quality and reduce its emission potential [93]. Currently, the H_2 feed is obtained by a local platformer or SMR plants that could be equipped with CC or replaced by electrolysis plants [94].

Iron and Steel Manufacturing: The use of hydrogen instead of coke for direct reduction of the iron ore is being explored to avoid the production of CO_2 that is emitted when using carbon monoxide as a reducing agent. Hydrogen DRI is assessed to increase the steel price by about one third [95].

Industrial Heat: Burning hydrogen avoids emitting CO_2 in the flue gases, the combustion process differs in terms of combustion temperatures, flame propagation velocity, and radiative properties and flammability limits [96], which could be exploited in processes that require local high-grade heat (e.g., clinker burning in cement plants), although thermal NO_x could be increased [7, 97–99].

Combined Heat and Power (CHP): Hydrogen or blended NG/H_2 can be fed to CHP systems (internal combustion engines, microgas turbines, or fuel cells) in order to provide both electricity and heat to industrial users. Correct sizing criteria and operational strategies (stationary base load and load following) should be assessed in function of the deployed technology to achieve a successful system integration.

Synthetic electrofuels (e-fuels) are produced by the reaction of renewable-based intermediate vectors (such as hydrogen produced via electrolysis), with CO_2 captured from emitting processes (or possibly extracted from the air), into gaseous and liquid products (methanol, dimethyl ether, diesel, gasoline, and jet fuel) with similar characteristics. Hydrogen-based e-fuels are considered attractive for multiple reasons: easier storage than hydrogen, easier integration with existing logistic infrastructure (e.g. use in gas pipelines, tankers, and refueling infrastructure), and ability to address various markets (e.g. aviation, shipping, freight, building heating, petrochemical, and feedstocks). One of the key technical challenges related to integrating variable electrolyzer operation with continuous industrial processes and/or making CC derived CO_2 available for these applications. In fact, synthetic fuels based on carbon-neutral CO_2 are completely neutral as CO_2 emission [100].

4.5.3 Hydrogen Mobility: A Complementary Solution to Battery Electric Vehicles

While for most light-duty mobility applications battery electric vehicles (BEVs) are the technology of choice, being modular, highly efficient, and with good dynamic response, heavy-duty mobility requires higher on-board energy volumes, longer driving ranges, and faster recharging times, which cannot be delivered successfully by a BEV. Fuel cell electric vehicles (FCEVs) are a promising option for mobility applications such as trucks, buses, ships, trains, large cars, and commercial vehicles (Table 4.4) as FCEV powertrains can provide sufficient power for long driving ranges under high payloads because of the increased energy density of its carrier, albeit a lower tank-to-wheel efficiency. In addition, hydrogen-based synthetic fuels can decarbonize other mobility sectors that are very far from electrification, without changing the powertrain configuration [101, 102].

Cost is surely the major barrier for hydrogen mobility both in terms of vehicle and infrastructure cost. Technological development and economy of scale will play a key role in generating cost reduction in the short and long term. Already today, FCEVs represent a competitive alternative for the heavy-duty applications in terms of total cost of ownership, surpassing the BEV alternatives. For smaller vehicles such as passenger vehicles and commercial vehicles, the operating conditions of the specific use cases (especially the average mileage per day) will determine the competitiveness of FCEV with respect to BEV [3, 103, 104], as shown in Figure 4.8.

Widespread refueling infrastructure is a crucial element for mass deployment of hydrogen mobility and poses a significant challenge as its current availability (in terms of capacity and spatial distribution) is limited. Hydrogen refueling stations (HRSs) present similar characteristics with respect to conventional refueling stations, enabling an easy integration in multi-fuel stations other than H_2-only stations, which reduces the upfront investment costs. Also, HRS are easily scalable while meaningful BEV penetration requires massive upgrades to power grids [105]. On the downside, HRS present an increased complexity (multiple compression, storage, and dispenser units operating up to 700 bar or more) increasing safety requirements and potentially posing a barrier from a permitting point of view [106–108].

Table 4.4 Summary of hydrogen use in mobility.

Mobility segment		Role of hydrogen-based FCEVs
Passenger cars	Competing with BEV	• FCEVs offer similar ranges and refueling times as ICE vehicles; • Hydrogen refueling station (HRS) infrastructure required; • Competition with BEVs depends from driving habits (average yearly mileage, variability of routes, etc.).
Trucks		• Long-range heavy duty trucks; • Weight comparable to ICE (comparable battery electric powertrain would weigh several times more because of lower energy density).
Ships	Highly advantageous with respect to BEV	• Large passenger/cargo ships (cruise ships and ferries); • Lower local emissions, less noise, and less water pollution; • Fuel cells can provide auxiliary power, replacing diesel-based units.
Trains		• Direct electrification of train lines is the preferred route for new tracks but upgrades of existing non-electrified tracks is costly. • No CO_2 emissions, reduce noise, and eliminate local particulates.
Industrial vehicles		• Material handling vehicles and internal port mobility vehicles (reach stackers, yard trucks, tow trucks, forklifts, and rubber-tyred gantry cranes) • Suitable for long shifts at constant base load power requirements
Aviation	To be defined	• Hydrogen and e-fuels present a strong potential in fuel cells or in dedicated turbines; • Hydrogen is seen as a potential key enabler of sustainable aviation; • FCs are increasingly used for auxiliary power units and ground power units but also propulsion in civil aircraft.

4.5.4 Fuel Cells, Flexible Electrochemical Conversion Systems for High-Efficiency Power, and/or CHP Applications

Fuel cells can convert the chemical energy of H_2-based fuel gases directly to electrical energy via an electrochemical process [109, 110]. Sharing the basic operating principle based on redox reactions in an electrode–electrolyte assembly, a vast range of available fuel cell technologies have been developed in terms of constituent materials (alkaline solutions, polymeric membranes, solid ceramic oxides, and molten carbonate electrolytes), operating temperatures (low-temperature fuel cells between 60 and 100 °C and high-temperature fuel cells between 600 and 900 °C), and operating fluids (pure H_2, methanol, hydrocarbon fuel reformate,

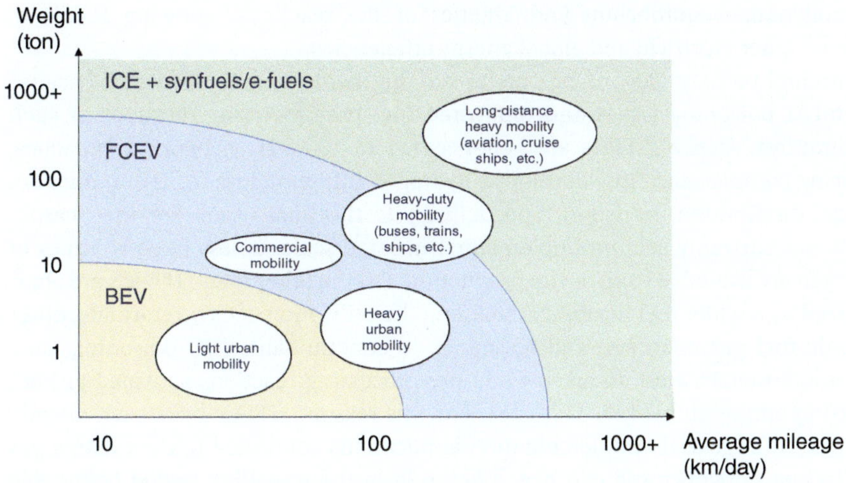

Figure 4.8 Mobility segments and characteristics addressed by EVs (FCEVs and BEVs). Source: Adapted from Hydrogen Council [102].

etc.). Different fuel cell technologies provide different operating envelopes in terms of operating temperature, current/power density ranges, dynamic response, fuel flexibility, among others [111, 112]. An overview of global fuel cell cumulative capacity is provided in Refs. [113, 114].

Given the electrochemical energy conversion pathway, very high chemical-to-electrical efficiencies can be achieved (exceeding the Carnot limit, which is an upper limit in conventional thermoelectric energy conversion pathways) in a highly modular system, resulting in a highly flexible and efficient conversion system for power production, which can play a key role in integrated energy networks. Despite that fuel cells are usually more advantageous for electricity generation, various heat streams can be recovered given the mostly exothermal reactions occurring in the stack, allowing to obtain a combined heat and power (CHP) unit with a heat interface at a temperature dependent on the fuel cell technology [78, 112, 115, 116]. A characteristic feature of fuel cell systems is that the electrical efficiency is also high at part loads and that heat can be recovered at all scales (mini-CHP and also μ-CHP levels) in nominal and off-design conditions, allowing to also meet smaller and modular thermal loads. On the contrary, other conventional CHP technologies (mostly combustion engines or gas turbines) present very poor conversion efficiencies at part loads and are not suitable for very small-scale CHP applications because of critical down-sizing issues. However, it must be noted that among other CHP technologies, fuel cells present a low heat/electricity ratio; therefore, the match with a suitable thermal/electrical demand is not straightforward and must be done with care [3].

Fuel cells can be categorized into low-temperature fuel cells (LTFCs) – operating between 60 and 200 °C – and high-temperature fuel cells (HTFCs) – operating between 600 and 1000 °C [117]. The temperature levels significantly affect the thermodynamic equilibrium and kinetics of the reactions, allowing HTFC to achieve higher electrical and global energy efficiencies.

From fuel point of view, LTFCs are mostly limited to pure H_2 feedstock because of catalyst poisoning issues and unfavored fuel pre-processing reactions at such temperatures. Hence, LTFCs are better suited to 100% H_2 network approaches, requiring complex and fully developed hydrogen infrastructure for H_2 production, storage, distribution, transport, and delivery to the final users. For this reason, LTFCs are currently seldom implemented in large scale, unless large volumes of hydrogen are available from onsite production. On the other hand, HTFCs are characterized by a wider fuel flexibility (enabling the utilization of CH_4 reformate, other biogenic fuel gas mixtures) and resistance to contaminants (CO poisoning, tars, H_2S, alkali-metals, etc.), thanks to fuel pre-processing reactions activated at high operating temperatures [57, 117, 118]. For this reason, HTFCs have been already deployed in large-scale implementation applications connected to the existing gas distribution networks and can play a key role in the transition period being able to increase the conversion efficiency without having to completely reinvent the energy vector distribution infrastructure paradigm. Nevertheless, research and development toward H_2-fed and H_2/NG blend-fed HTFC is ongoing where the FC manufacturers are redesigning stacks and systems based on H2 only fuel supply (or better, based on fuel flexibility supply). Several demonstrations are underway, especially evaluating HTFCs supplied with fuels with variable concentration in time, which might be a plausible long-term future projection.

Among the LTFC technologies, PEMFCs are the most suitable option for small- to medium-scale applications (from 1 to several 100 kW_e) [118]. Its compactness, construction simplicity, and high modulation characteristics can be effectively used in load-following mode to cover variable electrical demands (in households, micro-grids, and, especially, portable/automotive applications) with electrical efficiencies around 45–55% [111, 119]. PEMFCs as electricity generators are extremely compact, flexible, and modular, and the stack designs can be easily adapted to different application requirements. From a thermal output perspective, only low-grade heat is available (around 50–60 °C), limiting the thermal end uses to solely domestic hot water circuits and other low-temperature heat applications [116, 120]. In terms of technology penetration, PEMFCs are usually limited to locations where pure H_2 is available (either locally produced or supplied by a HRS). Another limitation in the use of PEMFCs is related to the high fuel cost (pure H_2), which must be stored onsite [79, 85, 121], increasing the total investment cost and facing permitting issues.

The HTFC technologies – SOFC and MCFC – present several commonalities if seen from a black-box point of view as electricity generators. As previously mentioned, the high temperatures allow high electrical conversion efficiency under different fuel supplies in highly compact systems. This can be obtained with a fairly simple and compact stack design, mainly composed of non-critical raw materials and non-noble metals as catalysts. On the downside, the system design can be more complex with respect to LTFC because of the higher operating temperatures, which are challenging to integrate in a simple and user-friendly way. In fact, the thermal inertia is a limiting factor for HTFC utilization in dynamic conditions, and the heat management is challenging in the design engineering. From a thermal output point of view, HTFCs present several advantages with respect to LTFC in relation to the exploitation of thermal energy, providing a high-grade heat interface that can be used for space heating or industrial heat purposes (beyond 80–90 °C) such as implementation in space heating or centralized district heating networks [115]. HTFCs, particularly SOFC, are a very promising option to decarbonize the residential sector [116, 120], providing an opportunity of substituting – at single household level – gas-fired boilers with small-scale SOFC μ-CHP units (<10 kW_e output) producing electricity and heat with extremely high efficiency (over 70% electrical efficiency and over 85% global efficiency [112]) with reasonable dynamic response while operating directly on natural gas, thus without having to modify the existing gas transmission and distribution infrastructure [81, 103, 104, 122]. Instead, MCFC units are more suitable for large-scale (MW-scale) stationary applications operated in stationary conditions [123, 124]. Moreover, MCFCs can be integrated in CC concepts, operating with CO_2 as a process gas [125, 126].

Another key aspect of HTFC technologies is the fact that the systems can be operated in the reversible mode, thanks to the favorable thermodynamic and kinetic conditions and the compatibility of the constituent materials and catalysts in both operation modes. Therefore, unitized electrolyzer/fuel cell systems can be engineered drastically reducing investment cost, with respect to LTFC schemes, which require one system operating in electrolysis mode and one system operating in fuel cell mode. Although there are still technological challenges in terms of operation dynamics (switching time and frequency between fuel cell and electrolysis mode, load modulation dynamics, etc.) and heat management, reversible high-temperature fuel cells are increasing their level of maturity, recently entering the market with several commercial-scale fully integrated systems. The development of reversible fuel cell systems is extremely promising in the perspective of local decentralized hydrogen production from RES and subsequent utilization in the same systems with highly efficient electrochemical conversion pathways.

4.6 Conclusions

The global energy system must complete a transition to a decarbonized system to reduce greenhouse gas emissions and mitigate climate change. Global greenhouse gas emissions continue to increase, and the urgency of moving to a secure low-carbon energy supply is evident. Nevertheless, energy security, fuel affordability, and availability are important aspects to be guaranteed.

Hydrogen has all these characteristics, and its versatility and potential for emissions reductions can allow it to have an important role in a low-carbon energy future. In fact, in addition to not emitting CO_2 when used as a fuel, hydrogen can be produced and used through a multiplicity of processes and systems, guaranteeing the principle of fairness and neutrality with respect to availability in the world. Hydrogen derived from fossil fuels is already widely used in some industries, but only a large-scale switch to green hydrogen can fully realize its decarbonization potential to support the clean energy transition.

Collaborative efforts from both the public and private sector are needed to stem ambitious, targeted, and expedite policy writing, which can enable all phases of the hydrogen value chain, reduce costs, and further overcome technical and non-technical barriers. Clear short-term goals for ready-accessible markets for hydrogen should be set, while strategically paving the way toward a zero carbon energy future, strongly rooted on green hydrogen.

The time is right to tap into hydrogen's potential to play a key role in a clean, secure, and affordable energy future.

List of Abbreviations

AEL	alkaline electrolyzers
AGC	automatic generation control
BEV	battery electric vehicle
CAPEX	capital expenditure
CC	carbon capture
CGH_2	compressed gaseous hydrogen
CHP	combined heat and power
DRI	direct reduced iron
EMS	energy management system
EV	electric vehicle
FC	fuel cell
FCEV	fuel cell electric vehicle
HRS	hydrogen refueling station
HTE	high-temperature electrolyzers
HTFC	high-temperature fuel cell
ICE	internal combustion engine
LH_2	liquid hydrogen
LHV	lower heating value
LOHC	liquid organic hydrogen carrier
LTFC	low-temperature fuel cell
MCFC	molten carbonate fuel cell
PEMEL	proton exchange membrane/polymer electrolyte membrane electrolysis
RES	renewable energy sources
SMR	steam methane reforming
SOFC	solid oxide fuel cell

References

1 European Commision (2020). Powering a climate-neutral economy: an EU Strategy for Energy System Integration. Communication from the Commission to the European Parliament, the Council, the European Economic and Social Committee and the Committee of the Regions, 5, 55.

2 European Commission (2020). EC COM (2020) 301 final. A hydrogen strategy for a climate-neutral Europe, Brussels.

3 Staffell, I., Scamman, D., Velazquez Abad, A. et al. (2019). The role of hydrogen and fuel cells in the global energy system. *Energy Environ. Sci.* 12: 463–491. https://doi.org/10.1039/c8ee01157e.

4 IPCC (2018). Summary for Policymakers - Global warming of 1.5°C, an IPCC special report.

5 IRENA (2018). Hydrogen From Renewable Power: Technology outlook for the energy transition.

6 Noussan, M., Raimondi, P.P., Scita, R., and Hafner, M. (2020). The role of green and blue hydrogen in the energy transition—a technological and geopolitical perspective. *Sustainability* 13: 298. https://doi.org/10.3390/su13010298.

7 IEA International Energy Agency (2019). The future of fuel: the future of hydrogen. *Report* https://doi.org/10.1016/S1464-2859(12)70027-5.

8 Kloess, M. and Zach, K. (2014). Bulk electricity storage technologies for load-leveling operation - an economic assessment for the Austrian and German power market. *Int. J. Electr. Power Energy Syst.* 59: 111–122. https://doi.org/10.1016/j.ijepes.2014.02.002.

9 Lund, P.D., Lindgren, J., Mikkola, J., and Salpakari, J. (2015). Review of energy system flexibility measures to enable high levels of variable renewable electricity. *Renew. Sust. Energ. Rev.* 45: 785–807. https://doi.org/10.1016/j.rser.2015.01.057.

10 Lew, D., Piwko, D., Miller, N. et al. (2010). How do high levels of wind and solar impact the grid? The Western wind and solar integration study. *Energy* 1–10. https://doi.org/10.2172/1001442.

11 Buffo, G., Marocco, P., Ferrero, D., and Lanzini, A. (2019). *Power-to-X and Power-to-Power Routes*. Elsevier Inc. https://doi.org/10.1016/B978-0-12-814853-2.00015-1.

12 Brinner, A., Schmidt, M., Schwarz, S. et al. (2017). Technologiebericht Querschnittstechnologien innerhalb des Forschungsprojekts TF_Energiewende, pp. 1–55.

13 IEA (2018). *Global CO_2 Emissions by Sector*, 2–5. IEA.

14 IRENA (2020). Green hydrogen: a guide to policy making.

15 The European House (2020). Ambrosetti, Snam. H_2 Italy 2050.

16 Lambert, M. and Schulte, S. (2021). *Contrasting European Hydrogen Pathways: An Analysis of Differing Approaches in Key Markets*. The Oxford Institute for Energy Studies.

17 International Renewable Energy Agency (2020). *Global Renewables Outlook: Energy Transformation 2050.*

18 Vittal, V. (2010). The impact of renewable resources on the performance and reliability of the electricity grid. In: National Academy Of Engineering, editor. *Bridg. Link. Eng. Soc.* 40: 5–13. https://doi.org/10.1038/138611c0.

19 Buffo, G., Ferrero, D., Santarelli, M., and Lanzini, A. (2020). Energy and environmental analysis of a flexible power-to-X plant based on reversible solid oxide cells (rSOCs) for an urban district. *J. Energy Storage* 29: 101314. https://doi.org/10.1016/j.est.2020.101314.

20 Kabouris, J. and Kanellos, F.D. (2009). Impacts of large scale wind penetration on energy supply industry. *Energies* 2: 1031–1041. https://doi.org/10.3390/en20401031.

21 Monforti Ferrario, A.M., Amoruso, C., Robles, R.V. et al. (2020). Power-to-Gas from curtailed RES electricity in Spain: potential and applications, pp. 1–6. https://doi.org/10.1109/eeeic/icpseurope49358.2020.9160820.

22 Tractebel Engineering and Hinicio (2017). Study on Early Business Cases for H_2 in Energy Storage and More Broadly Power To H_2 Applications, p. 228.

23 Saboya, I., Rouco, L., and Linares, P. (2017). Development perspectives of power to gas (P2G) in Spain review of past wind generation curtailments in the Mainland Spain power system The ROM model. *Renew. Energy Power Qual. J. (RE&PQJ)* 1: 1–6.

24 Gallardo, F.I., Monforti Ferrario, A., Lamagna, M. et al. (2020). A techno-economic analysis of solar hydrogen production by electrolysis in the north of Chile and the case of exportation from Atacama Desert to Japan. *Int. J. Hydrog. Energy* https://doi.org/10.1016/j.ijhydene.2020.07.050.

25 Heuser, I.P., Reuss, M., Grube, T. et al. (2018). Techno-Economic Analysis of a Global Hydrogen Supply Chain based on Wind-Generated Hydrogen • Results of Exemplary Hydrogen Supply Chain between.

26 Al-Sharafi, A., Sahin, A.Z., Ayar, T., and Yilbas, B.S. (2017). Techno-economic analysis and optimization of solar and wind energy systems for power generation and hydrogen production in Saudi Arabia. *Renew. Sust. Energ. Rev.* 69: 33–49. https://doi.org/10.1016/j.rser.2016.11.157.

27 IEA International Energy Agency (2015). Technology Roadmap: Hydrogen and Fuel Cells.

28 World Resources Institute (2021). CAIT Climate Data Explorer 2021. http://cait.wri.org (accessed April 18, 2021).

29 Ball, M. and Wietschel, M. (2009). The future of hydrogen–opportunities and challenges. *Int. J. Hydrog. Energy* 34: 615–627. https://doi.org/10.1016/j.ijhydene.2008.11.014.

30 Ball, M. and Weeda, M. (2015). The hydrogen economy - vision or reality? *Int. J. Hydrog. Energy* 40: https://doi.org/10.1016/j.ijhydene.2015.04.032.

31 European Commission (2003). Hydrogen Energy and Fuel Cells A vision of our future. Brussels.

32 Congressional Research Service (2020). Hydrogen in Electricity's Future Hydrogen in Electricity's Future.

33 IRENA IREA (2019). Hydrogen: A Renewable Energy Perspective - Report prepared for the 2nd Hydrogen Energy Ministerial Meeting in Tokyo, Japan.
34 IRENA (2020). Green Hydrogen Cost Reduction Scaling up Electrolysers to Meet The 1.5 °C Climate Goal.
35 International Renewable Energy Agency (IRENA) (2017). Renewable energy auctions: analysing 2016. IRENA, Abu Dhabi. ISBN 978-92-9260-008-2, https://www.irena.org/-/media/Files/IRENA/Agency/Publication/2017/Jun/IRENA_Renewable_Energy_Auctions_2017.pdf.
36 Ministerio de Energìa - Chile (2017). Acta Apertura Ofertas Economicas Licitacion 2017-01 n.d.
37 TaiyangNews (2017). Enel Offers Lowest Bid For Chilean Auction, pp. 1–2. http://taiyangnews.info/MARKETS/ENEL-OFFERS-LOWEST-BID-FOR-CHILEAN-AUCTION/ (accessed 19 October 2021).
38 Chapman, A.J., Fraser, T., and Itaoka, K. (2017). Hydrogen import pathway comparison framework incorporating cost and social preference: case studies from Australia to Japan. *Int. J. Energy Res.* 41: 2374–2391. https://doi.org/10.1002/er.3807.
39 ACIL Allen Consulting (2018). Opportunities for Australia from Hydrogen Exports 2018.
40 Arias, J. (2019). EU-Japan Centre for Industrial Cooperation. *Hydrogen and Fuel Cells in Japan*, pp. 1–145.
41 NEDO (2019). The effort to promote hydrogen in Japan. *Eur. Fuel Cell. Conf. Exhib.*, Naples Italy 2019.
42 van Renssen, S. (2020). The hydrogen solution? *Nat. Clim. Chang.* 10: 799–801. https://doi.org/10.1038/s41558-020-0891-0.
43 Miller, J. and Keohane, D. (2021). Car groups throw spanner in works of EU's hydrogen drive. *Financ Times*, pp. 1–10.
44 European Commission for Mobility and Transport (2017). Summary on national plans for alternative fuel infrastructure, p. 12.
45 Biebuyck, B. (2019). FCH-JU FC and HJU, FCH-JU making hydrogen and fuel cells an everyday reality 2019.
46 Gas for Climate (2020). Gas Decarbonisation Pathways 2020–2050.
47 Susmozas, A., Iribarren, D., and Dufour, J. (2015). Assessing the life-cycle performance of hydrogen production via biofuel reforming in Europe. *Resources* 4: 398–411. https://doi.org/10.3390/resources4020398.
48 Melideo, D., Ortiz-Cebolla, R., and Weidner, E. (2020). *Life Cycle Assessment of Hydrogen and Fuel Cell Technologies*. JRC Joint Research Centre.
49 Kolmetz, K. (2017). Refinery Catalytic Reforming Unit Selection, Sizing and Troubleshooting (Engineering Design Guidelines) 2017. *Handbook*, p. 94.
50 Keçebaş, A., Kayfeci, M., and Bayat, M. (2019). Electrochemical hydrogen generation. *Sol. Hydrogen Prod. Process Syst. Technol.* 299–317. https://doi.org/10.1016/B978-0-12-814853-2.00009-6.
51 McKenna, R.C., Bchini, Q., Weinand, J.M. et al. (2018). The future role of power-to-gas in the energy transition: regional and local techno-economic analyses in Baden-Württemberg. *Appl. Energy* 212: 386–400. https://doi.org/10.1016/j.apenergy.2017.12.017.

52 Bertuccioli, L., Chan, A., Hart, D. et al. (2014). Fuel cells and hydrogen Joint undertaking Development of Water Electrolysis in the European Union.

53 Schiebahn, S., Grube, T., Robinius, M. et al. (2015). Power to gas: technological overview, systems analysis and economic assessment for a case study in Germany. *Int. J. Hydrog. Energy* 40: 4285–4294. https://doi.org/10.1016/j.ijhydene.2015.01.123.

54 Buttler, A. and Spliethoff, H. (2018). Current status of water electrolysis for energy storage, grid balancing and sector coupling via power-to-gas and power-to-liquids: a review. *Renew. Sust. Energ. Rev.* 82: 2440–2454. https://doi.org/10.1016/j.rser.2017.09.003.

55 RSE (2021). SpA Ricerca Sistema Energetico. Idrogeno Un vettore energetico per la decarbonizzazione, Milano.

56 Schalenbach, M., Zeradjanin, A.R., Kasian, O. et al. (2018). A perspective on low-temperature water electrolysis - challenges in alkaline and acidic technology. *Int. J. Electrochem. Sci.* 13: 1173–1226. https://doi.org/10.20964/2018.02.26.

57 Del Zotto, L., Monforti Ferrario, A., Hatunoglu, A. et al. (2021). Experimental procedures & first results of an innovative solid oxide fuel cell test rig: parametric analysis and stability test. *Energies* 14: 2038. https://doi.org/10.3390/en14082038.

58 EERA European Energy Research Alliance, Joint Research Programme on Fuel Cells and Hydrogen Technologies (JP FCH) & Hydrogen Europe Research (HER), Research Grouping of the Fuel Cells and Hydrogen Joint Undertaking (FCH JU), (2020) KEY performance indicators (KPIS) for FCH research and innovation, 2020 – 2030 Version: 5.0.

59 Laurencin, J. and Mougin, J. (2015). High-temperature steam electrolysis. In: *Hydrogen Production by Electrolysis* (ed. A. Godula-Jopek), 191–272. https://doi.org/10.1002/9783527676507.ch6.

60 Hansen, J.B. (2015). Solid oxide electrolysis - a key enabling technology for sustainable energy scenarios. *Faraday Discuss.* 182: 9–48. https://doi.org/10.1039/c5fd90071a.

61 Barelli, L., Bidini, G., Cinti, G., and Milewski, J. (2020). High temperature electrolysis using Molten Carbonate Electrolyzer. *Int. J. Hydrog. Energy* https://doi.org/10.1016/j.ijhydene.2020.07.220.

62 Dincer, I. and Acar, C. (2014). Review and evaluation of hydrogen production methods for better sustainability. *Int. J. Hydrog. Energy* 40: 11094–11111. https://doi.org/10.1016/j.ijhydene.2014.12.035.

63 NREL (2011). Hydrogen production cost estimate using biomass gasification independent review. *Advances.* https://doi.org/10.2172/1028523.

64 Bocci, E., Di Carlo, A., McPhail, S.J. et al. (2014). Biomass to fuel cells state of the art: a review of the most innovative technology solutions. *Int. J. Hydrog. Energy* 39: 21876–21895. https://doi.org/10.1016/j.ijhydene.2014.09.022.

65 Parker, N., Fan, Y., and Ogden, J. (2010). From waste to hydrogen: an optimal design of energy production and distribution network. *Transp. Res. Part E Logist. Transp. Rev.* 46: 534–545. https://doi.org/10.1016/j.tre.2009.04.002.

66 McPhail, S.J., Cigolotti, V., and Moreno, A. (2012). Fuel cells in the waste-to-energy Chain. *Green Energy Technol.* 45: https://doi.org/10.1007/978-1-4471-2369-9.

67 Stetson, N.T., McWhorter, S., and Ahn, C.C. (2016). *Compendium of Hydrogen Energy: Introduction to Hydrogen Storage*, 1e, vol. 1 (ed. A.B. Subramani and T.N. Veziroglu). Elsevier, 2006, https://www.elsevier.com/books/compendium-of-hydrogen-energy/unknown/978-1-78242-361-4.

68 LeGault, M. (2012). CompositesWorld. Pressure vessel tank types, pp. 1–2. https://www.compositesworld.com/articles/pressure-vessel-tank-types (accessed 19 October 2021).

69 EIHP (2021). European Integrated Hydrogen Project 2021. http://www.eihp.org/ (accessed 19 October 2021).

70 Peschel, A. (2020). Industrial perspective on hydrogen purification, compression, storage, and distribution. *Fuel Cells* 20: 385–393. https://doi.org/10.1002/fuce.201900235.

71 Andersson, J. and Grönkvist, S. (2019). Large-scale storage of hydrogen. *Int. J. Hydrog. Energy* 44: 11901–11919. https://doi.org/10.1016/j.ijhydene.2019.03.063.

72 Hydroville (2021). How is hydrogen transported? | Hydroville, 1–5. http://hydroville.be/en/waterstof/hoe-transporteer-je-waterstof/ (accessed 19 October 2021).

73 McKinsey (2021). Hydrogen Insights 2021.

74 Quarton, C.J. and Samsatli, S. (2020). Should we inject hydrogen into gas grids ? Practicalities and whole-system value chain optimisation. *Appl. Energy* 275: 115172. https://doi.org/10.1016/j.apenergy.2020.115172.

75 Lo Basso, G., Nastasi, B., Astiaso Garcia, D., and Cumo, F. (2017). How to handle the Hydrogen enriched Natural Gas blends in combustion efficiency measurement procedure of conventional and condensing boilers. *Energy* 123: 615–636. https://doi.org/10.1016/j.energy.2017.02.042.

76 Melaina, M.W., Antonia, O., and Penev, M. (2013). Blending Hydrogen into Natural Gas Pipeline Networks: A Review of Key Issues Blending Hydrogen into Natural Gas Pipeline Networks: A Review of Key Issues 2013.

77 Bloomberg New Energy Finance (2020). Hydrogen Economy Outlook 2020, p. 12.

78 Steilen, M. and Jörissen, L. (2015). Hydrogen conversion into electricity and thermal energy by fuel cells: use of H_2-systems and batteries. *Electrochem. Energy Storage Renew. Sources Grid Balanc.* 143–158. https://doi.org/10.1016/B978-0-444-62616-5.00010-3.

79 Kotowicz, J., Węcel, D., and Jurczyk, M. (2018). Analysis of component operation in power-to-gas-to-power installations. *Appl. Energy* 216: 45–59. https://doi.org/10.1016/j.apenergy.2018.02.050.

80 Mori, D. and Hirose, K. (2009). Recent challenges of hydrogen storage technologies for fuel cell vehicles. *Int. J. Hydrog. Energy* 34: 4569–4574. https://doi.org/10.1016/j.ijhydene.2008.07.115.

81 Bartolini, A., Carducci, F., Muñoz, C.B., and Comodi, G. (2020). Energy storage and multi energy systems in local energy communities with high renewable energy penetration. *Renew. Energy* 159: 595–609. https://doi.org/10.1016/j.renene.2020.05.131.

82 Blanco, H. and Faaij, A. (2018). A review at the role of storage in energy systems with a focus on Power to Gas and long-term storage. *Renew. Sust. Energ. Rev.* 81: 1049–1086. https://doi.org/10.1016/j.rser.2017.07.062.

83 Kharel, S. and Shabani, B. (2018). Hydrogen as a long-term large-scale energy storage solution to support renewables. *Energies* 11: https://doi.org/10.3390/en11102825.

84 Monforti Ferrario, A., Bartolini, A., Comodi, G. et al. (2021). Optimal sizing of battery and hydrogen energy storage systems configurations in a hybrid renewable microgrid, 09002.

85 Monforti Ferrario, A., Vivas, F.J., Segura Manzano, F. et al. (2020). Hydrogen vs. battery in the long-term operation. A comparative between energy management strategies for hybrid renewable microgrids. *Electronics* 9: 1–27.

86 Berrada, A., Loudiyi, K., and Zorkani, I. (2016). Valuation of energy storage in energy and regulation markets. *Energy* 115: 1109–1118. https://doi.org/10.1016/j.energy.2016.09.093.

87 Mohanpurkar, M., Luo, Y., Terlip, D. et al. (2017). Electrolyzers enhancing flexibility in electric grids. *Energies* 10: 1DUMMU. https://doi.org/10.3390/en10111836.

88 Vivas, F.J., De las Heras, A., Segura, F., and Andújar, J.M. (2018). A review of energy management strategies for renewable hybrid energy systems with hydrogen backup. *Renew. Sust. Energ. Rev.* 82: 126–155. https://doi.org/10.1016/j.rser.2017.09.014.

89 Cau, G., Cocco, D., Petrollese, M. et al. (2014). Energy management strategy based on short-term generation scheduling for a renewable microgrid using a hydrogen storage system. *Energy Convers. Manag.* 87: 820–831. https://doi.org/10.1016/j.enconman.2014.07.078.

90 Torreglosa, J.P., García, P., Fernández, L.M., and Jurado, F. (2014). Hierarchical energy management system for stand-alone hybrid system based on generation costs and cascade control. *Energy Convers. Manag.* 77: 514–526. https://doi.org/10.1016/j.enconman.2013.10.031.

91 Fertilizers Europe (2000). Best Available Techniques (BAT) for Pollution Prevention and Control in the European Fertilizer Industry.

92 Armijo, J. and Philibert, C. (2019). Flexible production of green hydrogen and ammonia from variable solar and wind energy: case study of Chile and Argentina. *Int. J. Hydrog. Energy* https://doi.org/10.1016/j.ijhydene.2019.11.028.

93 Baharudin, L. and Watson, M.J. (2017). Hydrogen applications and research activities in its production routes through catalytic hydrocarbon conversion. *Rev. Chem. Eng.* 34: 43–72. https://doi.org/10.1515/revce-2016-0040.

94 Monforti Ferrario, A., Santoni, F., Della Pietra, M. et al. (2021). A system integration analysis of a molten carbonate electrolysis cell as an off-gas recovery system in a steam-reforming process of an oil refiner. *Front. Energy Res.* https://doi.org/10.3389/fenrg.2021.655915.

95 EPRS European Parliamentary Research Service (2020). The potential of hydrogen for decarbonising steel production.

96 COAG Energy Council Hydrogen Working Group (2019). Hydrogen for industrial users.

97 Association Mineral Products (2019). Options for switching UK cement production sites to near zero CO_2 emission fuel: technical and financial feasibility.

98 Spinelli, M., Romano, M.C., Consonni, S. et al. (2014). Application of molten carbonate fuel cells in cement plants for CO_2 capture and clean power generation. *Energy Procedia* 63: 6517–6526. https://doi.org/10.1016/j.egypro.2014.11.687.

99 De Silvestri, A., Stendardo, S., Della Pietra, M., and Borello, D. (2021). Decarbonizing cement plants via a fully integrated calcium looping-molten carbonate fuel cell process: assessment of a model for fuel cell performance predictions under different operating conditions. *Int. J. Hydrog. Energy* https://doi.org/10.1016/j.ijhydene.2020.12.024.

100 Tremel, A., Wasserscheid, P., Baldauf, M., and Hammer, T. (2015). Techno-economic analysis for the synthesis of liquid and gaseous fuels based on hydrogen production via electrolysis. *Int. J. Hydrog. Energy* 1–8. https://doi.org/10.1016/j.ijhydene.2015.01.097.

101 IEA International Energy Agency (2020). Global EV Outlook 2020. https://doi.org/10.1787/d394399e-en.

102 Hydrogen Council (2017). How hydrogen empowers the energy transition, pp. 1–28. https://hydrogencouncil.com/wp-content/uploads/2017/06/Hydrogen-Council-Vision-Document.pdf.

103 FCH-JU FC and HJU (2019). Hydrogen Roadmap Europe. https://doi.org/10.2843/249013.

104 FCH-JU FC and HJU (2018). Fuel Cells and Hydrogen for Green Energy in European Cities and Regions 2018.

105 Reddi, K., Elgowainy, A., Rustagi, N., and Gupta, E. (2017). Impact of hydrogen refueling configurations and market parameters on the refueling cost of hydrogen. *Int. J. Hydrog. Energy* 42: 21855–21865. https://doi.org/10.1016/j.ijhydene.2017.05.122.

106 Monforti-Ferrario, A., Hamedani, R., Zotto, D. et al. (2018). Techno-economic analysis of in-situ production by electrolysis and biomass gasification and delivery systems for hydrogen refuelling stations: rome case study. *Energy Procedia* 148C: 82–89.

107 Caponi, R., Monforti Ferrario, A., Bocci, E. et al. (2021). Thermodynamic modeling of hydrogen refueling for heavy-duty fuel cell buses and comparison with aggregated real data. *Int. J. Hydrog. Energy* https://doi.org/10.1016/j.ijhydene.2021.02.224.

108 Minutillo, M., Perna, A., Forcina, A. et al. (2020). Analyzing the levelized cost of hydrogen in refueling stations with on-site hydrogen production via water electrolysis in the Italian scenario. *Int. J. Hydrog. Energy* 46: 13667–13677. https://doi.org/10.1016/j.ijhydene.2020.11.110.

109 Khotseng, L. (2020). Fuel cell thermodynamics. *Thermodyn. Energy Eng.* 1–17. https://doi.org/10.5772/intechopen.90141.

110 Cassir, M., Jones, D., Ringuedé, A., and Lair, V. (2013). 15 - Electrochemical devices for energy: fuel cells and electrolytic cells. In: *Handbook of Membrane Reactors, Reactor Types and Industrial Applications*, vol. 2 (ed. A. Basile), 553–606. Woodhead Publishing, ISBN 9780857094155, https://doi.org/10.1533/9780857097347.3.553.

111 Pachauri, R.K. and Chauhan, Y.K. (2015). A study, analysis and power management schemes for fuel cells. *Renew. Sust. Energ. Rev.* 43: 1301–1319. https://doi.org/10.1016/j.rser.2014.11.098.

112 Energy USD (2015). Fuel Cells Factsheet. Fuel Cell Technol Off Fuel 2015, pp. 23–54. https://doi.org/10.1007/978-1-349-04829-8_3.

113 Weidner, E., Cebolla Ortiz, R., and Davies, J. (2019). *Global Deployment of Large Capacity Stationary Fuel Cells*. JRC Joint Research Centre https://doi.org/10.2760/372263.

114 Hart, D., Jones, S., Lewis, J. (2020). The fuel cell industry review. *E4Tech*.

115 Dincer, I. and Rosen, M.A. (2013). Chapter 18 - Exergy analysis of fuel cell systems. *Exergy Energy, Environ. Sustain. Dev.* 978- 0- 08– 0: 363–382. https://doi.org/10.1016/C2010-0-68369-6.

116 Hawkes, A., Staffell, I., Brett, D., and Brandon, N. (2009). Fuel cells for micro-combined heat and power generation. *Energy Environ. Sci.* 2: 729–744. https://doi.org/10.1039/b902222h.

117 Mcphail, S.J., Aarva, A., Devianto, H. et al. (2010). SOFC and MCFC : Commonalities and opportunities for integrated research. *Int. J. Hydrog. Energy* 36: 10337–10345. https://doi.org/10.1016/j.ijhydene.2010.09.071.

118 U.S. Department of Energy (2004). *Fuel Cell Handbook*, 7e. https://doi.org/10.1016/s0031-9422(00)82398-5.

119 Garcia, P., Fernandez, L., Garcia, C.A., and Jurado, F. (2010). Comparative study of PEM fuel cell models for integration in propulsion systems of urban public transport. *Fuel Cells* 10: 1024.

120 Arsalis, A. (2019). A comprehensive review of fuel cell-based micro-combined-heat-and-power systems. *Renew. Sust. Energ. Rev.* 105: 391–414. https://doi.org/10.1016/j.rser.2019.02.013.

121 Bartolucci, L., Cordiner, S., Mulone, V., and Pasquale, S. (2019). Fuel cell based hybrid renewable energy systems for o ff -grid telecom stations: data analysis and system optimization. *Appl. Energy* 252: 113386. https://doi.org/10.1016/j.apenergy.2019.113386.

122 Comodi, G., Cioccolanti, L., and Renzi, M. (2014). Modelling the Italian household sector at the municipal scale: micro-CHP, renewables and energy efficiency. *Energy* 68: 92–103. https://doi.org/10.1016/j.energy.2014.02.055.

123 Mcphail, S.J., Leto, L., Della Pietra, M. et al. (2015). International status of molten carbonate fuel cells 2015. *Advanced Fuel Cells Implementing Agreement*, Annex 23 - MCFC 2015, pp. 1–34.

124 Della Pietra, M., McPhail, S.J., Prabhakar, S. et al. (2016). Accelerated test for MCFC button cells: first findings. *Int. J. Hydrog. Energy* 41: 18807–18814. https://doi.org/10.1016/j.ijhydene.2016.07.021.

125 Rinaldi, G., McLarty, D., Brouwer, J. et al. (2015). Study of CO_2 recovery in a carbonate fuel cell tri-generation plant. *J. Power Sources* 284: 16–26. https://doi.org/10.1016/j.jpowsour.2015.02.147.

126 Desideri, U., Proietti, S., Sdringola, P. et al. (2012). MCFC-based CO_2 capture system for small scale CHP plants. *Int. J. Hydrog. Energy* 37: 19295–19303. https://doi.org/10.1016/j.ijhydene.2012.05.048.

Annex 1: Relevant European and International Projects 2000–2020

List of relevant projects (non-exhaustive).

Region	Title and description	Funding program and grant ID	Duration dates budget (financed)	Relevant partners
EU Spain	**Green Hysland**[a)]: Deployment of a H_2 ecosystem on the island of Mallorca **Main contribution:** H_2 islands by deploying a fully integrated and functioning H_2 ecosystem in the island of Mallorca, Spain	H2020 FCH-JU ID:101007201	5 yr 2021–2025 20 M€ (10 M€)	Enagas, Acciona, Transports Urbans de Palma de Mallorca
EU Germany, Scandi-navia, France, and the UK	**Hydrogen Mobility Europe**[b)]: Flagship project giving fuel cell electric vehicle (FCEV) drivers access to the first truly pan-European network of hydrogen refueling stations. **Main contribution**: The H2ME Project will significantly expand the European hydrogen vehicles' fleet and aims to confirm the technical and commercial readiness of vehicles, fueling stations, and hydrogen production techniques	H2020 ID: 671438	5 yr 2015–2020 62 M€ (32 M€)	40 Partners from 9 countries, drawing together their expertise from across the transport, hydrogen, and energy industries
EU	**JIVE and JIVE 2**[c)]: Deployment of 142 fuel cell buses (fleets 20–30 buses) across nine locations, more than doubling the number of FC buses operating in Europe. **Main contribution**: Deploy large-scale fuel cell bus mobility and refueling infrastructure throughout Europe	H2020, FCH-JU ID: 735582	5 yr + 5 yr 2017–2023 228 M€ (57 M€)	50 Partners between bus operators, municipalities, suppliers, RO, engineering, and associations
EU Spain	**H2PORTS**[d)]: Implementing fuel cells and hydrogen technologies in ports. **Main contribution**: Deployment of port equipment equipped with FC technologies and use of hydrogen as zero-emission fuel through innovative market-sided solutions ready for market adoption by the end of the project	H2020 ID: 826339	3 yr 2019–2022 4 M€ (4 M€)	FVP, ATENA, Hyster Yale, Grimaldi Group, MSC, Ballard, CNH2

EU The Netherlands	**Hydrogen Energy Applications for Valley Environments in Northern Netherlands**[e]**:** Large-scale deployment and integration of six existing and planned project clusters in the Netherlands. **Main contribution**: Production, distribution, storage, and local end use of H_2 into a fully integrated and functioning "H_2 valley" (H2V), which can serve as a blueprint for replication across Europe and beyond	H2020 ID: 875090	5 yr 2020–2025 96 M€ (20 M€)	Total NL, Shell NL, Nouryon Industrial Chemicals B.V.
EU The Netherlands	**Port of Amsterdam, H2ermes**[f]**; Port of Amsterdam**: Establishment of a 100 MW hydrogen plant on the Tata Steel site in IJmuiden, together with Nouryon and Tata Steel; Establishment of a 250 MW hydrogen plant BPs refinery **Main contribution:** Port of Amsterdam focuses on the infrastructure for the further distribution of green hydrogen. Port of Rotterdam frontrunner in the energy transition, which is an important differentiator for the port industry	Private investment		Nouryon and Tata Steel, Port of Amsterdam BP, Nouryon and Port of Rotterdam
EU Belgium	**Port of Antwerp North-C-Methanol project**[g]: 65 MW electrolyzer to convert the collected CO_2 emissions of local industrial players into green methanol **Main contribution:** New circular economy in the North Sea Port area	Private investment	2024–2030	Engie, HELM Proman Methanol, ArcelorMittal, Alco Biofuel, and Nippon Gases
EU Germany	**REFHYNE**[h]**:** Integration of clean hydrogen into refinery processes including the desulphurization of conventional fuels **Main contribution:** It lays the foundation for future large-scale, commercial 100 MW industrial plants	H2020, FCH JU ID: 779579	4 yr 2018–2022 20 M€ (10 M€)	Shell's Rheinland Refinery ITM Power, SINTEF, Element Energy, and Sphera
EU Sweden	**HYBRIT**[i] **fossil-free steel**: Developing the technology and the value chain for hydrogen-based iron and steel production for a fossil-free future **Main contribution:** Direct reduction of iron ore using renewable energy and hydrogen; the hydrogen reacts with the oxygen in the iron ore, thus creating metallic iron and water	Swedish Energy Agency	5 yr 2016–2021 250 MSEK	SSAB, LKAB, and Vattenfall

(*continued*)

Annex 1: (Continued)

List of relevant projects (non-exhaustive).				
Region	**Title and description**	**Funding program and grant ID**	**Duration dates budget (financed)**	**Relevant partners**
EU Austria	**H2FUTURE**[j]: 6 MW PEM electrolyzer provided by Siemens is working on a Voestalpine steel site using power from Verbund's almost entirely renewable-based portfolio in Austria **Main contribution:** Advance in the decarbonization of the steel industry	H2020, FCH JU Hydrogen Europe and N.ERGHY ID: 735503	4 yr 2017–2021 18 M€ (12 M€)	Verbund, Voestalpine, Siemens, TNO, and K1-MET
EU Spain	**Puertollano green hydrogen plant**[k]: Including 100 MW PV plant, storage system, and 20 MW electrolyzer for producing green ammonia **Main contribution:** First in Europe producing green fertilizers	Private investors	Operation planned by 2021 150 M€	Iberdrola, Fertiberia
EU Spain	**Talgo Vittal-One:** Development of a hybrid traction powertrain (H_2 FC+ batteries) on an existing train and refueling station **Main contribution:** Transition of non-electrified railway line modular design for future-generation trains extended to other rail vehicles, including the conversion of diesel vehicles to hydrogen	Private investors	2021–2023	Talgo TRAVCA loco plus TPH unpowered coaches
Extra-EU KSA	**NEOM**[l]: Self-sustaining smart city, including hydrogen as one of the most important energy carriers of the future (1, 2 Mt/yr)	Private investment	7 BUSD	NEOM (KSA), ACWA Power, Airproducts
Extra-EU Chile	**HIF – Highly Innovative Fuels**[m]: Integrated and commercial large-scale plant for climate-neutral e-Methanol (0.75 ML/yr by 2022 to over 550 ML/yr by 2026) **Main contribution:** Producing e-fuels from Patagonian wind resources and carbon capture technology at commercial costs toward the decarbonization of the global transport sector	BMWi Germany	2 yr 2021–2022 38 MUSD	Mabanaft, Siemens, Porsche, AME, EGP Chile, ENAP

Extra-EU Chile	**HYDRA**[n)]: Design and supply of a new powertrain for off road mining electric drive trucks (>2 MW) to run on renewable hydrogen instead of diesel **Main contribution:** Optimize the design that could replace the traditional diesel powertrain for Chilean copper mining operations. Establish safety protocols for hydrogen use at scale in the copper mining industry	Chilean Economic Development Agency (CORFO)	2 yr 2021–2022 1.3 MUSD	ENGIE, CSIRO, Mining3
Extra-EU Australia	**HYSTRA**[o)]: Demonstration project for the establishment of mass hydrogen marine transportation supply chain derived from coal gasification **Main contribution:** Development of technology for long-distance transportation of liquid H_2 and hydrogen loading and unloading technologies	NEDO, Australian Government	2020–2030	Iwatani Corporation, Kawasaki Heavy Industries, Shell JP, J-POWER, Marubeni Corporation, ENEOS Corporation
Extra-EU Japan	**Fukushima Hydrogen Energy Research Field**[p)]: Establish a hydrogen usage business model and hydrogen sales business model for demand response **Main contribution:** Adjust supply/demand power grid in order to maximize utilization to establish low-cost green hydrogen production technology			NEDO, Toshiba ESS, Tohoku Electric Power Co., Inc., and Iwatani Co.
Extra-EU Canada	**Markham Energy Storage, Ontario**[q)]: Demonstrate H_2 grid service to enhance the flexibility and reliability of operating the grid for ISO of Ontario (±1.05 MW regulation service, <2 s response time, 2 MW/s ramp rate) **Main contribution:** Link the energy sector through the production of renewable H_2 for zero-emission fuel cell electric vehicles (train, bus, and truck fleets) or other applications while providing grid services to system operators			Enbridge Gas Distribution and Hydrogenics Corporation

(*continued*)

Annex 1: (Continued)

List of relevant projects (non-exhaustive).

Region	Title and description	Funding program and grant ID	Duration dates budget (financed)	Relevant partners
Extra-EU Argentina	**Hychico**[r)]: Production of green hydrogen from wind in the south of Argentina **Main contribution:** Gain experience to carry H_2 through polymeric 2.3 km long pipeline to underground hydrogen storage in depleted oil and gas reservoirs	Cont. of HyUnder 2012–2014 EU 303417		Shell, Eon, ECN, Ludwig-Bölkow Systemtechnik GmbH

a) https://cordis.europa.eu/project/id/101007201/it
b) https://h2me.eu
c) https://cordis.europa.eu/project/id/735582/it; https://cordis.europa.eu/project/id/779563
d) https://h2ports.eu
e) https://cordis.europa.eu/project/id/875090/it
f) https://www.portofamsterdam.com/en/business/cargo-flows/liquid-bulk/h2-hydrogen
g) https://www.proman.org/news/proman-to-build-worlds-largest-green-methanol-plant-at-north-sea-renewables-hub
h) https://refhyne.eu
i) https://www.hybritdevelopment.se/en
j) https://www.h2future-project.eu
k) https://www.iberdrola.com/conocenos/lineas-negocio/proyectos-emblematicos/puertollano-planta-hidrogeno-verde
l) https://www.neom.com/en-us
m) https://www.hif.cl
n) https://www.mining3.com/engie-and-mining3s-renewable-hydrogen-powertrain-project-receives-funding-support-from-chilean-economic-development-agency
o) http://www.hystra.or.jp/en
p) https://www.toshiba-energy.com/en/info/info2020_0307.htm
q) http://www.h2gta.ca/wp-content/uploads/2018/06/AMurray-Hydrogenics-Markham-Energy-Storage-Facility-061318.pdf
r) www.hychico.com.ar/eng/hydrogen-plant.html

5

Review on the Energy Storage Technologies with the Focus on Multi-Energy Systems

Morteza Vahid-Ghavidel[1,2], Sara Javadi[3], Matthew Gough[1,2], Mohammad S. Javadi[2], Sérgio F. Santos[2,4], Miadreza Shafie-khah[3], and João P.S. Catalão[1,2]

[1] *University of Porto (FEUP), Faculty of Engineering, R. Dr. Roberto Frias, Porto 4200-465, Portugal*
[2] *Institute for Systems and Computer Engineering, Technology and Science (INESC TEC), Campus of FEUP, Porto 4200-465, Portugal*
[3] *University of Vaasa, School of Technology and Innovations, Wolffintie 34, Vaasa 65200, Finland*
[4] *Portucalense University Infante D. Henrique (UPT), R. Dr. António Bernardino de Almeida 541, Porto 4200-072, Portugal*

5.1 Introduction

Energy storage systems (ESSs) play an essential role in multi-energy systems. These storage systems not only allow for the balancing between fluctuations in energy supply and demand but can also offer important means to convert energy from one form to another. This ability of energy storage systems to store energy across time, location, and energy type greatly increases the flexibility of the integrated energy systems [1]. This chapter provides a comprehensive overview of energy storage technologies being applied to multi-energy system and shows how these emerging technologies and systems play a critical role in any future energy system. Expectations are that needs for energy storage systems will triple by 2030 [1].

As the energy system evolves into one dominated by intermittent renewable energy sources, energy storage systems have experienced a massive increase in research and development from both academic and commercial developers [2]. This has led to immense reductions in cost and improvements in system efficiency, and this is expected to continue in the near-term future. Despite these improvements, there still needs to be further development in this sector. This can be done through a combination of deployment led innovation and active policies and regulation, which shape research and development [3].

The breadth of energy storage applications is rapidly accelerating and is shown in the emerging sector of hybrid or multi-energy systems. These are systems that combine various renewable energies, traditional energy sources, and storage systems, which complement each other to develop energy systems that take advantages of each of the component systems [2].

Within multi-energy systems, energy storage technologies can be applied at nearly all scales and time frames [4]. Each of the different energy storage technologies has

Technologies for Integrated Energy Systems and Networks, First Edition.
Edited by Giorgio Graditi and Marialaura Di Somma.

its own advantages and disadvantages, and the exact combination of technologies for a given application should be carefully studied to ensure that the full potential of energy storage systems in multi-energy systems is harnessed [5].

This chapter introduces the concept of energy storage and discusses the various types of energy storage systems. Recent projects using energy storage systems are highlighted to show the diversity of applications of energy storage systems. Then, the concept of multi-energy systems is discussed briefly, with a detailed focus on the application of energy storage technologies to multi-energy systems.

5.2 Energy Storage

In the energy system, an important component is energy storage. Within the power system, the energy storage can be defined as a component that can be employed to generate a form of energy or storing energy for use at a different time or location.

5.2.1 Main Concept of Energy Storage in the Power System

Applications of renewable energy resources around the world have developed and increased exponentially because of their advantages over traditional energy resources such as power plants that use fossil fuels. Despite these advantages, the fast growth of renewable energy resources has bought some challenges to the power system as well. One of the main issues of the renewable energy resources is the intermittent generation, which is dependent on many factors such as solar irradiation, wind speed, and direction, among others [6, 7]. These factors lead to fluctuations in electricity generation from renewable energy resources. Utilization of ESSs can address this issue and play a complementary role for renewable energy resources in order to create a reliable and sustainable energy system.

The top countries in terms of installed ESS projects or those to be built are shown in Figure 5.1. The total capacity of ESS that these 20 countries have is approximately 175.45 GW [8]. According to Figure 5.1, the United States is the country with the highest capacity of ESS. The United States has around 33.5 GW of ESS capacity [8, 9]. China and Japan are at the second and third places with 31.7 and 28.1 GW of capacity, respectively. If we take a deeper look at Figure 5.1, there is a major difference in the capacities of these three countries relative to the remaining 17 countries. For instance, Germany's ESS capacity is 7.8 GW, which is around four times lower than the ESS capacity of the United States. This can indicate that ESSs are considered as an important component of the energy system in the United States, China, and Japan. Where the dimensions of Germany and Japan are almost the same, however, the ESS capacity of Japan is much greater than Germany. However, the observed data show that the dimensions of the country have a positive relation on the ESS capacity on most of the cases. Another important observation from Figure 5.1 is related to the share of ESSs in each continent. For instance, in Asia and Australia, considering China, Japan, India, South Korea, Taiwan, and Australia, the ESS share

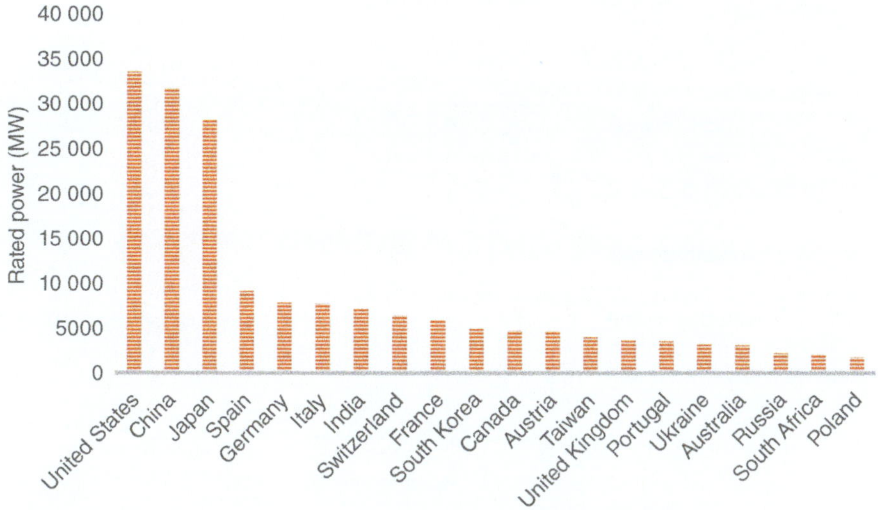

Figure 5.1 Top countries in ESS capacities. Source: Based on Nguyen [8].

is equal to 79 GW. In European counties, Figure 5.1 is 56 GW of ESS capacity, which includes Spain, Germany, Italy, Switzerland, France, Austria, United Kingdom, Portugal, Ukraine, Russia, and Poland. North America has 38.3 GW of capacity that includes the United States and Canada. Finally, South Africa is the only country from Africa that is listed in the top 20 countries with an ESS capacity of 2 GW. Therefore, Asia and Australia are leading, followed by Europe and North America. This number of ESS capacity for Europe indicates that if the energy network of European countries is connected together, their real ESS capacity is higher than that of the three top countries in the list (Figure 5.2).

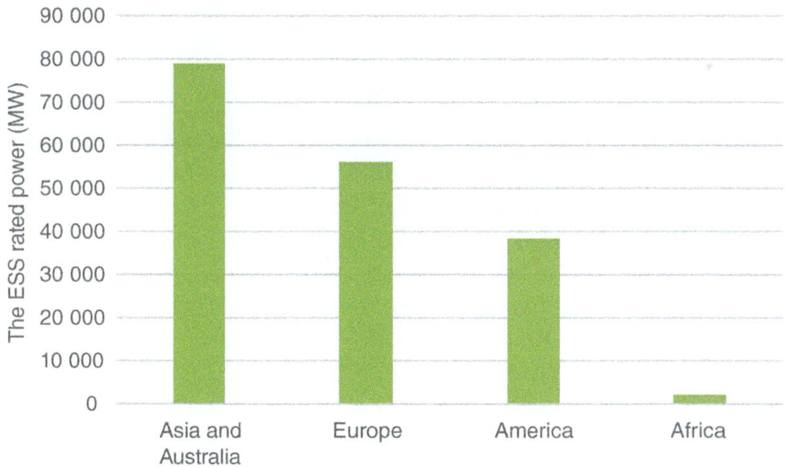

Figure 5.2 The cumulative capacity of ESSs of each continent. Source: Based on Nguyen [8].

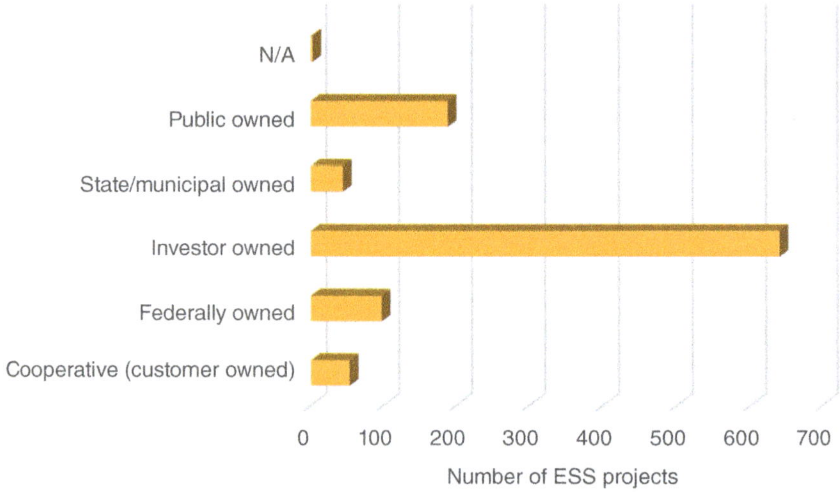

Figure 5.3 Type of the ESS ownership. Source: Based on Nguyen [8].

The ownership type of the current ESS projects is given in Figure 5.3. As shown in Figure 5.3, there are five main categories of ESS ownership. Most of the ESS projects are investor-owned projects, which indicate that they belong to the investment companies that are developing the project. There are 640 projects that are being implemented and managed by their investors. Public-owned ESS projects are in the second place and then federally owned and state/municipal-owned are next. There are 188 projects considered to be publicly owned projects and 53 projects that are owned by cooperatives. In Figure 5.3, it is shown that there are two projects that fall outside of the above-mentioned categories. In other words, their ownership is not belonged to the public, state, investor, or other above-mentioned categories.

5.2.2 Different Types of Energy Storage Systems

Many different technologies are being utilized for current ESS projects. The classification of these technologies is dependent on many factors such as the purpose of energy storage. For instance, they can be classified according to their operation duration or type of function [10, 11]. Electrical and thermal energies are the main types of energies that are being stored. The different energy storage technologies are listed in Figure 5.4. In this list, the main storage systems are as follows: mechanical storage, pumped hydro storage, lithium-ion battery, liquid air energy storage, lead–carbon, hydrogen storage, electromechanical, electrochemical, and thermal storage systems [12–19]. According to the information provided in the literature, electrochemical energy storage systems are the most popular and common storage technology [10]. There are currently at least 998 projects around the world that are categorized as electrochemical energy storage systems [8]. Pumped hydro storage technology is another common type, and there are more than 350 projects that use pumped hydro storage

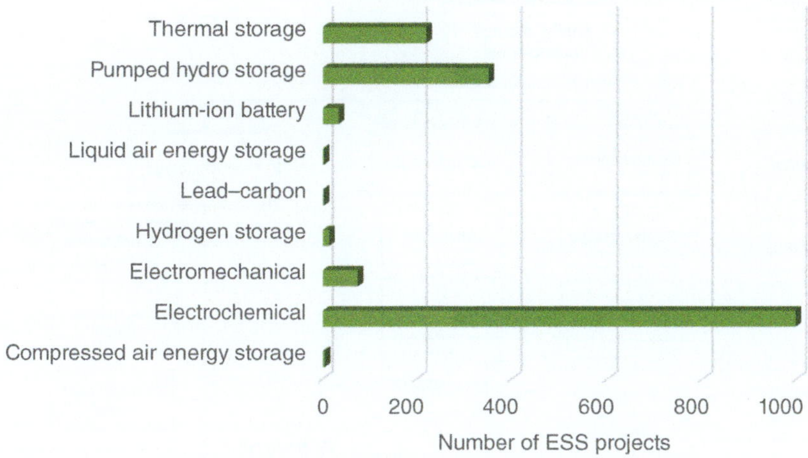

Figure 5.4 The various ESS technologies. Source: Based on Nguyen [8].

technology. There are 220 projects that use thermal storage technology according to the information shown in Figure 5.4. However, there are some technologies that are less common and popular for companies that are designing and implementing energy storage technologies for the energy network. These less-common technologies are liquid air energy storage and lead–carbon technologies [15, 16].

Based on a report from the US Department of Energy, the global capacity of energy storage systems is equal to 191.2 GW in 2020, and this is a 12% increase compared to 2017 [20]. Table 5.1 shows the cumulative capacity of energy storage systems for each type of energy storage technology. According to this table, electrochemical systems have the greatest capacity of rated power among all the currently available storage technologies at 118.2 GW. Lithium-ion batteries are also becoming an important source of energy storage because of their application in the electric vehicles section.

Table 5.1 The rated power of each ESS technology.

ESS technology	Rated power (GW)
Electrochemical storage	118.2
Compressed air energy storage	1.6
Electromechanical	4.5
Hydrogen storage	1.3
Lead–carbon	0.02
Liquid air energy storage	0.33
Lithium-ion battery	1.7
Pumped hydro storage	51.8
Thermal storage	12.5

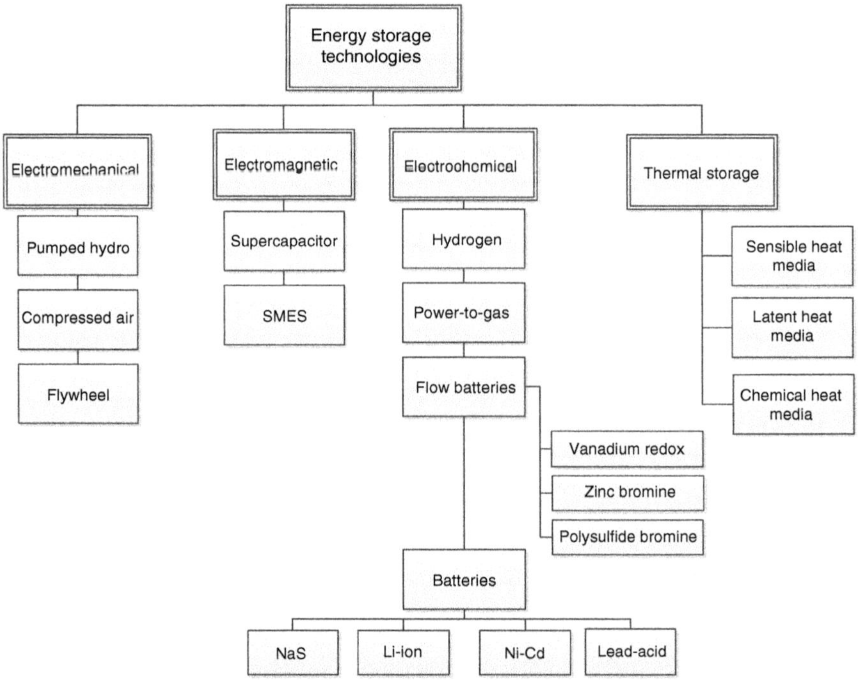

Figure 5.5 The main classification of energy storage systems. Source: Moseley and Garche [22] / With permission of Elsevier.

Then, the pumped hydro storage with 51.8 GW captures the second greatest cumulative rated power capacity. The lithium-ion battery is also becoming an important source of energy storage because of its application in the electric vehicles sector [21]. Next in terms of installed capacity, there is pumped hydro storage with 51.8 GW of rated power capacity.

It should be noted that in this chapter, energy storage systems are classified into four main categories, which are electrochemical energy storage, electromechanical energy storage, electromagnetic energy storage, and thermal energy storage (TES) as depicted in Figure 5.5. Each category will be introduced and explained in the following sections.

5.2.2.1 Electromechanical Energy Storage Systems

The most established ESS in high power application is pumped hydroelectric storage (PHS), which has been used since the 1890s. PHS is a sustainable energy source, with the flexibility and storage capacity to improve grid stability [23]. PHS is operated in low-demand periods; extra energy is used from the grid to pump water from a lower to an upper reservoir. Low-cost surplus off-peak electric power is normally used to run the pump. In high-demand periods, the opposite occurs with water flowing from the upper reservoir to the lower one and turning a turbine to generate electricity to export to the grid. The gravitational potential energy of the stored water determines the energy storage potential [22]. PHS allows energy from renewable sources such as solar and wind, or excess electricity from sources such as coal or nuclear, to be saved for periods of higher demand. PHS is a suitable technology for small autonomous island grids and large-scale energy storage. The energy efficiency of PHS is approximately 70–80% [23].

Another type of mechanical energy storage is compressed air energy storage (CAES). It also has a relatively simple operating principle. Air is compressed by an electrical compressor, and this compressed air can be stored in suitable storage vessels. In fact, electrical energy is changed to potential energy of compressed air. An air turbine expands the air, and it releases back the energy to the grid [24]. In comparison with other energy storage systems, CAES has a large storage capacity, low self-discharge, and a long lifetime [25]. These characteristics make CAES very suitable and cost-effective for bulk energy storage systems. In advanced CAES projects, the efficiency has been improved (around 70–80% efficiency) [26]. A vast amount of compressed air can be stored underground, so CAES can provide a large amount of the world's future energy storage demands [25].

Another common type of electromechanical storage technology is flywheels [27]. Flywheels consist of a massive rotating cylinder, attached to a shaft, which is supported on a stator. The cylinder rotates and stores kinetic energy. The flywheel is connected to a motor–generator that interacts with the grid through advanced power electronics. When the system is utilized as a motor and a generator, it is being charged and discharged, respectively. Nowadays, some magnetic bearings are used in order to decrease friction and shear. To maintain efficiency, the flywheel system is operated under vacuum to reduce drag.

Low maintenance, long lifetimes, and low environmental impacts are some of the advantages of flywheel energy storage systems. Flywheels are more applicable to short-term storage systems as the self-discharge rate is nearly 20% of the hourly stored energy [22]. Flywheel energy storage systems are good choices for various applications in power systems such as power quality improvements, power smoothing, renewable energy integration support, and stability improvements [28].

5.2.2.2 Electromagnetic Energy Storage Systems

One of the systems used to store the energy electromagnetically is supercapacitor. It is made from electrochemical cells containing two electrodes, an electrolyte, and a membrane. The porous membrane provides an area for the ions to transfer between the electrodes. No chemical reaction occurs in supercapacitors in contrast to what happens in batteries. Supercapacitors store the energy in the cells electrostatically. The anode contains negative charges, the cathode contains positive charges, and the electrolyte contains both. By applying a voltage to the electrodes, an electrical double layer forms in the vicinity of the anode and cathode. In fact, the electric field created by these double layers is where the energy is stored [22].

Because of the fast charge/discharge and high power density, supercapacitors are applicable as supplementary energy sources in electric vehicles, consumer electronics, and industrial fields. However, because of their fast self-discharge and low-energy densities, supercapacitors are not suitable as primary power sources [29]. To overcome the issues, some improvements are needed in configuration, electrode material, and electrolyte.

Another system for storing energy in a magnetic field is superconducting magnetic energy storage (SMES). SMES system stores energy in a magnetic field. This magnetic field is generated by a DC current traveling through a superconducting coil [30]. The wire is made of a superconducting material that is cryogenically kept cold, so the electric current passes through the coil with almost zero resistance. This allows the energy to be stored in the system for a longer period. Normally, the superconducting material can be mercury, vanadium, and niobium–titanium. To discharge the stored energy in an SMES, the conductive coil is connected to an AC power convertor. SMES systems are very efficient storage systems (around 90% efficiency), but they have very low energy densities and they are still far from being economically lasting [30, 31].

5.2.2.3 Electrochemical Energy Storage Systems

Hydrogen energy storage is a form of electrochemical energy storage in which electrical power is converted into hydrogen by an electrolyzer [32]. Later, this stored energy can be released by using gas as fuel in a combustion engine or a fuel cell [33]. Electrolysis of water is a simple process to produce hydrogen. The efficiency of water electrolysis depends on the technology, hydrogen production rate, and pressure level [22]. Most commonly, hydrogen is stored as a compressed gas in a container. Also, it can be stored in very low temperature as a cryogenic liquid. Some other methods such as metal hydride materials or chemical hydrides can be used to store hydrogen. In this method, the hydrogen is bonded to a material, and it can be released as required. Hydrogen can be utilized as fuel in a gas turbine, piston engines, and hydrogen fuel cells. Hydrogen energy storage systems can provide much longer duration storage compared to batteries [33].

Battery energy storage systems (BESS) are a family of technologies developed for storing electric charge by using batteries. In most of the energy storage systems with batteries, electrical energy is converted into chemical energy and vice versa. Redox, reduction, and oxidation reactions occur in the battery cell. Each battery consists of two electrodes, an electrolyte, a separator, and a container. The electrolyte is a material in which the ions can be transferred between the anode and cathode, and the redox reaction can take place. This electrolyte is an electronic insulation material. The separator prevents internal short circuits of the battery from occurring and the container is needed to enclose and protect the battery cell [22].

Battery energy storage systems have the advantages of small footprint and no restrictions on geographical locations where they could be located. Other storage technologies such as PHS and CAES are only suitable for a limited number of locations. For instance, topological conditions, long development time, and large land use are the main constraints in the development of PHS projects [34]. Batteries are of various types such as lithium-ion, lead-acid, sodium sulfur, zinc bromine, and flow.

5.2.2.4 Thermal Energy Storage Systems

Another form of energy storage is TES. In TES systems, thermal energy is stored by heating or cooling a storage medium [35, 36]. The stored energy can be released later for power generation and other demands where it can generate steam for electricity production [3]. In this storage system, different materials with different thermal properties can be used and various results can be achieved. TES systems are commonly used in buildings and industrial processes. Solar thermal systems have the most common application in TES systems. There should be a heat-sensitive material in a solar power plant such as molten salt. The solar field gathers the energy from the sun and heats up the molten salt. A heat transfer fluid is heated up by the hot salt through a heat exchanger and then a turbine (connected to a generator) is spun using this fluid. Even if there is no sun, the turbine can be run with the heat stored in the molten salt [22].

5.2.3 Advantages of Storage in the Energy System

The energy storage technologies are employed in power grids for various reasons [37–39]. The most common advantages of the application of energy storage systems are given in Table 5.2. In this table, the various services that the storage technologies are being used for are listed in the first column. In the first row, the different storage technologies are given. The different services for the storage are electric supply capacity, electric energy time shift, on-site power, electric supply reserve capacity, frequency regulation, voltage support, and electricity bill management. According to Table 5.2, it is shown that electrochemical energy storage is broadly employed for electric energy time shift, frequency regulation, and renewable capacity firming. For example, there are 267 projects employed as the electrochemical energy storage system for electric energy time shift service. There are also 225 electrochemical energy storage systems designed or operating for the purpose of frequency regulation.

Furthermore, through analysis of Table 5.2, it can be seen that the electric energy time shift service is the only service that uses all types of energy storage technologies. Besides that, electrochemical storage is also being used for all of the power grid services. As there are no services that do not use the electrochemical type of energy storage. The electrochemical energy storage with 998 projects worldwide is the most popular storage technology used to supply one of the services to power grids.

According to the data published by the US Department of Energy [8], there are 1698 projects based on the development of ESSs until 2020. Thus, the number of ESS projects in the top countries regarding implementation of this technology is depicted in Figure 5.6. As illustrated in Figure 5.6, the United States by working on more than 740 projects is the world's leading country in this regard. China and Germany are in the next place by running 101 and 97 ESS projects, respectively. However, all of the projects are not in the operational phase. In order to present a better view regarding the applied ESS projects around the world, Figure 5.7 represents the

Table 5.2 The energy storage system projects based on their storage type and services.

	Electro-chemical	Electro-mechanical	Thermal storage	Hydrogen storage	Lead–carbon	Liquid air energy storage	Lithium-ion battery	Pumped hydro storage	Compressed air energy storage
Electric supply capacity	100	4	0	1	0	0	0	302	2
Electric energy time shift	267	16	84	1	2	1	1	325	3
On-site power	121	5	1	0	0	0	27	1	0
Electric supply reserve capacity – spinning	63	22	3	0	0	0	0	83	1
Frequency regulation	225	34	6	3	0	1	1	66	0
Voltage support	157	18	1	0	0	1	1	37	1
Load following (tertiary balancing)	62	4	0	2	0	0	0	27	0
Black start	51	2	0	0	0	0	0	15	0
Electric bill management with renewables	105	3	2	1	0	0	3	1	0
Stationary transmission/distribution upgrade deferral	44	0	4	1	0	0	0	0	0
Transmission support	27	1	3	0	0	1	0	2	0
Renewable capacity firming	278	11	55	8	0	0	1	15	3
Renewable energy time shift	202	17	62	6	0	0	28	18	3
Grid-connected commercial (reliability and quality)	81	4	3	0	0	0	2	2	0
Transportation services	51	1	0	2	0	0	0	0	0
Distribution upgrade due to solar	48	0	0	0	0	0	0	0	0
Ramping	54	4	1	1	0	0	0	5	0
Grid-connected residential (reliability)	47	0	3	0	0	0	1	0	0
Microgrid capability	170	9	1	0	0	0	26	0	0
Transmission congestion relief	20	1	3	1	0	1	0	3	0
Transmission support	27	1	3	0	0	1	0	2	0

Source: Based on Nguyen [8].

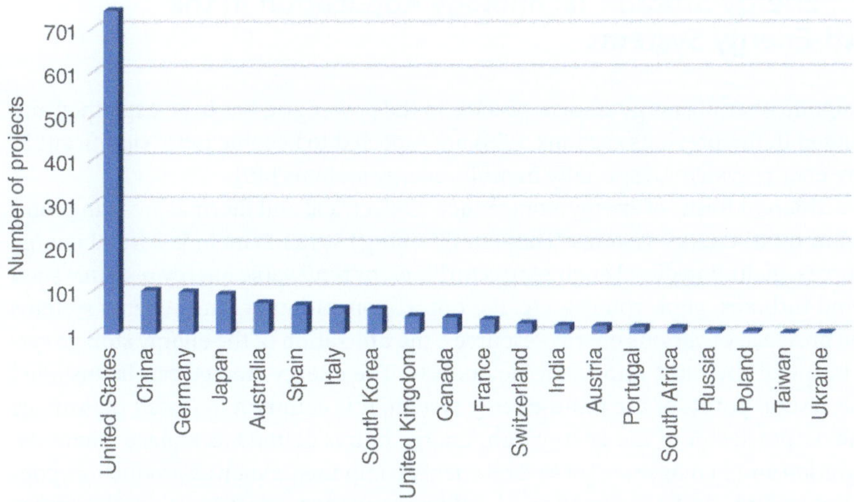

Figure 5.6 Number of ESS projects in the top countries. Source: Based on Nguyen [8].

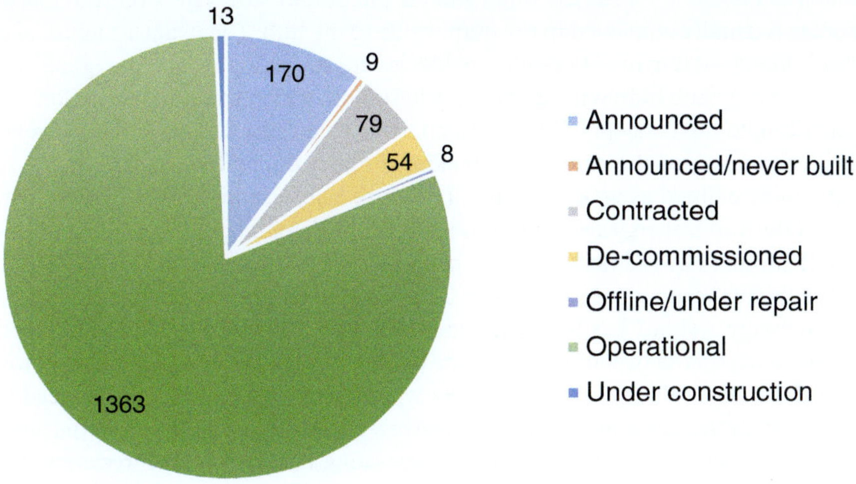

Figure 5.7 The current situation of the total ESS applications around the world. Source: Based on Nguyen [8].

current situation of the total ESS applications around the world. Figure 5.7 indicates that there are 1363 projects out of 1698 whose design and construction are done and they are in the operational mode. This number is equal to almost 80% of all of the currently published ESS projects. Hundred and eighty projects are also in the announcing phase, and from this number, nine ESSs were announced, but they were never built. More details about the current situations of the ESS programs can be found in Figure 5.7.

5.3 Energy Storage Technology Application in the Multi-Energy Systems

The significance of energy storage systems in the power grid has been explained and discussed in the previous sections. ESSs are expected to become more significant in future energy systems, especially in multi-energy systems [40].

The different forms of energy storage such as electrical and thermal are being combined in multi-energy systems. These multi-energy systems include several energy resources, including diesel engines, gas turbines, or renewable energy resource such as wind turbines, photovoltaics, etc. For optimal operation of multi-energy systems in the presence of various energy resources, the utilization of the energy storage system is one of the most important factors [41]. The energy storage can be installed in the several points of the multi-energy system. It is common to install the storage in the output sector of the energy hub. Energy hub is defined as a place where the integration and management of several energy components such as production, conversion, storage, and consumption of different energy carriers in the multi-energy systems occur [42]. While instalment of the storage in the input side of it is also proposed in some cases. Thus, hydrogen and electrical storage can be installed in both sides of the hub, i.e. the input side or the output side. However, the thermal storage is usually employed in the output side of the hub. The usual structure of the multi-energy system in the presence of ESS is depicted in Figure 5.8. In Figure 5.8a, a simple energy hub is drawn based on the definition, which was provided in Ref. [43]. According to the concept of energy hub in this study, any structure that correlates the generation and the consumption sides through transmission, conversion, and ESS can be defined as an energy hub. In Figure 5.8b, a more comprehensive structure of the multi-energy system is presented where demand can be supplied through electrical, cooling, and heating forms of energy.

The impact of storage size and forecasting period in the optimal operation of the multi-energy systems has been studied in Ref. [44]. This study proves that there is a reverse relation between the size of the energy storage system and the operational cost of the multi-energy system. In other words, larger energy storage will lead to a lower cost in the multi-energy system. However, the impact of the size of the energy storage is lower than the length of the forecasting horizon. Long forecasting horizon for the energy storage can lead to a reduction in the costs of the energy hub. Moreover, Ivalin Petkov et al. proved that the application of ESS has the capability to reduce emissions by 90% in multi-energy systems, which include renewable energy resources [45]. To better highlight the advantages of ESS in multi-energy systems, several applications of the energy storage systems are summarized in the following part of this section.

With the aim of optimization of the total operation cost of the energy hub and to consider the uncertainty posed from the distribution system including electricity, heating, and cooling loads, a power-to-gas storage with tri-state CAES system is proposed in Ref. [46]. Authors in Ref. [45] implemented a conditional value-at-risk approach for managing the uncertainties originating from wind power generation, electrical and thermal loads in a multi-energy system, which utilizes a CAES system

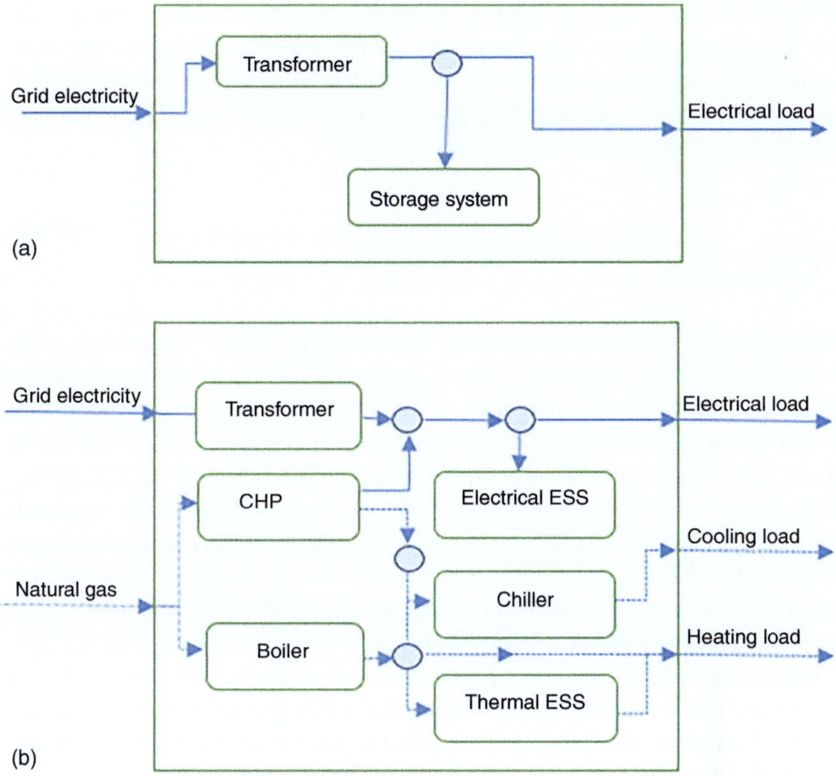

Figure 5.8 Multi-energy structure in the presence of ESS. Source: Modified from Mohammadi et al. [42].

in order to decrease the fluctuations caused by the renewable energy resources as well as to increase the freedom of the multi-energy system's operator. An underground hydrogen storage system is proposed in Ref. [47] in order to minimize the CO_2 emissions in the context of an integrated energy system by developing a mixed-integer linear program optimization model, which focuses on the dynamics of the stored energy during the hydrogen injection and withdraw processes.

The ESS is also a complementary component in the multi-energy systems for the demand response programs [48]. Since the goal of the operator from employment of the demand response programs is to meet the amount of the generation with the required load especially during the peak period. The energy storage technology can also support this goal by providing a percentage of the demand to the consumers when there is a lack in the supplying side, and it is not possible for the consumers to participate in the demand response programs. Therefore, the storage can be charged in the multi-energy systems through the acquired demand response during the off-peak period. This aggregated demand response can be discharged in the multi-energy system during the peak period to meet the consumers' demand and reduce the pressure from the generation side. For instance, the authors in Ref. [49] presented an optimal model for the operation of an energy hub by utilization of renewable energy resources, demand response programs, and energy storage systems. In order to provide more flexibility for the operation of the energy hub, several demand response programs for the residential sector of the consumers such as shifting program and curtailing program are proposed as a complementary component of the energy storage system in Ref. [50].

5.4 Conclusion

Energy storage is a necessary component in the energy system in order to store the generated energy to reuse whenever it is required. There are various regimes that could be defined for the status of the ESS such as the charging regime, storing regime, and discharging regime. The energy storage is considered to be one of the most important components in the energy system that can be employed to generate a form of energy or use the stored energy in the moment or place that is required. The application of ESSs can lead to a grid with enhanced stability, increased reliability in the integrated systems that include renewable energy resources, and improve the efficiency of the energy system and reduce the environmental emissions.

Many storage technologies are being employed in the current energy storage systems. The most recent information of the current ESS projects around the world has been presented in this chapter by emphasizing the leading countries that are developing the ESS projects. The United States, China, and Germany are the top three countries in the world with the highest number of ESS projects.

The classification of these technologies is dependent on many factors, including the purpose of storing energy. Electrochemical, electromechanical, electromagnetic, and thermal storage are considered as the main four categories for the ESS technologies, and they are presented and explained in detail. The advantages of the energy storage technologies are presented as well. The implementation of ESS in the power system is done to meet different services. The most common services for the application of energy storage systems are discussed in detail in this chapter such as electric supply capacity, electric energy time shift, on-site power, electric supply reserve capacity, frequency regulation, voltage support, and electricity bill management. Additionally, by integrating the various energy forms and developing the concept of the multi-energy systems, one of the key components of multi-energy systems is ESSs. The main role of ESSs in multi-energy systems is to compensate for the fluctuations introduced by renewable energy resources. In this chapter, ESS technologies in the context of the multi-energy systems are presented and explained.

Furthermore, in the context of a multi-energy system, the storage unit can be installed in both sides of the input or output of the system as hydrogen and electrical storage can be installed in both sides, while the thermal storage is usually employed in the output side of the system. Moreover, it is shown that the ESS can also be a complementary component for the demand response actions to provide more flexibility for the operation of the energy hub, especially during the high consumption periods.

List of Abbreviations

BESS	battery energy storage systems
CAES	compressed air energy storage
ESS	energy storage system
PHS	pumped hydroelectric storage
SMES	superconducting magnetic energy storage
TES	thermal energy storage

References

1 Shaqsi, A.Z.A.L., Sopian, K., and Al-Hinai, A. (2020). Review of energy storage services, applications, limitations, and benefits. *Energy Rep.* 6: 288–306. https://doi.org/10.1016/j.egyr.2020.07.028.

2 Murphy, C.A., Schleifer, A., and Eurek, K. (2021). A taxonomy of systems that combine utility-scale renewable energy and energy storage technologies. *Renew. Sustain. Energy Rev.* 139: 110711. https://doi.org/10.1016/j.rser.2021.110711.

3 Sarbu, I. and Sebarchievici, C. (2018). A comprehensive review of thermal energy storage. *Sustainability* 10 (1): 191.

4 Lai, C.S., Locatelli, G., Pimm, A. et al. (2021). A review on long-term electrical power system modeling with energy storage. *J. Cleaner Prod.* 280: 124298. https://doi.org/10.1016/j.jclepro.2020.124298.

5 Sani, S.B., Celvakumaran, P., Ramachandaramurthy, V.K. et al. (2020). Energy storage system policies: way forward and opportunities for emerging economies. *J. Energy Storage* 32: 101902. https://doi.org/10.1016/j.est.2020.101902.

6 Lund, H. (2007). Renewable energy strategies for sustainable development. *Energy* 32 (6): 912–919. https://doi.org/10.1016/j.energy.2006.10.017.

7 Østergaard, P.A., Johannsen, R.M., and Duic, N. (2020). Sustainable development using renewable energy systems. *Int. J. Sustain. Energy Plann. Manage.* 29: 1–6. https://doi.org/10.5278/ijsepm.4302.

8 T. Nguyen (2020) "DOE Global Energy Storage Database". [Online]. Available at: https://sandia.gov/ess-ssl/gesdb/public/ (accessed 16 December 2021).

9 Koohi-Fayegh, S. and Rosen, M.A. (2020). A review of energy storage types, applications and recent developments. *J. Energy Storage* 27: 101047. https://doi.org/10.1016/j.est.2019.101047.

10 Luo, X., Wang, J., Dooner, M., and Clarke, J. (2015). Overview of current development in electrical energy storage technologies and the application potential in power system operation. *Appl. Energy* 137: 511–536. https://doi.org/10.1016/j.apenergy.2014.09.081.

11 Ghosh, S. and Kamalasadan, S. (2017). An energy function-based optimal control strategy for output stabilization of integrated DFIG-flywheel energy storage system. *IEEE Trans. Smart Grid* 8 (4): 1922–1931. https://doi.org/10.1109/TSG.2015.2510866.

12 Mahmoud, M., Ramadan, M., Olabi, A.G. et al. (2020). A review of mechanical energy storage systems combined with wind and solar applications. *Energy Convers. Manage.* 210: 112670. https://doi.org/10.1016/j.enconman.2020.112670.

13 Barbour, E., Wilson, I.A.G., Radcliffe, J. et al. (2016). A review of pumped hydro energy storage development in significant international electricity markets. *Renew. Sustain. Energy Rev.* 61: 421–432. https://doi.org/10.1016/j.rser.2016.04.019.

14 Zubi, G., Dufo-López, R., Carvalho, M., and Pasaoglu, G. (2018). The lithium-ion battery: state of the art and future perspectives. *Renew. Sustain. Energy Rev.* 89: 292–308. https://doi.org/10.1016/j.rser.2018.03.002.

15 Antonelli, M., Barsali, S., Desideri, U. et al. (2017). Liquid air energy storage: potential and challenges of hybrid power plants. *Appl. Energy* 194: 522–529. https://doi.org/10.1016/j.apenergy.2016.11.091.

16 Zhang, W.-L., Yin, J., Lin, Z.-Q. et al. (2017). Lead-carbon electrode designed for renewable energy storage with superior performance in partial state of charge operation. *J. Power Sources* 342: 183–191. https://doi.org/10.1016/j.jpowsour.2016.12.061.

17 Zhang, F., Zhao, P., Niu, M., and Maddy, J. (2016). The survey of key technologies in hydrogen energy storage. *Int. J. Hydrogen Energy* 41 (33): 14535–14552. https://doi.org/10.1016/j.ijhydene.2016.05.293.

18 Yan, W., Wang, X., Gao, W., and Gevorgian, V. (2020). Electro-mechanical modeling of wind turbine and energy storage systems with enhanced inertial response. *J. Mod. Power Syst. Clean Energy* 8 (5): 820–830. https://doi.org/10.35833/MPCE.2020.000272.

19 Goodenough, J.B. (2014). Electrochemical energy storage in a sustainable modern society. *Energy Environ. Sci.* 7 (1): 14–18. https://doi.org/10.1039/c3ee42613k.

20 Rahman, M.M., Oni, A.O., Gemechu, E., and Kumar, A. (2020). Assessment of energy storage technologies: a review. *Energy Convers. Manage.* 223: 113295. https://doi.org/10.1016/j.enconman.2020.113295.

21 Fathabadi, H. (2019). Combining a proton exchange membrane fuel cell (PEMFC) stack with a Li-ion battery to supply the power needs of a hybrid electric vehicle. *Renew. Energy* 130: 714–724. https://doi.org/10.1016/j.renene.2018.06.104.

22 Moseley, P.T. and Garche, J. (2014). *Electrochemical Energy Storage for Renewable Sources and Grid Balancing*. Newnes.

23 Rehman, S., Al-Hadhrami, L.M., and Alam, M.M. (2015). Pumped hydro energy storage system: a technological review. *Renew. Sustain. Energy Rev.* 44: 586–598. https://doi.org/10.1016/j.rser.2014.12.040.

24 Budt, M., Wolf, D., Span, R., and Yan, J. (2016). A review on compressed air energy storage: basic principles, past milestones and recent developments. *Appl. Energy* 170: 250–268. https://doi.org/10.1016/j.apenergy.2016.02.108.

25 Dooner, M. and Wang, J. (2020). Compressed-air energy storage. In: *Future Energy: Improved, Sustainable and Clean Options for Our Planet* (ed. T. Letcher), 279–312. Elsevier.

26 Koohi-Kamali, S., Tyagi, V.V., Rahim, N.A. et al. (2013). Emergence of energy storage technologies as the solution for reliable operation of smart power systems: a review. *Renew. Sustain. Energy Rev.* 25: 135–165. https://doi.org/10.1016/J.RSER.2013.03.056.

27 Olabi, A.G., Wilberforce, T., Abdelkareem, M.A., and Ramadan, M. (2021). Critical review of flywheel energy storage system. *Energies* 14 (8): 2159.

28 Arani, A.A.K., Karami, H., Gharehpetian, G.B., and Hejazi, M.S.A. (2017). Review of flywheel energy storage systems structures and applications in power systems and microgrids. *Renew. Sustain. Energy Rev.* 69: 9–18. https://doi.org/10.1016/j.rser.2016.11.166.

29 Dincer, I. (2018). *Comprehensive Energy Systems*. Elsevier.

30 Hall, P.J. and Bain, E.J. (2008). Energy-storage technologies and electricity generation. *Energy Policy* 36 (12): 4352–4355. https://doi.org/10.1016/j.enpol.2008.09.037.

31 Semadeni, M. (2004). Storage of energy, overview. *Encycl. Energy* 719–738. https://doi.org/10.1016/B0-12-176480-X/00104-2.

32 Breeze, P. (2018). *Power System Energy Storage Technologies*. Academic Press.

33 Agbossou, K., Kolhe, M., Hamelin, J., and Bose, T.K. (2004). Performance of a stand-alone renewable energy system based on energy storage as hydrogen. *IEEE Trans. Energy Convers.* 19 (3): 633–640. https://doi.org/10.1109/TEC.2004.827719.

34 Behabtu, H.A., Messagie, M., Coosemans, T. et al. (2020). A review of energy storage technologies' application potentials in renewable energy sources grid integration. *Sustainability* 12 (24): 10511. https://doi.org/10.3390/SU122410511.

35 Aydin, D., Casey, S.P., and Riffat, S. (2015). The latest advancements on thermochemical heat storage systems. *Renew. Sustain. Energy Rev.* 41: 356–367. https://doi.org/10.1016/j.rser.2014.08.054.

36 Song, X., Zhu, T., Liu, L., and Cao, Z. (2018). Study on optimal ice storage capacity of ice thermal storage system and its influence factors. *Energy Convers. Manage.* 164: 288–300. https://doi.org/10.1016/j.enconman.2018.03.007.

37 Zhang, D., Li, J., and Hui, D. (2018). Coordinated control for voltage regulation of distribution network voltage regulation by distributed energy storage systems. *Prot. Control Mod. Power Syst.* 3 (1): 1–8. https://doi.org/10.1186/s41601-018-0077-1.

38 Stroe, D.I., Knap, V., Swierczynski, M. et al. (2017). Operation of a grid-connected lithium-ion battery energy storage system for primary frequency regulation: a battery lifetime perspective. *IEEE Trans. Ind. Appl.* 53 (1): 430–438. https://doi.org/10.1109/TIA.2016.2616319.

39 Parra, D., Gillott, M., Norman, S.A., and Walker, G.S. (2015). Optimum community energy storage system for PV energy time-shift. *Appl. Energy* 137: 576–587. https://doi.org/10.1016/j.apenergy.2014.08.060.

40 Gabrielli, P., Fürer, F., Mavromatidis, G., and Mazzotti, M. (2019). Robust and optimal design of multi-energy systems with seasonal storage through uncertainty analysis. *Appl. Energy* 238: 1192–1210. https://doi.org/10.1016/j.apenergy.2019.01.064.

41 Simeoni, P., Nardin, G., and Ciotti, G. (2018). Planning and design of sustainable smart multi energy systems. The case of a food industrial district in Italy. *Energy* 163: 443–456. https://doi.org/10.1016/j.energy.2018.08.125.

42 Mohammadi, M., Noorollahi, Y., Mohammadi-ivatloo, B., and Yousefi, H. (2017). Energy hub: from a model to a concept – a review. *Renew. Sustain. Energy Rev.* 80: 1512–1527. https://doi.org/10.1016/j.rser.2017.07.030.

43 Favre-Perrod, P., Geidl, M., Klöckl, B., and Koeppel, G. (2005). A vision of future energy networks. *Proceedings of the Inaugural IEEE PES 2005 Conference and Exposition in Africa*, 2005, vol. 2005, pp. 13–17, https://doi.org/10.1109/pesafr.2005.1611778.

44 Adamek, F., Arnold, M., and Andersson, G. (2014). On decisive storage parameters for minimizing energy supply costs in multicarrier energy systems. *IEEE Trans. Sustain. Energy* 5 (1): 102–109. https://doi.org/10.1109/TSTE.2013.2267235.

45 Petkov, I. and Gabrielli, P. (2020). Power-to-hydrogen as seasonal energy storage: an uncertainty analysis for optimal design of low-carbon multi-energy systems. *Appl. Energy* 274: 115197. https://doi.org/10.1016/j.apenergy.2020.115197.

46 Mirzaei, M.A., Oskouei, M.Z., Mohammadi-Ivatloo, B., and Loni, A. (2020). Integrated energy hub system based on power-to-gas and compressed air energy storage technologies in the presence of multiple shiftable loads. *IET Gener. Transm. Distrib.* 14 (13): 2510–2519. https://doi.org/10.1049/iet-gtd.2019.1163.

47 Gabrielli, P., Poluzzi, A., Kramer, G.J. et al. (2020). Seasonal energy storage for zero-emissions multi-energy systems via underground hydrogen storage. *Renew. Sustain. Energy Rev.* 121: 109629. https://doi.org/10.1016/j.rser.2019.109629.

48 Vahid-Ghavidel, M., Javadi, M.S., Santos, S.F. et al. (2020). Demand response based trading framework in the presence of fuel cells using information-gap decision theory. *SEST 2020 - 3rd Int. Conf. Smart Energy Syst. Technol.* https://doi.org/10.1109/SEST48500.2020.9203313.

49 Pazouki, S., Haghifam, M.R., and Moser, A. (2014). Uncertainty modeling in optimal operation of energy hub in presence of wind, storage and demand response. *Int. J. Electr. Power Energy Syst.* 61: 335–345. https://doi.org/10.1016/j.ijepes.2014.03.038.

50 Brahman, F., Honarmand, M., and Jadid, S. (2015). Optimal electrical and thermal energy management of a residential energy hub, integrating demand response and energy storage system. *Energy Build.* 90: 65–75. https://doi.org/10.1016/j.enbuild.2014.12.039.

6

Digitalization and Smart Energy Devices

Maher Chebbo

CTS (Chebbo Technology Services), Paris, France

6.1 Introduction

Nowadays, electricity operates the world. Without electricity, the quality of life decreases, the industries stop functioning, and the overall development, economic progress, and social evolution decelerate. This is the reason why digitalization of power grid is rightly referred to as a revolution – it affects everyone, from individual families and communities to large companies, nations, and even the cross-border economy.

From plant to plug and across demand applications, digital transformation leads to multiple benefits for all players in the power ecosystem. In addition to enhanced grid reliability through access to real-time information and remote management, utilities provide support to increase savings through operational efficiency and earn income through new revenue streams. Digitalization also transforms the grid from unidirectional to multi-directional flow, enabling distributed generation and encouraging the integration and consumption of renewable energy. In addition, across the power value chain, one of the most important developments that digital technology has facilitated is transparency across the value chain, and through this, access to actionable intelligence is achieved. More information on assets, behavior, and demand is now available, and it can be leveraged to create value for everyone [1].

In the context of power industry, digital transformation is found across the energy value chain from electricity generation through transmission and distribution and all the way to end-use electricity delivery in homes, factories, and businesses. The potential applications range from asset performance management, grid optimization and aggregation, and integrated customer services.

The information and communications technology (ICT) sector reports at least 2% of global emissions. This is comparable with the civil aviation. The real benefits from green ICT will be gained from developing energy-efficient solutions that affect the remaining 98% of global emissions. The ICT sector is a contributor to global warming, but more importantly, it is key to monitoring and mitigating its effects.

The challenges deriving from the integration in the network of large-scale renewable sources, electric vehicles, systems for multi-directional flow management, new

Technologies for Integrated Energy Systems and Networks, First Edition.
Edited by Giorgio Graditi and Marialaura Di Somma.

equipment and power electronics, and real-time information devices can be faced only with a strong, reliable, and fast digital infrastructure.

The sustainable development of the energy value chain in the future will be increasingly based on and supported by digital improvements and innovations.

The International Telecommunications Union (ITU) emphasized that the ICT sector is also part of the solution to climate change and could help control emissions by approximately 15–40%. The Smart 2020 study has estimated that smart technology could reduce global emissions by 15%.

If directed to sustainable uses, the ICT sector could increase the energy efficiency in all areas of the economy while continuing to account for 40% of Europe's productivity growth.

Global energy demand is predicted to increase by 60% over the next 30 years. The European Union (EU) energy dependency could rise by 50–70% by 2030.

Getting from the current non-efficient energy to the future optimized-energy will be a major challenge.

The cost, performance, and deployment of many clean energy technologies have dramatically increased in recent years, accelerating transitions toward cleaner energy systems around the world. Digital technologies already play a vital role in accelerating decarbonization efforts, but for digitalization to reach its full decarbonization potential, we need good policies implemented by rigorous analysis. The International Energy Agency (IEA) report, Digitalization and Energy, states that driven by advancing technology, falling costs, and ubiquitous connectivity, the energy sector is on the edge of a new digital era, with a wide range of impacts on all energy sector stakeholders, from manufacturers and utilities to producers and consumers.

According to the IEA, €1000 billion will be spent to deploy smart grids by 2030 (average of €45 billion/yr, representing a capital expenditure [CAPEX] of 11% based on €400 billion annual turnover of the power sector), of which 50% will be spent on generation and the remaining 50% will be spent on transmission and distribution. Utilities allocate 2–6% of their turnover to information technology (IT) spending (average of 4% when investing intensively in market liberalization and smart grid deployment), representing an additional investment of €176 billion by 2030 to reach €352 billion of total ICT spending by 2030.

Digital technologies have been used within energy systems for decades. The energy sector was one of the early adopters of large ICT systems. In the 1970s, electric utilities used ICT to help manage the transmission & distribution (T&D) system. Many electricity markets around the world are monitored and controlled in real time across large customer bases and geographic areas. Similarly, oil and gas companies have a long history of using digital technologies to aid exploration and production efforts. Similarly, a variety of industries have used process controls and automation to optimize energy use. Digital technologies have long been used across transportation modes to improve safety and increase energy efficiency [2].

Energy generation and distribution uses one-third of all the primary energy. Electricity generation could be made more efficient by 40% and its transport and distribution by 10%. ICT could not only make the management of power grids more efficient but also facilitate the integration of renewable energy sources (RES).

Digital technologies could not only make the management of power grids more efficient but also facilitate the integration of RES.

Blockchain, artificial intelligence (AI), and machine learning (ML) are three digital technologies that are undergoing rapid development today and have the potential to bring disruptive changes to the energy landscape. Blockchain technology presents an exciting opportunity for decentralized energy environments to enable, validate, record, and settle energy transactions in real time. Blockchain is a distributed digital ledger built on a decentralized transaction verification system; this framework could enable peer-to-peer transactions, where neighbors can make transactions directly with each other and trade energy generated from their rooftop solar panels and electric vehicles through the grid.

If the digital power generation solutions were installed across the global fleet of coal and gas-fired power plants, carbon dioxide emissions from power plants would be reduced by up to 10%, which is roughly equivalent to taking all the cars in the United States (US) off the road. This single digital application represents a US$200 billion opportunity for the global power industry. The potential impact of the digital grid is also vast. Recent analysis suggests that if digital grid technologies were fully deployed globally, electricity consumption could be reduced by as much as 12% and carbon dioxide emissions could be reduced by up to 2 billion metric tons by 2030 [3].

The Global e-Sustainability Initiative (GESI) has recently found that an Industrial Internet of Things (IIoT)-enabled world by 2030 can be cleaner, smarter, and more prosperous. These findings indicate that ICT can bring about a 20% reduction in global carbon dioxide emissions by 2030 through the application of Internet-enabled solutions. This would also reduce energy costs by US$4.9 trillion by 2030, with US$1.2 trillion reduction in electricity expenditures and US$1.1 trillion reduction in fuel expenses [4].

In the near future, new applications can be easily added into home and building automation to save energy. A bioclimatic house may provide energy savings up to 80–90% compared to a conventional building.

Heating, cooling, and lighting of buildings account for more than 40% of European energy consumption. ICT could provide consumers with real-time updates on their energy consumption to stimulate behavioral changes.

The Green Deal and the digitalization of the European economy will be important new priorities of the European Commission. Moreover, by 2050, renewables could reach as much as 87% of the electricity mix, with wind and solar playing a dominant role. Cheap renewables, flexible demand, and battery storage will be digitally combined to shift the European power system away from fossil fuels and nuclear power to a cleaner society around variable renewables and emission-free energy.

At the Paris Climate Conference in 2015, 195 countries agreed to cut carbon dioxide emissions and review progress every five years. Beyond climate-specific policies, a broad set of associated policies have been implemented in the past two decades, which have already led to sustained decarbonization of the global electricity system, which produces 42% of global carbon dioxide emissions. Among the most notable policies are renewable portfolio standards in the United States, feed-in tariffs in Europe, and a potpourri of other subsidies and tax incentives for low-carbon generation sources across the globe.

This shift in energy transition will be enabled by smart digital technologies. Digital technologies will optimize the value that battery storage systems can bring to the energy markets, thereby enabling opportunities for new energy stakeholders, creating a new generation of jobs for the circular economy, and bring Europe to the forefront of leadership in the fight against climate change.

Data strategy and security are subject to the largest degree of change and the most significant impact on investments and returns for all parts of the value chain. Data security, privacy, collection, storage, and manipulation will all determine the utility's ability to stay competitive.

New digital technologies are unlocking opportunities across the electricity value chain (Figure 6.1), generation, networks, and retail, leading to cost reduction, efficiency, and new revenue streams. While we are focusing on digital grids, it is important to specify the digital focus for every stakeholder across the value chain:

- Digital generation covering predictive maintenance, asset performance management, asset process automation within operational technology (OT), digital twins, digital worker, and field service automation.
- Digital grids covering network predictive maintenance, network asset performance management, network digital twins, process automation (OT), substation automation, and network inspection using unmanned aerial vehicles (UAVs) such as drones for vegetation detection, which is a critical need for grid operators. While grids in Europe are unbundled, first legal separation of transmission grids (transmission system operators [TSOs]) from distribution grids (distribution

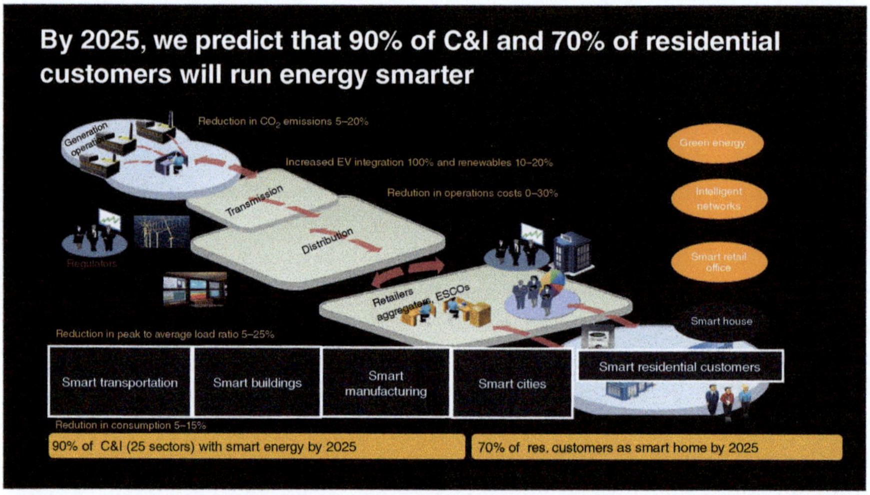

Figure 6.1 By 2025, 90% of business and 70% of residential customers will run smarter energy. Dream or reality?. Source: Based on Ref. [5].

system operators [DSOs]) and distribution grids from retailers, still DSOs in Europe have the responsibility of deploying and maintaining the smart meters enabling the smart metering services to be provided to the end customers by retailers. Technologies driving digitalization include distributed energy capacity, behind-the-meter generation, improved computer processing, cloud storage and computing, widespread communications networks, ML algorithms, cyber security, advanced distributed energy management systems, and trading technologies such as blockchain.

Following the European Technology Platform (ETP) SmartGrids Strategic Research Agenda (SRA) and Strategic Deployment Document activities between 2005 and 2010, the European Electricity Grid Initiative (EEGI), which was setup in 2010, has described the requirements in research, development, and large-scale demonstration (RD&D) per stakeholder and layer (Figure 6.2): smart customers, smart energy management, smart integration, smart distribution network and processes, smart pan-European transmission network, and new-generation technologies. It was estimated that more than €2 billion of RD&D were estimated covering 37 projects for TSOs, DSOs, and retailers across the value chain.

- Digital retail is focusing exclusively on services as retailers do not own or operate assets. It covers services using AI, ML, and chatbots to perform better customer services and multi-channel communication with customers. Retailers are focusing on providing a wide range of energy services and sometimes offering non-energy services as well. Retailers are leveraging digital technologies to offer more efficient and innovative customer services, managing customers' lifecycle and prosumers' expectations. Prosumers are motivated to digitalize in order to look for cost reductions and new revenue streams. Regulated grid operators will use digitalization to meet government mandates and flexibility requirements.
- Energy consumers are the ones who consume energy products and services. Consumers could be residential, households, and communities; small and midsize enterprises (SMEs), industries; and electricity-intensive industries. A specific example of a consumer category is the set of users with specialized mobility requirements for plug-in hybrid and pure electric vehicles. These users need digital mobility interfaces with quality and security of supply of the electricity system. Energy consumers will reverse the value chain and "dictate" their requirements, their priorities, and the products and services they will be willing to acquire. There is a revolutionary change in the current value chain where consumers become centric after being historically just passive consumers. Energy prosumers are consumers with the additional role of self-provided (owned) electricity generation and/or storage for private, daily life needs, comfort, and SME business needs.

Everything behind the electricity meter in an average consumer's house has so far been mysterious to them. However, the consumer profile is now changing. Many consumers are now "prosumers"; that is, they have invested in solar, wind, and other independent systems and want to sell power back to the grid. With the grid getting "smarter," utilities are now looking to deepen their relationship with customers, and this change is a win–win for both. Through mobile apps,

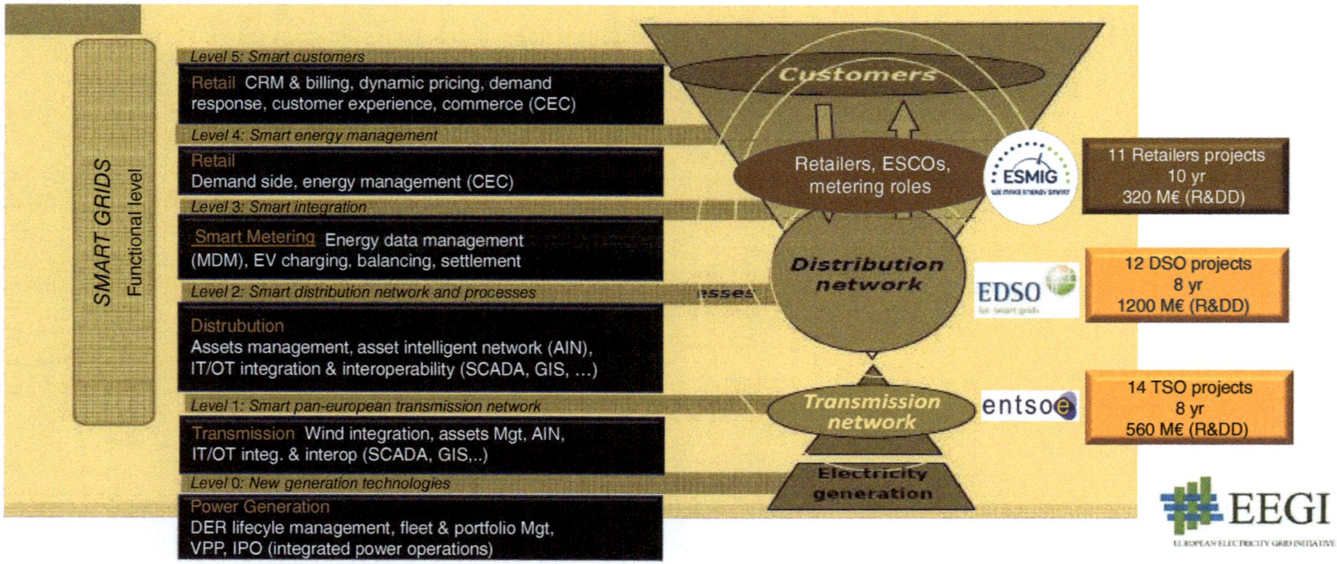

Figure 6.2 Smart grid requirements per stakeholder, mapped to the technology solutions [6].

consumers have control and choice over how they consume power and can exert environment-friendly decisions by choosing renewable power. They can also take advantage of lower tariffs by planning activities that consume more power during off-peak hours [7].

Retailers have a particular place within the energy value chain. Retailers are the first contact for the household customer regarding billing, house-moves, switching requests, and energy supply. They are also the last value adding party before energy is delivered. They are also serving business customers. Although market structures vary, the main function of the retailer in the value chain is similar.

With the coming of smart grids, the world of energy will become more complex, with a bidirectional value chain (Figure 6.3) coming together in a smart energy ecosystem. Figure 6.4 also shows the distinction between parties in the market domain and those in the pure regulated domain (transmission and distribution), with metering responsibilities varying from country to country.

Retailers and system operators face pressure from customers, government policy, and regulation to improve their business and leverage digitalization to streamline operations and improve customer services.

There is a lot to learn from the telecommunications sector where the revolution happened already more than 20 years ago. Energy and telecom consumer management is converging to flexible, real-time, and value-added products and services expected to be more and more competitive.

The digitalization of energy empowered by the digital multi-channel interaction with the customers leads to innovative e-transactions operated through business

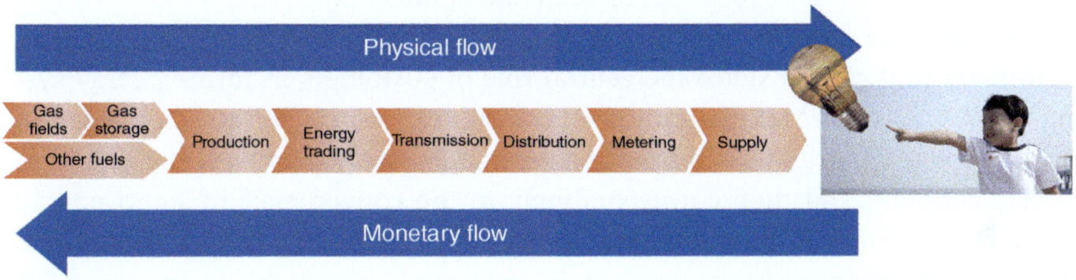

Figure 6.3 Traditional energy value chain with one-way direction physical and monetary [8]. Source: Jeroen Scheer.

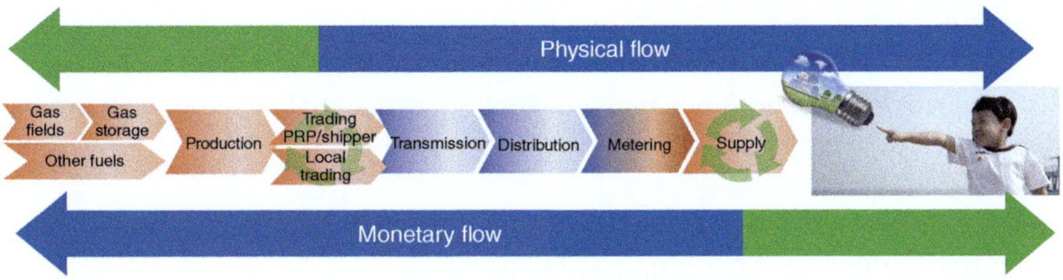

Figure 6.4 Reversing the energy value chain with two-way interaction of physical, monetary, and other flows [9]. DERMS, distributed energy resources management system. Source: Jeroen Scheer.

networks such as buying and selling on electronic marketplaces and the digital management of information collected and operated by the workforce.

Digitalizing the power sector in 2025 will generate $64 billion in revenue across the value chain, according to Bloomberg New Energy Finance (BNEF). Within this total, grid automation is forecast to be a $10 billion market by 2025, with connected home systems reaching $11 billion [10].

Digitalization is the transformation to a digital business, that is, using digital technologies to change business models and provide new revenue streams and value producing opportunities.

Digital technologies have been in use for more than 50 years and have become the dominant form of electronics and communications. Initially, the term was used to describe the shift from continuously varying signals to those made up of step changes, of zeroes and ones. However, as rapid data processing and transfer have become simpler, cheaper, and more widespread, the term "digitalization" has taken on a new meaning, representing increasing connectivity and data sharing in industrial, consumer, and energy sectors.

The digital technologies are being installed in our society in a record time. The communication technologies, such as mobile phones, web, and smartphones, have been adopted in 12, 7, and 5 years, while electricity and telephone took 47 and 35 years, respectively, to get adopted.

The digitalization of the energy system is not a recent task, but it is a process that has been ongoing since at least 10 years. In reality, digitalization is even older if you consider the installation of components such as remote terminal unit (RTU). The main focus so far has been on infrastructure operation, and, coherently, the concept of smart grid has been the focus of research and applications.

A key reference in this sense is given by the Winter Package of the European Commission that clearly states the central role of customers in future energy systems. Thus, with respect to the traditional concept of a smart grid, the digitalization process involves other new factors such as customer involvement, greater attention to sector coupling and then correspondingly to the convergence of smart energy and smart cities and communities, and new concepts and technologies that are also emerging at the physical layers thanks to a greater role played by electronics in the new digital energy system. The convergence of physical and digital happening in the smart networks for energy transition will be like the convergence that happened in the telecom and automotive industry.

Hence, the digital energy network paradigm is a broader concept than smart grid with significant social components and focused on service. The final goal is to enable a flexible open, transparent trade market of energy with equal possibility of participation of every player as envisioned by the Winter Package.

6.2 Our Vision of the Digital Networks

The ETP SmartGrids SRA 2035 focused on research necessary for the further development of the electricity system beyond 2020. It takes a perspective of the year 2035 and beyond. Clearly, SmartGrids research in a 2035 perspective must contribute to go beyond a reduction of EU greenhouse gases (GHGs) of 20% by 2020: a factor of 4 is envisioned by 2050. SmartGrids research for 2035 must consider the increases of renewable generation well beyond the expected 34% of the final consumption by 2020.

Energy systems are vital infrastructures that meet essential societal needs and without which our modern life would be impossible. They drive industrial processes, help exploit natural resources, make agriculture more efficient, enable most services, distribution of goods and movement of people, treat and pump water, and keep us comfortable with heating and cooling. They are the basis for all industrial revolutions, including the Internet, which is also an example of the increasingly crucial role of electricity. They are a central foundation for prosperity, and by making life comfortable and productive for all, they support societal cohesion.

In 2050, energy network operators work within a legal framework to ensure overall security, reliability, resilience, and quality of supply to satisfy citizen and business needs at affordable costs. To realize the energy transition, network operators design and operate robust and smart electricity, gas, and heat and cooling grids to become more active players, for example, by purchasing flexibility from prosumers in order to adapt consumption and generation behaviors depending on local needs.

The digitalization of the grid includes smart operations of the grid at all voltage levels to reduce losses and outage times, retailers that optimize their portfolio by balancing based on forecasting algorithms, aggregators that control flexible consumption for various business cases, and new market platforms, such as EnOS of Envision Digital, that provide a suitable interaction between all these actors to optimize the overall efficiency.

Digitalization is central to transmission system operators (TSOs) as they have to manage a fast-evolving power system 24/7 while keeping the lights on. Today, they are required to overcome a variety of new challenges, such as the increased amount of variable generation, sector coupling, power and transport connected through e-mobility, increasing electrification and, in particular, heating and cooling, as well as the rise of the Energy Internet of Things (IoT) and the many flexibility opportunities and needs. Digitalization also contributes to further market facilitation and enables new actors and new roles, centered around prosumers and active system management [11].

Traditionally, DSOs have operated as asset-centric companies, physically managing electricity distribution infrastructure assets, such as electrical lines and cables, substations, and transformers. With the upgrade of this infrastructure into "smart grids," however, their asset base is being expanded to include intelligent monitors and sensors and smart meters. This results in DSOs becoming data-centric companies, using digital technologies to optimize asset management, integrate distributed RES, and improve network stability and security. They are also able to leverage consumer and network data to deliver better quality of service and engagement and to serve better as a neutral facilitator among market players.

Figure 6.5 illustrates the typical digital technologies and digital applications or use cases accelerating the energy transition while generating cost savings, increased

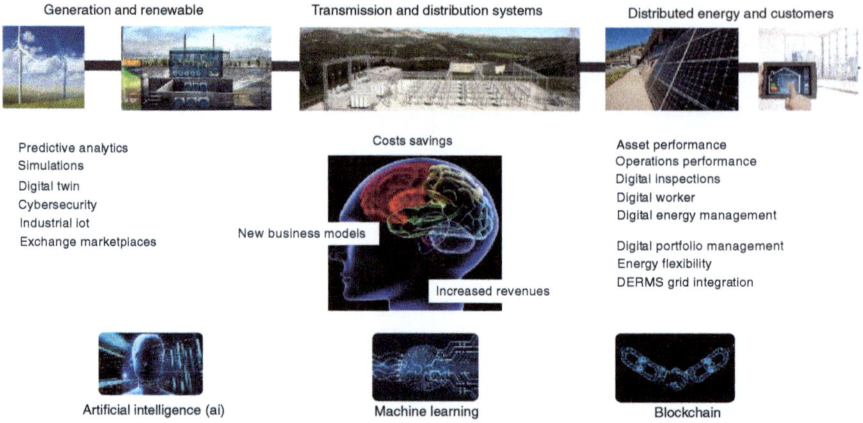

Figure 6.5 Digitalization across the energy value chain: technologies (e.g. digital twin, AI, and ML) and applications (asset performance and flexibility). Digital energy across the value chain.

savings, and new business models targeting socio-economic benefits across the value chain.

A number of interdependent "digital" trends are prompting DSOs to refocus their business and strategy on "digitalization." The most notable trends include [12] the following:

1) **Connectivity:** Every asset, every device, and every person is now connected and interconnected via electronic communications.
2) **Collaboration:** This connectivity allows for multiple forms of collaboration, increasing efficiency and improving performance, e.g. consumers and prosumers to network operator via smart meters; machine-to-machine (M2M) smart appliance and meter exchanges in consumer premises.
3) **Personal data innovation:** Data management innovations are emerging even while respecting the important EU data protection legal framework and the fundamental right of privacy.
4) **Big data:** With the deployment of intelligent sensors and smart meters and the proliferation of customer-owned connected devices (i.e. the IoT), the volume of data is expanding by orders of magnitude.
5) New forms of collecting, storing, processing, and exploiting such data can lead to improved operations and enable new business opportunities.
6) **Industry 4.0:** A redesign of the industrial value chain is underway, embracing automation, IoT, data exchange, and modern manufacturing/industrial technologies.
7) **Open data:** The expectation that all government and other public data be published and made openly reusable, encouraging the development of innovative applications and services.
8) **Cyber attacks:** Connected, collaborative energy networks and systems and the use of cloud computing are potential target for cyber attacks.

9) **Digital DSO:** to which European Distribution System Operator (EDSO) members are aspiring and already are transitioning today toward an upgraded network and systems.
10) **Global Cyber security:** for digitalization in all domains.

Digitalization in **network management and operation** is predictive analytics intensive:

1) **Ability to predict** and handle power infeed with bidirectional power flow to manage intermittent and decentralized power production.
2) **Evaluation of energy data** to predict grid loads and anticipate bottlenecks. This allows for the optimization of network investments.
3) **Real-time processing** of load data and generation, enabling the integration with demand/supply balancing service to optimize grid utilization.
4) **New capabilities in predictive** maintenance and self-healing concepts help to further reduce operational costs.
5) **Predictive analytics**, using AI (Figure 6.6) based on sensor data, enabling smarter asset management with a fully digital allocation of spare parts, work, and logistics.
6) **Long-term system planning and integration** with other (regional) grids.
7) **Hardware-in-the-loop** testing that can use **complex simulation** coupled with actual hardware testing.

Digitalization in **mobile field operations**, for example:

1) **Digital support for grid operations**, allowing very fast response and/or allowing very thorough analysis of contingencies and their consequences.
2) **Digital support for field technicians**.

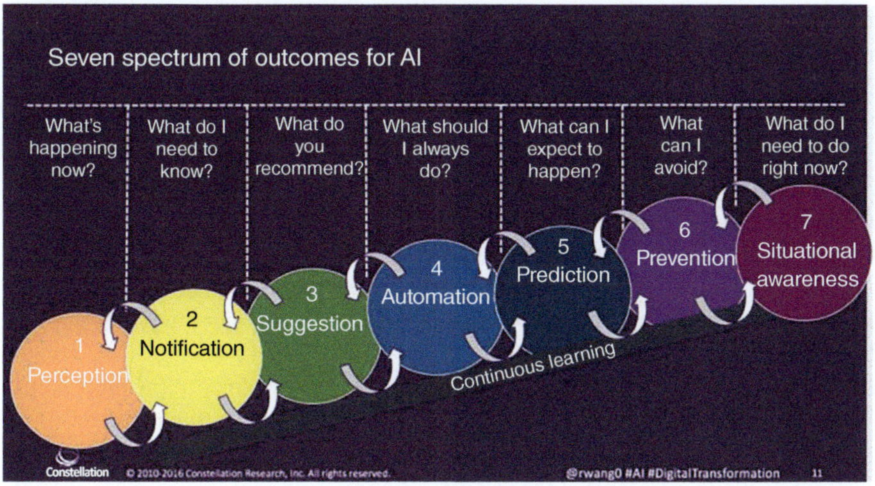

Figure 6.6 The spectrum of seven artificial intelligence outcomes: from perception and reaction to prediction and situational awareness [13].

3) **Digital business processes**, replacing manual transactions in procurement, inventory management, invoicing, and payment processing.

Digitalization for **market facilitation**, for example:

1) **Meter-point operations** to digitally connect to the consumer and enable value-adding services.
2) **Collaboration with consumer and prosumer**, consumers who produce their own energy, to reduce consumption and optimize network management.
3) **Data-enabled transactions** among DSOs, aggregators and supplier, aggregator and consumers, and energy start-ups.
4) **Fast transactions allowing close-to-real-time** intra-day market closure for better integration of variable renewables in the wholesale market.
5) **Standardized and secure data exchange** to support market communications (e.g. supplier switching, meter data exchange, billing data exchange, and nominations).
6) **Provision of anonymized and/or aggregated data** to public administrations and market parties – as mandated by regulation or motivated by market facilitation – to enable market innovation.

When talking about smart grids, customer engagement will be essential if the new technology is to be used to further energy efficiency [14].

Smart grids are energy networks that can automatically monitor energy flows and adjust to changes in energy supply and demand accordingly. When coupled with smart metering systems, smart grids reach consumers and suppliers by providing information on real-time consumption.

To date, following a cost–benefit analysis (CBA) (Figure 6.7), EU Member States have committed to rolling out close to 200 million smart meters for electricity and 45 million for gas by 2020 at a total potential investment of €45 billion. By 2020, almost 72% of European consumers are equipped with a smart meter for electricity while 40% will have one for gas [15].

The implementation of smart metering is crucial to the development of the European energy efficiency programs. The drive for lower carbon emissions combined with greatly improved efficiency on the demand side will empower consumers to select automatically the suppliers that provide the best value-added services capable of offering outstanding consumer-focused programs (Figure 6.8). Member states shall ensure the implementation of smart metering in order not only to meet the carbon emissions and renewable targets but also to meet the energy efficiency target in which consumers will participate actively reversing the overall production to supply value chain.

However, smart meters are only a first step toward smart grids – many more steps and devices will be required to ensure a successful smart grid deployment.

Smart meters should allow consumers to reap the benefits of the progressive digitalization of the energy market via several different functions. Consumers should also be able to timely access their energy consumption data and dynamic electricity price contracts.

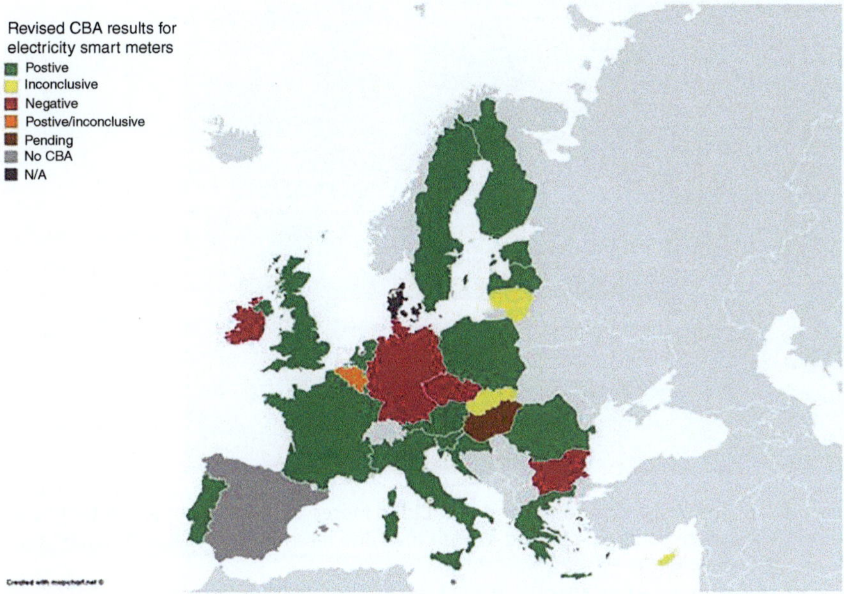

Figure 6.7 Revised CBA (cost–benefit analysis) results electricity smart meters, considering a large-scale rollout to at least 80% by 2020 (as of July 2018) [16].

Figure 6.8 Ranking of the considered benefits in the electricity CBA vs. number of member states [16].

Traditionally, energy actors, typically DSOs, were deploying software on-premise meter data management solutions. However, with the development of the Cloud technology and the benefits provided by the continuous innovation and improvement of the solution being hosted on the Cloud (Figure 6.9) compared to on-premise solutions, new solution providers specialized in Cloud meter data management became key players meeting market expectations.

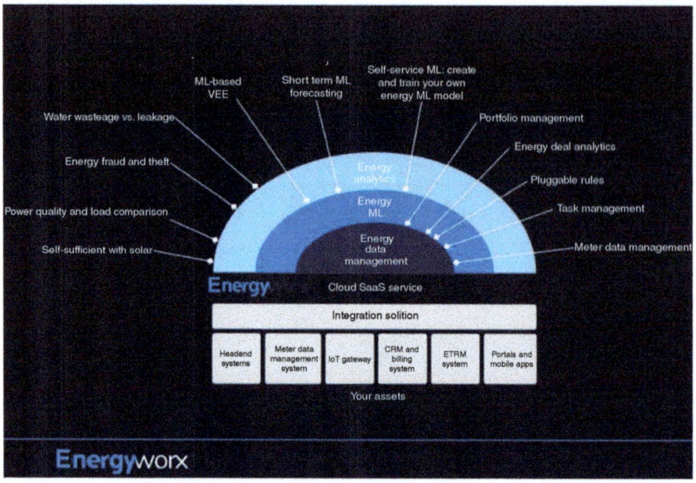

Figure 6.9 Energyworx, solution provider of Cloud Meter Data Management (MDM) as a software as a service solution, offering as well machine learning and customer analytics.

A study from December 2019 on the deployment of smart meters in the EU found that nearly 225 million smart meters for electricity and 51 million for gas will be rolled out in the EU by 2024. This represents a potential investment of €47 billion by 2024, and it is expected that almost 77% of European consumers will have a smart meter for electricity. About 44% will have a smart meter for gas. The cost of installing a smart meter in the EU is on average between €180 and €200 on average, smart meters provide savings of €230 for gas and €270 for electricity per metering point (distributed amongst consumers, suppliers, distribution system operators, etc.) as well as an average energy saving of at least 2% and as high as 10% based on data coming from pilot projects [17].

The digital transformation of electricity is one of the most exciting areas to work in either in Europe, the United States, or Asia.

Beyond Europe, we have a huge opportunity to improve the reliability and availability of electricity worldwide:

- Unplanned power outages leave homes and businesses in many emerging countries without electricity for many days each month.
- In South Asia, consumers experience an outage every day. In the Middle East and North Africa, businesses experience more than 15 days of outages each month.
- These outages have a huge economic impact equal to 6–11% of annual sales.
- Wind and solar energy create new stresses that the grid was not designed to handle.
- North American Electric Reliability Corporation (NERC) reports that even US power systems are unable to meet the demand 6% of the time.
- Combined with an uptick in extreme weather, the United States is experiencing a steady rise in blackouts. If this continues, the US Society of Civil Engineers estimates that US gross domestic product will fall by a total of $819 billion by 2025 and $1.9 trillion by 2040.

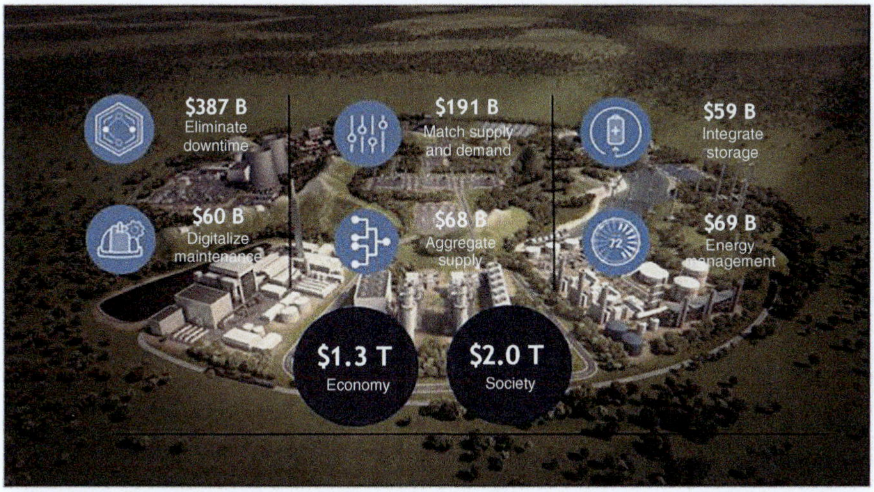

Figure 6.10 Value generated from the digitalization of the energy system across the value chain. Source: WEF [18].

- World Economic Forum (WEF) estimates the economic impact of digitalization at $1.3 T over 10 years 2025 (Figure 6.10). Digital solutions in the power industry can create over US$2 trillion from the reduction in GHG emissions, new job creation, and value for consumers.

The new (11 May 2017) International Energy Agency report "Digitalization and Energy" highlights that the power industry could save US$80B/yr or 5% of total annual power gen costs:

- Primary benefits from **reducing operations & maintenance (O&M) costs**, improving power plant and network efficiency, reduced unplanned downtime, and extending the operational lifetime of assets.
- Greater value from **breaking down boundaries between energy sectors** increasing flexibility and enabling integration across entire systems. e.g.:
  - **Smart demand response:** Digitalization could provide 185 GW of system flexibility, offsetting US$ 270 billion of investment in new electricity infrastructure that would have otherwise been needed, which is equivalent to that of Australia and Italy together. Demand response management (DRM) is enabled by connecting 1 billion households and 11 billion smart appliances.
    Fifteen organizations representing the EU have agreed on a joint approach to deliver **demand side flexibility and energy efficiency** with a joint declaration [19].
    - Providing market access for all demand side resources.
    - Promoting scarcity pricing and dynamic tariffs.
    - Establishing the right for consumers to self-generate and consume.
    - Establishing system-wide adequacy assessments.
    - Ensuring "efficiency first" in system operation.
    - Promoting the deployment of smart and integrated energy solutions.
    - Ensuring better market surveillance and governance.

Flexibility is also a key enabler of the decarbonization of buildings and is vital to deliver on the EU's climate and energy objectives, given that buildings are responsible for 40% of total energy consumption. The system needs to be designed for consumers to enable them to participate in the energy market and to protect their rights and privacy [20]:

- **More Efficient Integration of Renewables:** Digitalization can enable grids to better match the energy demand to intermittent power sources. In the EU alone, storage + smart demand response could reduce curtailment of solar and wind power from 7% to 1.6% in 2040, avoiding 30 million tons of CO_2 in 2040.
- **Smart Charging for Electric Vehicles:** They could help shift charging to periods when electricity demand is low and supply is abundant. This would provide further flexibility to the grid while saving between US$ 100 billion and US$ 280 billion in avoided investment in new electricity infrastructure between 2016 and 2040.
- **Facilitating distributed energy resources (DER) growth:** Digitalization can facilitate the development of DER by creating better incentives and making it easier for producers to store and sell surplus electricity to the grid. New tools such as blockchain could help to facilitate peer-to-peer electricity trade within local energy communities as it is digitally promoted by GreenCom Networks.

6.3 Enabling State-of-the-Art Digital Technologies

A broad range of technologies are available today. What was science fiction yesterday is a reality today (Figure 6.11): fully electric planes and UAVs, autonomous driving, additive manufacturing and 3D, augmented reality (A/R) and virtual reality (V/R), and sensors everywhere including body skins and navigation in smart cities.

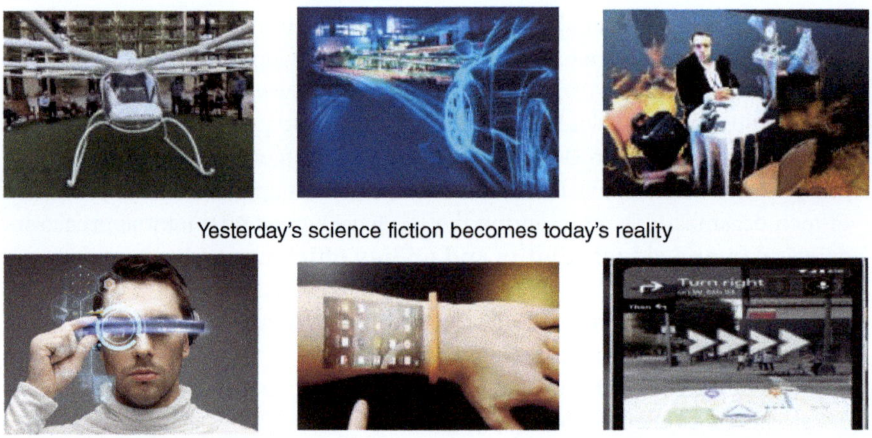

Figure 6.11 Science fiction technologies of yesterday become today's reality. Source: Syda Productions/Adobe, Maher Chebbo.

What makes yesterday's science fiction a reality today? It is the increasing power of technology, computing power, internet traffic, fast communications (now 5G is already on the market!), IoT, business intelligence (BI), ML, predictive analytics, data science, blockchain, and a continuous accelerated innovation of technology providers.

Some background about the computer and the evolution of the computing power:

The von Neumann computer concept (Figure 6.12) was developed in the 1940s when the first electronic computers were built. Nearly all modern computers are based on this stored program scheme, in which both machine instructions and program data are stored in computer memory in the same manner. After the 1940s, the computer industry began a rapid development with the speed and cost of computer systems improving by a factor of 2 every two years.

Today, the power of computers has no limitation: Summit, the US new supercomputer, is more than twice as powerful as the current world leader. The machine can process 200 000 trillion calculations/s or 200 petaflops.

IoT is the internetworking of physical devices, vehicles, connected devices and smart devices, buildings, and other items, embedded with electronics, software, sensors, actuators, and network connectivity that enable these objects to collect and exchange data without requiring human-to-human or human-to-computer interaction. The worldwide IoT market spending has been growing from $592 billion in 2014 to $1.3 trillion in 2019 according to International Data Corporation (IDC), while the installed base of IoT end points have been growing from $9.7 billion in 2014 to $30 billion in 2020 where 40% of all data in the world will be data resulting from M2M communication. IIoT market is estimated at $60 trillion by 2030.

Gartner survey showed that 43% of organizations are using or plan to implement the IoT in 2016. Gartner estimates today that $2.5 M/min in IoT is being spent and 1 M new IoT devices sold every hour.

From 2019, Global Business Intelligence market is exceeding $23 billion and Global Predictive Analytics market has reached $3.6 billion in 2020, driven by the growing need to replace uncertainty in business forecasting with probability and the increasing popularity of prediction as a key toward improved decision making. Predictive analytics is the branch of advanced analytics, which is used to

Figure 6.12 Evolution of the spiops AQ computers power since Von Neuman launched the machines called computers in 1946. Twenty-five billion instructions/s can be handled now in a microprocessor chipof 100 g. Source: Péter Molnár/behance/Adobe.

Figure 6.13 Historical detailed data gathered is crucial, even not exhaustive, to predict the future.

Figure 6.14 The majority of data collected has not been smartly analyzed. Only 2%.

make predictions about unknown future events using historical data (Figure 6.13). Predictive analytics use many techniques from data mining, statistics, modeling, ML, and AI to analyze current data – where only 2% have been analyzed so far (Figure 6.14) – to make predictions about future.

Big Data investments have been growing from $46 billion in 2016 to $72 billion by the end of 2020.

A new brand of analysts called "data scientists" are introducing data science courses into degrees ranging from computer science to business. Data scientists (Figure 6.15) usually require a mix of skills such as mathematics, statistics, computer science, algorithmic, ML, and most importantly business knowledge using digital models (Figure 6.16). If data scientists are lacking business knowledge, they will definitely fail. They also need to communicate the findings to C-Level management to be able to take the right strategic decisions.

Data science needs to be a fundamental component of any digital transformation effort. All sectors, including the energy sector and electricity networks, will have to hire and educate a significant number of data scientists.

Figure 6.15 Data scientists and modeling are highly needed to unlock the potential of data.

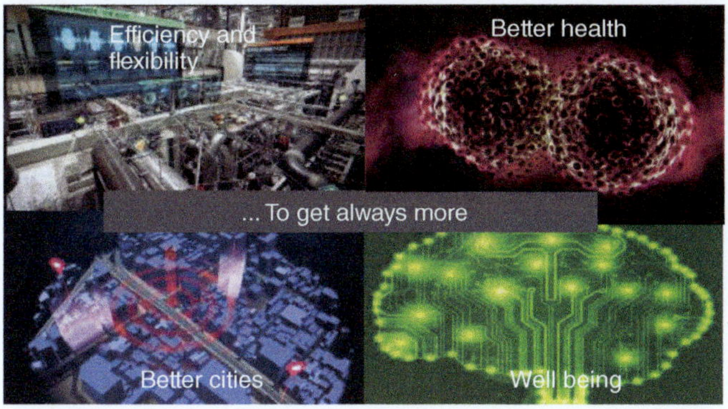

Figure 6.16 Digital models aim always for more, such as efficiency and flexibility of assets, better health, better cities, and well-being.

By 2022, a blockchain-based business will be worth $10 B, with blockchain being a digital platform that records and verifies transactions in a tamper and revision-proof way that is public to all.

In the energy sector where the digital transformation is playing a crucial role to reach global and European energy targets, European spending on digital transformation is estimated to be around 50 b€ with 330 b€ of annual economic benefits expected by 2020.

Energy players are seeking to find the upside of disruption to manage risks and capture new opportunities across the rapidly evolving value chain. We are moving toward a future energy system, augmented and interconnected by digital technologies, where both power and information flows circulate in both directions.

The work done under the SmartGrids mandate M/490 of the European Commission resulted in the definition of a Smart Grid Architecture Model (SGAM) based on a Reference model initially defined by the National Institute of Standards and Technology (NIST) and evolved by the European Committee for Standardization (CEN)–European Committee for Electrotechnical Standardization (CENELEC)–European Telecommunications Standards Institute (ETSI).

Consisting of the five interoperability layers, the SGAM framework allows the representation of entities and their relationships in the context of smart grid domains, information management hierarchies, and in consideration of interoperability aspects.

A specific layer-focused analysis has shown that the European TSOs are already widely using digitalization technologies. Of the 100 projects, more than half address the physical layer and data management layer [21].

The digitalization of the energy system Working Group 4 (WG4) within the European Technology and Innovation Platform Smart Networks for Energy Transition (ETIP SNET) has defined the technology and use cases requirements needed to make the smart networks for energy transition prepared for a future where physical assets and digital are converging, with digital twins (Figure 6.17) being the best dynamic representation.

The development of digital technologies is required to improve the industrialization of new batteries and shorten the time to market. The design of ML algorithms will accelerate the discovery of materials and the development of AI orchestrated characterization of battery materials and battery cells. Combining computer-aided engineering (CAE) tools and experimental measurements will help to understand and predict the battery performance.

The European Technology and Innovation Platform (ETIP) Batteries Digital Taskforce leveraged the activities of the ETIP Batteries Working Groups to define the digitalization agenda for Batteries in Europe that was included in the Batteries SRA.

Digital technology is not an objective but an important enabler to achieve the development of innovative new services, what we call applications or use cases that can deliver significant measurable benefits: saving costs, increasing revenues, and setting up new business models, all contributing to achieve socio-economic benefits.

The digital technologies that are listed in Table 6.1 are enablers to the definition of the use cases described in Table 6.2. Some of these technologies are typically

Figure 6.17 Digital twin technology provided by large energy equipment providers (Source: Siemens, GE, OECD/IEA 2018).

Table 6.1 Digital technologies summary table [22].

Technology	Readiness	EU competitive advantage	Research & innovation (R&I) investment required	Budget	Key performance indicator (KPI) example
CAE for modeling and simulation	++	++	++	40 M€	% Model accuracy vs. measurements, reduction in computational time, and reduction in product time to market
Design of experiments	++	++	+	15 M€	#Relationships established between parameters
ML algorithms/AI	+	+++	++	40 M€	% Level of prediction, reduction in time
Data infrastructure	+	+++	++	25 M€	#Test data, #organizations sharing information
(Big) Data analytics	++	++	++	25 M€	#Analytics programs, amount of data proceeded
Digital twin[a]	+	+++	+++	60 M€	% Accuracy of digital twins vs. real batteries/production lines
Wireless communication	++	+	+	15 M€	#Data flow through wireless communication

a) Combining all above technologies.

simulation technologies, digital twins, IIoT, wireless communications, AI, (ML), CAE for modeling and simulation, etc.

6.4 Key Digital Use Cases and Associated Benefits

Table 6.2 describes important use cases and their benefits for the different actors of the digital energy system.

We define "digitalization of the energy system" as "the process of implementing and operating a set of assets by monitoring, transferring, and analyzing data, which have been generated by one of the actors in the energy system."

This includes smart operation of the grid at all voltage levels to reduce losses and outage times, retailers that optimize their portfolio by balancing based on forecasting algorithms, aggregators that control flexible consumption for various business cases, and new market platforms, such as EnOS of Envision Digital, that provide suitable interaction between all these actors to optimize the overall efficiency.

An overview of practical use cases is given per stakeholder. The objective is not to present an exhaustive list of use cases, rather than providing practical information on actual field trials and projects, to build a vision on how digital technologies are changing the energy landscape.

We list and describe some promising use cases through projects in Europe, based on which we can build a vision on the future power system. Most of the use cases described are still at the innovation stage, which is in line with the ETIP SNET ambition to discuss the future technologies and applications.

Several use cases are illustrated by practical examples in the field. In particular, for each use case, it was emphasized what digitization contributed and what the benefits (Table 6.2) or business case was for the technologies introduced.

Digital grid is the digitization of electricity, gas, and water networks supported by a two-way communications network. It provides insight and control into the operations, empowering various energy stakeholders, producers, TSOs, DSOs, and retailers or aggregators to solve today and tomorrow's challenges.

A number of use cases are listed in Table 6.3, number of use cases related projects and field trials, most of them EU-funded projects, per energy stakeholder and system operator.

Figure 6.18 illustrates the project "Flexiciency," a Horizon 2020 project demonstrating a DSO facilitator platform connecting market players, such as energy services company (ESCO), aggregators, and retailers.

Digital interconnection of the centralized and decentralized storage systems by hybridization and multi-use of battery electric storage systems (BESS) into flexible portfolios will be an important step toward democratized energy systems (Table 6.4).

Electric storage systems for industrial vehicles have even tighter and application customized requirements rendering them ideal test cases for digital approaches. Long lifetime and a fast return on investment even under heavy duty require precise state estimation. Digital twins of the battery developed with large data sets

Table 6.2 Description and benefits of selected representative digital use case across the value chain [23].

General uses	Description	Practical use cases in this report	Expected benefits
Forecasting generation	Improved forecasting tools can allow a more efficient operation of the grid, in combination with curtailment, reactive power injection, and dynamic line rating	The SWIFT project, discussing a practical case where forecasting, dynamic line rating, curtailment, and reactive power injection allow connection of a wind farm without a costly grid upgrade	Increase reliability of supply
			Increase renewable energy penetration
			Increase the reliability of supply, reduce the cost of operations and CAPEX, and improve the quality of service
Network planning and operations	Improved digital options for digital network planning. Smart operation of the network using IT and OT integration, Big Data, and predictive services (a large Dutch DSO collects 1.5 billion grid sensor measurements to forecast the required operations in real time). Transmission and distribution networks and power generators' assets new models: smart products, data-driven business models, technology-driven customer engagement, and new alliances	The STAR grid project: grid management at LV and MV level	Increase the reliability of supply, reduce the cost of operations and CAPEX, and improve the quality of service
		The iTesla project, electrical system security within large areas	Reduce CAPEX investments
		The GRID4EU German demonstrator, autonomous grid reconfiguration, and forecasting in the MV grid	Improve the quality of service
		The GRID4eu Swedish demonstrator: meter data management for network operation in the LV grid	Provide real-time asset cockpits reducing incidents and outages
		Collaborative asset management, the SAP view	Defer grid upgrades
		NOBEL GRID: advanced tools and ICT services for DSOs	Enable flexible demand
		The Servo Platform: interface demand side management with DSO needs	Increase the renewable energy penetration
Digital use cases for retailers and aggregators	The data that market facilitator provides (e.g. DSO) can be offered to other commercial parties to facilitate market operation	The Smarter EMC2 project: empowering market actors through ICT technologies	Integrate flexible demand in the market
		IDE4L: digital tools for the technical and the commercial aggregator	Take grid constraints into account in market operation

(continued)

Table 6.2 (Continued)

General uses	Description	Practical use cases in this report	Expected benefits
Customer participation in the market	Full customer energy management using Big Data and IoT. Smart devices to understand customer behavior by utilities. Storage assets and EVs can become a part of the solution to integrate DER	The linear project, dynamic pricing, and residential demand response	Peak shifting and portfolio optimization
		SmartHouse/SmartGrid	Integrate flexible demand
		FINESCE: smart charging of electric vehicles	Defer grid upgrades and peak shifting
		COOPERATE: neighborhood energy management	Reduce losses and ensure safe grid operation
Balancing	Digital tools allow the TSO to balance the network more efficiently	IoT big data collection from end customers and data aggregation balancing	Better matching of demand and supply across the energy mix
Integration of EV and storage	Storage assets and EVs can become a part of the solution to integrate DER (estimated total world energy storage market by the year 2020 is $50 billion)	Decentralized operation using DER	Increase reliability of supply
			Decreasing grid upgrade costs
			Avoid over-voltages
		EV coordination schemes	Peak shaving for the consumer
Enable flexibility by leveraging electronic market places	The data that market facilitator provides (e.g. DSO) can be offered to other commercial parties to facilitate market operation	FINESCE local energy markets	Enable flexibility to support the local grid
		eBadge: ICT tools for cross-border markets	Improved international market model
		Use cases defined by the "Flexiciency" H2020 project	New roles of the DSO as a market facilitator
		The Universal Smart Energy Framework	Developing an adequate market model for stakeholder interactions

Table 6.3 Overview of digital use cases per stakeholder and related projects or field trials [23] such as Flexiciency.

Stakeholder	Practical use case	Project/field trial
3.1 Power generator	3.1.1 Probabilistic forecasting of wind generation, extremes, and optimal use in the system	Anemos/Safewind
	3.1.2 Smart curtailment, dynamic line rating, and improved forecasting tools to maximize integration of wind	SWIFT
3.2 Transmission and distribution networks	3.2.1 Innovative tools for electrical system security within large area	iTesla
	3.2.2 Autonomous grid reconfiguration and forecasting in the MV grid	FP7 GRID4EU
	3.2.3 Meter data management for network operation in the LV grid	FP7 GRID4EU
	3.2.4 Collaborative asset management	SAP Asset Intelligence Network
	3.2.5 Advanced tools and ICT servicers for distribution system operators	NOBEL GRID
	3.2.6 A Platform to interface demand side management with DSO needs	SERVO
3.3 Retailers and Aggregators	3.3.1 Empowering SmartGrids (SG) market actors through information and communication technologies	SmarterEMC2
	3.3.2 IDE4L use cases on technical and commercial aggregators	IDE4L
3.4 Consumers and prosumers	3.4.1 Dynamic pricing and demand response management	Linear
	3.4.2 Smart houses in a smart grid environment	SmartHouse/SmartGrid
	3.4.3 Smart charging of electric vehicles	FINESCE
	3.4.4 Neighborhood energy management	FP7 COOPERATE
3.5 New market platforms	3.5.1 Local energy markets	FINESCE
	3.5.2 ICT tools for cross-border markets	eBadge
	3.5.3 The DSO as market facilitator	FLEXCIENCY
	3.5.4 The Universal Smart Energy Framework	USEF Foundation

of interconnected batteries will allow for statistical backed maintenance requests from the batteries.

Electrical vehicle (EV) smart charging vs. EV charging would provide further flexibility to the grid saving between US$ 100 and 280 billion investments in new electricity infrastructure.

Table 6.4 Digital battery use cases' summary table, feasibility, impact, investment effort, budget, and KPIs [25].

Use case	Feasibility	Impact	R&I investment required	Budget	KPIs example
Automated materials discovery	+	++	+++	100 M€	– Number of new battery chemistries discovered – Cost of discovery and development of new battery materials – Number of materials developed in a time period – Lead time for material development
Green battery passport and digital referential	+++	+++	++	45 M€	– Percentage of batteries that go to the second life market – Percentage of savings in the design and development cycles – Percentage of cost avoided related to health and safety along the whole battery value chain
Advanced methods for SoX now and forecasting (sensors & big data)	++	++	++	20 M€	– Number and models of batteries with initialization problems – Percentage of batteries with maintenance needs above the normal maintenance range
Hybridization of battery energy storage systems (BESS) into flexible portfolios	+++	+++	++	50M€	– Percentage of use of renewable energy in BESS – Percentage of battery users who monetize their energy storage
Accurate datasheet generator based on application-specific Big Data Simulation Platform	++	++	++	30 M€	– Percentage of elements changed over original design in production phase – Percentage of batteries with a lower life than the forecasted one
Digitalization of the battery cell production	+++	+++	++	50 M€	– Percentage savings in producing batteries with digital twin developed in the design and development phase vs. without digital twin
Lighthouse project: Battery Ecosystem Marketplace Platform for Management, Services, and Supplies	++	+++	+	20 M€	– Percentage of market participants using the platform to identify and purchase services – Sustained use of the platform by market participants – Year-to-year growth of the number of active vendors on the platform

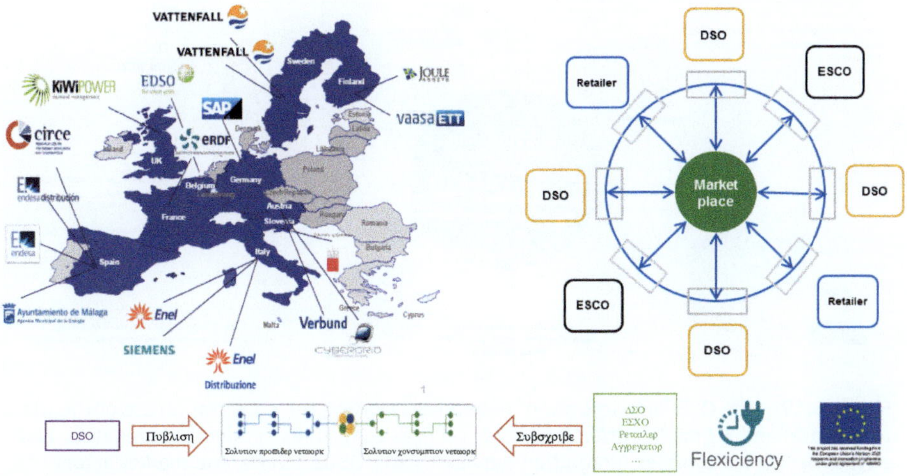

Figure 6.18 H2020 project "FLEXICIENCY" where SAP, ENEL, ERDF, Vattenfal, Verbund, Siemens EDSO, etc., participate. SAP HANA Cloud Platform used for the marketplace connecting DSOs (publishers) with retailers, ESCOs, and aggregators (subscribers). Source: Adapted from Flexiciency H2020 project [24].

The automated materials discovery aims to increase the pace of development of new battery chemistries. Others will leverage the flexibility and competitiveness of the European battery cell production by making use of advanced Industry 4.0 concepts. All stakeholders in the supply chain will benefit from a transparent and fraud-proof traceability of the full life cycle, from the raw material to the end-of-life, in the form of a digital battery passport and advanced monitoring based on sensors and the use of big data analytics.

Digital interconnection of centralized and decentralized storage systems by hybridization and multi-use of BESS into flexible portfolios will be an important step toward democratized energy systems.

6.5 Integrated Digital Platform Across Stakeholders

The European Technology and Innovation Platform Smart Networks for Energy Transition Working Group 4 (ETIP SNET WG4) Digital Energy group worked on a value proposition as a one-stop-shop (OSS) integrated digital platform across stakeholders [26]:

Customer-Centric OSS (One-Stop Shop) Universal, Democratized, and Simplified Digital Access to the available Market Energy Services (Figure 6.19).

One major goal of the value proposition or program is to provide a simplified and intuitive customer access leveraging heavily digital technologies, new approaches to big data, IoT, Cloud, data models, data integration, system interoperability, data management, predictive analytics approaches, and cyber security. The development of digital technologies represents 70% of the program.

Figure 6.19 The OSS "one-stop shop" network supporting the universal access of the users to the local and EU market offers of energy services companies, retailers, aggregators, and others. Backoffice complex integration between TSOs, DSOs, ESCOs, and market actors leads to simple front-office access for end users.

Another major objective of the program is to change as needed the operation of Grid Operators to create from the planning an approach that is customer oriented. Technically speaking, this change indicates an integration of planning, operation, and market that is not possible with current approaches.

It has been recognized that the urgent need for a customer-centric, multi-stakeholder, multi-process-scalable ICT architecture enables a country-specific integration of planning vs. operation vs. market to connect the dots, manage the complexity using digital advanced technologies, and make it transparent, simple, and intuitive for the customers.

The OSS process [26] allows the customer willing to have a new connection or to choose any energy service in Europe to get a unique point of access and request, where choosing among a conventional connection or a dynamic flexible connection or a service is offered. Cost savings for service activation would lead to about 25% cost savings.

Empowering the consumer will be a vital aspect of the new energy economy. By democratizing and expanding the battery storage sector, new demands will be generated within the market and will give rise to a new generation of innovative services and companies. However, easy accessibility to these resources via a common platform will be the key to unlocking this potential. The Batteries Digital Task Force proposes a lighthouse project (Figure 6.20) to provide a model for an OSS platform to meet consumers' needs for batteries. This type of platform would address the needs of both EVs and stationary applications and inform consumers about how to optimize the use of their battery storage resource.

6.6 Key Digital Recommendations

The digital grid of the future, part of the value chain, connected and digitalized, has been described in previous chapters. The digital technologies and digital use

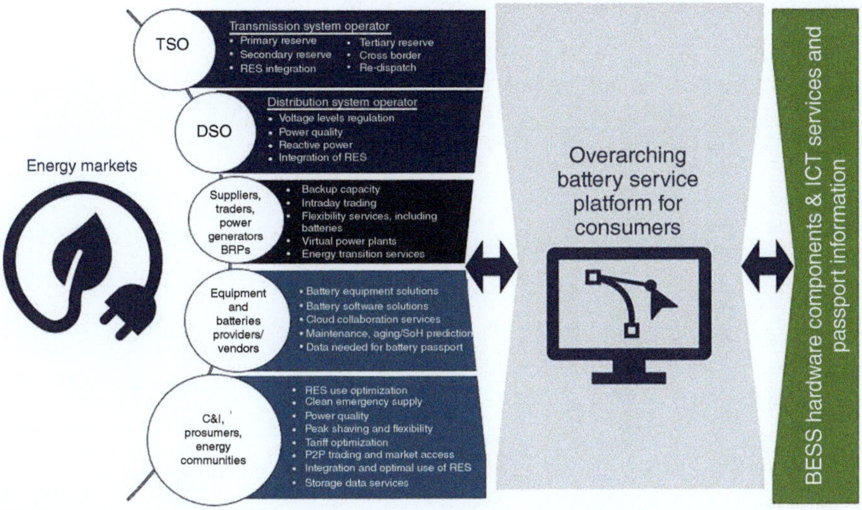

Figure 6.20 Usage cycle and services of the batteries integrated with the rest of the energy system illustrating the role of energy market stakeholders.

cases across networks and the rest of the energy system, generation and distribution generation, virtual power plants, storage, retailers, aggregators and ESCO (energy services companies), and customers have been illustrated as well.

In this chapter are summarized some of the recommendations issued by the ETIP SNET WG4 Digital Energy working group [27], the ETIP Batteries Digital task force [28], as well as the additional strategic recommendations supporting an accelerated energy and networks digitalization toward a decarbonized Europe in 2050.

Recommendations for a digital energy and network transition:

Digitalization is affecting the energy system at every level. In particular, the transformation from an electromechanical system to an electronic system is a fundamental change that will transform the fundamental principles around which the energy system is operating. The main digital facts and recommendations are summarized in Figure 6.21.

Recommendations for digital technologies' wide adoption for accelerating the energy and network transition:

On the technology's layer side, we provide hereafter the digital recommendations of the network energy transition system:

1) Need for new principles of operation.
2) Using analog AC current (AC) vs. direct current (DC).
3) Sharing infrastructure investments.
4) Need for overall covering architectures.
5) Need for open API's that will support interaction with other business sectors.
6) Need for a data economy based on open platforms.
7) Need for trust raising technologies.
8) Need for adequate service management and operations.

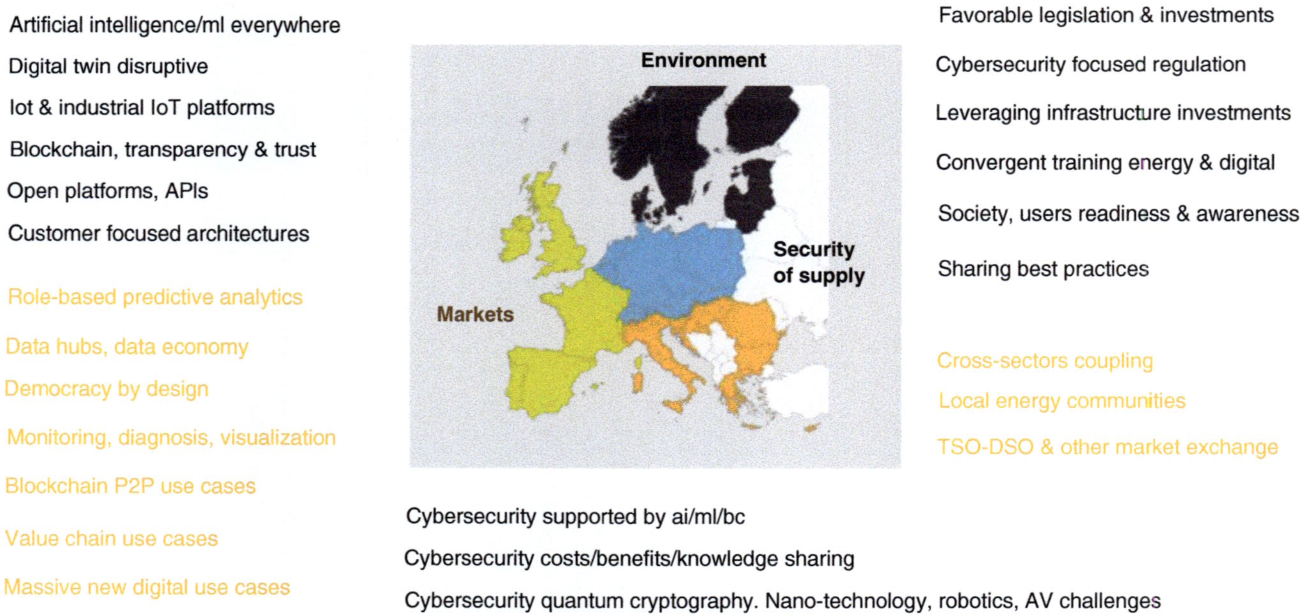

Figure 6.21 Smart networks for energy transition digital facts and recommendations [29].

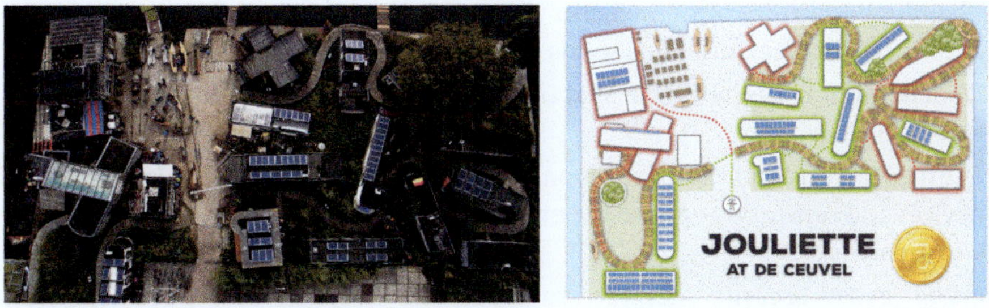

Figure 6.22 Spectral and Alliander have launched a new blockchain-based energy sharing token at De Ceuvel in Amsterdam. Named the "Jouliette," the new token aims to empower individuals and communities to easily manage and share their locally produced renewable energy [30].

9) Need for adequate education.
10) Need for adaptation of legislation.

Recommendations for digital use cases' wide adoption for accelerating the energy and network transition:

1) **Customers** or **prosumers** need to allow the connection to their assets.
2) **Data exchange platforms (DEPs)** seek to improve data exchange processes.
3) **Projects involving cross-sector coupling** offer an accessible market.
4) **Local energy communities (LECs**) offer many benefits.
5) **A strong collaboration between industry leaders and utilities** is needed.
6) **Existing infrastructure such as smart metering** should be further exploited.
7) **New consumer demands aim to accelerate penetration of** EVs.
8) **The Digital DSOs, TSOs, and other energy stakeholders** using digital twins.
9) **Blockchain is certainly one of the biggest trends in the** energy industry (Figure 6.22).

Recommendations for digital cyber security's wide adoption for accelerating the energy and network transition:

Digitization and customer participation is an interdisciplinary issue that affects many different verticals and also entails potentially new directions of necessary research in basically all topic areas previously listed in clusters. Despite the (r)evolution of new cyber security topics, it is evident that know-how and expertise from of related domains information security, telecommunications, automotive, healthcare, is essential for the development of cyber security and resilience in the future. The summarized takeaway messages are recommendations for research in the three clusters:

Technology: Research in the following topic areas is recommended because of the following:

1) **AI** helps the cyber security industry to monitor sophisticated threats efficiently.

2) The **blockchain** is considered as a promising technology to address **authentication, authorization, consensus**, and **immutability**.
3) **Decentralized distributed systems** efficiency needs to be measured and understood.
4) **Digitalization** enables and relies on data of massive deployment of **IoT**-enabled devices and sensors that make the energy system more transparent and efficient with analytics.
5) OT/IT cyber security architecture raises the question of on-**premise vs. cloud**-based.
6) For highly networked components, **safety** is **not reachable without cyber security**.
7) ML enables **predictive analytics** that help in detecting cyber attacks.

Policy: Research in the following topic areas are recommended because of the following:

1) **Metrics** and frameworks should be developed for decision-making tools on cyber risks.
2) A cyber security **communication platform needed among stakeholders** (IT, TSOs, DSOs, ESCOs, and policy).
3) Transparency of data flows and standardized data models are required for **General Data Protection Regulation (GDPR)**.
4) To lower burden on society, **CBAs** shall be considered
5) Opposing demands of **anonymization** and **aggregation** need research to allow both.
6) Investigate **privacy layer** design principles and techniques beyond cryptography.
7) The EU should go beyond network and information system security (NIS), following USA NERC example, organizing research of large-scale interdisciplinary attack scenarios.
8) **Knowledge databases** are used to share and access known vulnerabilities.
9) Regular **trainings** are vital to make our critical infrastructure resilient to cyber attacks.

Future challenges: Research in the following topic areas are recommended because of the following:

1) Technological **progress** is ongoing and predicting research needs to include variations.
2) **Society** and energy users need awareness about cyber security in the energy system.
3) **New crypto-environments** should include field demonstrations with open protocols.
4) New communication technologies, e.g. 5G, need new methods to guarantee service level agreements (SLAs).

5) **Bio- and nanotechnologies** raise the number of cyber threats that require research.
6) **Robotics** introduces new threats together with opportunities, e.g. physical unclonable functions (PUF) for robot identification.
7) Investigate **autonomous vehicles**, such as drones and cars, introducing new threats.

Recommendations for digital batteries' wide adoption by stakeholders:

To European Commission/national/regional funding and policy makers:

1) **Digitalization, decarbonization, decentralization, and democratization** tasks are paving the way for the "platformization" of the energy sector. Europe should invest in building a battery- and customer-centric digital leading platform
2) **A comprehensive "resilient, digitally democratized, and sustainable battery by design"** concept – including technical and regulatory dimensions – should be translated into tailored for member states plans but still leveraging global best practices.

Definition of policies and requirements to ensure quality and sustainability (i.e. processes, materials, and stakeholders involved) along the battery life cycle and value chain (i.e. design and development to production, setup, O&M, usage, decommission, and recycling) in alignment with the quality and sustainability policies in the EU. Procedures and sources of information and enforced measuring and monitoring through indicators should also be considered.

To industrial stakeholders:

1) **Enhance cross-sector partnerships along the battery value chain.**
2) **Co-innovation pilots and prototypes should drive decisions.**
3) **Technology innovation must be open to everyone.**

Many large energy players will appoint Chief Digital Officers to drive the digital transformation of their processes and create new businesses.

To Chief Digital Officers' stakeholders:

Customer-centric innovations will require the digital transformation roadmap to be adopted:

1) **Accelerate customer innovations** by making data available for market participants.
2) **Build massive energy services** as downloadable apps through energy exchange platforms business to business (B2B), business to consumer (B2C), and consumer to consumer (C2C).
3) **Full customer participation** by making customer usability as simple as one click.
4) **Build the pan-European Energy Union** of Customer Services by extending to cross-border energy management.

6.7 Conclusion

Digitalization is impacting all economic and social sectors in such a way that all of them are being transformed. The energy sector, in particular, strongly needs to go through a specific transformation with two major targets: on the one hand, establishing a clean and sustainable energy system; on the other hand, allowing energy users to participate in the entire value chain of the energy system that will enable the achievement of our targets to fully decarbonize Europe by 2050.

Innovative research work on identified issues has already begun, for example, on blockchain in the Jouliette token project, to gain experience with this emerging technology, to include cyber security by design, and to implement LEC's digital solutions, allowing us to receive a real-time flow view of electricity within the microgrid. However, many challenges remain to be explored to meet the vision 2050 timeframe.

Within this strategic frame, digitalization must be seen as a very powerful enabler and accelerator, rather than an objective in itself, to accelerate the achievement of our energy targets, especially in Europe where clean energy is being socially accepted almost everywhere.

Digitalization is already happening. It is time to scale up: Several use cases have been presented and their associated benefits were discussed. For some use cases, the CBA is not yet positive; however, the costs are rapidly decreasing and with increasing distributed generation and introduction of appropriate market models (e.g. including demand response and dynamic pricing), the digitalization of the entire energy system is definitely happening in the coming years.

Actors need to adapt their strategy to the market structure: The actors that have been involved in the energy system for many years are challenged to adapt their way of operating and incorporate new technologies that are adopted from other sectors such as the mobile communication sector.

Regulation plays an important role to develop digitalization: It is clear that regulators play an important role providing the correct incentives to develop the required technologies. As an example, smart metering functionality integration of flexible demand and dynamic pricing.

Funding research agencies will be necessary: Funding agencies are recommended to keep investing in research, as large challenges continue to exist even with the current available technologies and decreasing cost of communication. For example, the increased connectivity and digital evolution to tackle a major challenge in sustainable energy, i.e. the increased penetration of distributed and intermittent generation. A variety of use cases is being demonstrated in innovation projects.

Do not miss the non-reversible digital transformation reality today. Alternatively, you might be out of business: Digitalization for grid operations is ongoing. Currently, the DSOs are investigating which technologies to roll out for a smarter operation of their grid, allowing a higher penetration of distributed generation, a massive integration of storage, smart metering, and network big data at minimum costs and high flexibility.

Currently innovative smart grid management projects are usually performed by larger DSOs. Some smaller DSOs (e.g. more than 880 DSOs exist in Germany) may not be able to adopt these technologies without a clear articulation of the value proposition, time to market, industrialization, shared platforms, and accelerated deployment. A crucial aspect is that smaller DSOs who do not have the financial resources to perform innovation projects should join the smart grid operation community of practices and join the electronic marketplaces connecting distributors and retailers.

Meaning the management of SmartGrids: The European and global experts that discussed digitalization of the grid, either at medium voltage (MV) or low voltage (LV) level, did put a considerable effort in setting up communication infrastructure and coupling with the data platform. There is not yet an end-to-end solution on the market that provides supervisory control and data acquisition (SCADA) and ICT services, which are fully interoperable and easy to implement. However, the emergence of the IT/OT integrated or convergent platforms running on Big Data with real-time predictive analytics services can play a key role in making implementations more efficient.

Empower ICT infrastructures using digital simulation models: As a DSO, simulation models can be a great tool to identify weak spots in the grid and the most cost-effective way to operate the grid. With the availability of sensor data, a combination with grid analysis tools can effectively increase the vision of the operator on the flows in the grid, reducing the amount of locations where hardware implementations are needed.

Open electronic marketplaces boost digital energy: In most EU member states that proceed with a smart meter rollout, the DSO is responsible for the collection and the management of the data. In the United Kingdom, retailers play an active role in smart metering data collection. DSOs and retailers should be connected through electronic marketplaces and exchange B2B or B2C digital services (for example, Horizon 2020 program of the European Commission [H2020] project "Flexiciency"). The access to these marketplaces should be open as well to technology and energy service providers.

Well-guided data confidentiality, privacy, and security accelerates the digital transformation: Traditionally, grid operators have a rather strict policy on data confidentiality with regard to their grid topologies and other grid-related data. However, with the digital transformation and the increased possibilities for smart network control, cooperation with third parties such as research institutes in large innovation projects is much more efficient if the actual field data can be used.

Data privacy concerns need to be fully addressed: A major barrier to the increased use of digital technologies in the energy arena is industry and public concerns about data privacy and security. Energy use, cost, and production data are highly sensitive, and the risk of data loss or theft as a result of increased connectivity is a disincentive for companies as they attempt to embrace digitalization. Increased interoperability and standardization will enable energy data systems to interact with one another in a reliable, safe, secure, and user-friendly manner.

Smart management is the only way to successfully integrate a massive number of renewables: As one of the contributions to this document is showing that increased forecasting possibilities, in combination with measures such as reactive power control and curtailment, are effectively able to increase the hosting capacity of the grid and connecting renewables at an overall lower cost for the society.

Leveraging digital technologies frees up creativity and enables open and transparent flex markets: Several initiatives are ongoing to integrate flexibility in the market and create a simple cross-border system for flex trading. However, in some European member states, industrial flexibility is already integrated in the market, and the challenge is to increase the share of flexible demand in the system and develop a transparent market model that can take into account the needs of all actors. This requires leveraging on digital technologies such as data handling, IoT and predictive analytics platforms.

Automated technologies can shift residential consumption: Demand response is already active in several member states in Europe and flexible consumption should be integrated in the market of the remaining countries as soon as possible, by working out appropriate market products.

Keep investing in disruptive digital technologies: Even with the planned reduction of GHGs, global warming is still a problem that presents enormous challenges to the power sector.

Despite the long-term potential and early success of energy digitalization efforts, there remain several challenges that must be overcome in order to unlock the full potential of our digital energy future:

- Digital energy adoption needs to be encouraged.
- Industry, government, and universities must advance digital energy R&D.
- Cyber security needs to be improved.
- There is a need for standards and specifications.
- Talent should be cultivated.

What will change in the energy and networks, transportation, technology, consumers, and cities (Figure 6.23)? Changes are disruptive:

		Will change to...	Become	With...
1	Energy	No more a commodity	Distributed	DER, SMARTGRIDS
2	Transportation	More integrated and cleaner	Interoperable	SMART TRANSPORTATION
3	Technology	Connect citizens to all city services in real time	Anywhere	SMART TECHNOLOGIES
4	Consumers	Be empowered in their daily activities & producers	Prosumers	DER, FLEXIBILITY, STORAGE
5	Cities	Live, work, innovate and sustain better	Sustainable	SMART CITIES

Figure 6.23 What will change in energy, transportation, technology, citizens, and cities?

As the industrial revolution was the transition to new manufacturing processes between 1760 and 1840, the digital revolution will be the disruptive transformation of the twenty-first century to a new economy, to a new society, and to a new era of low-emission energy across the energy and transportation sectors as well as cities.

List of Abbreviations

AC	analog AC current
AI	artificial intelligence
B2B	business to business
B2C	business to consumer
C2C	consumer to consumer
BESS	battery electric storage systems
BI	business intelligence
BNEF	Bloomberg New Energy Finance
CAE	computer-aided engineering
CAPEX	capital expenditure
CBA	cost–benefit analysis
CEN	European Committee for Standardization
CENELEC	European Committee for Electrotechnical Standardization
DC	direct current
DEP	data exchange platform
DER	distributed energy resources
DRM	demand response management
DSO	distribution system operator
EDSO	European Distribution System Operators
EEGI	European Electricity Grid Initiative
ESCO	energy services company
ETIP Batteries	European Technology and Innovation Platform for Batteries
ETIP SNET	European Technology and Innovation Platform Smart Networks for Energy Transition
ETIP SNET WG4	European Technology and Innovation Platform Smart Networks for Energy Transition Working Group 4
ETP	European technology Platform
SRA	Strategic Research Agenda
ETSI	European Telecommunications Standards Institute
EV	Electrical Vehicles
GDPR	General Data Protection Regulation
GESI	Global e-Sustainability Initiative
GHG	Greenhouse Gas
H2020	Horizon 2020 program of the European Commission
ICT	information and communication technologies
ITU	International Telecommunications Union
IEA	International Energy Agency

IIoT	Industrial Internet of Things
Industry 4.0	Fourth Industrial Revolution
IoT	Internet of Things
KPI	key performance indicators
LEC	local energy community
M/490	Standardisation Mandate Smart Grids – European Commission
M2M	machine-to-machine
ML	machine learning
MV	Medium Voltage
LV	Low Voltage
NERC	North American Electric Reliability Corporation
NIST	National Institute of Standards and Technology
O&M	operations & maintenance
OT	operational technology
IT	information technology
RD&D	research, development and large-scale demonstration
RES	renewable energy sources
RTU	remote terminal unit
SCADA	supervisory control and data acquisition
SGAM	smart grid architecture model
SLA	service level agreement
SME	small and midsize enterprise
TSO	transmission system operator
UAV	unmanned aerial vehicle
WEF	World Economic Forum

References

1. Economic Times (2018). Opinion: How power grid digitalization is unlocking new possibilities EnergyWorld.com (accessed 2 November 2021).
2. GE Renewables & Power (2018). The Digital Energy Transformation, white paper coordinated by M. Chebbo.
3. GE Renewables & Power (2018). The Digital Energy Transformation, white paper coordinated by M. Chebbo.
4. Global e-Sustainability Initiative/Accenture (2015). *SMARTer2030: ICT Solution for the 21st Century*. GeSI.
5. The Digitalisation of the Energy system 4.0 (2016). Value Chain of the unbundled market, generation, TSOs, DSOs, retailers illustrated by M. Chebbo and included in the ETP SmartGrids WG3 chaired by M. Chebbo.
6. SmartGrids R&DD agenda described by the ETP SmartGrids and EEGI (2010) and Technologies mapped illustrated by M. Chebbo (former SAP General Manager Energy for EMEA) (2017).

7 Economic Times (2018). Opinion: How power grid digitalization is unlocking new possibilities. EnergyWorld.com (accessed 2 November 2021).

8 ETP SmartGrids WG3 chaired by M. Chebbo "Demand and Metering/Task Force Retail": Energy Retailers' perspective on the deployment of SmartGrids in Europe (2011).

9 ETP SmartGrids WG3 by M. Chebbo "Demand and Metering/Task Force Retail": Energy Retailers' perspective on the deployment of SmartGrids in Europe (2011).

10 Bloomberg New Energy Finance (2017). Digitalization of Energy Systems, a white paper.

11 ENTSOE (2019). The Cyber Physical System for the Energy Transition. Digital Challenges, Opportunities and Projects from TSOs and ENTSO-E.

12 European Distribution System Operators (EDSO) for SmartGrids (2016). "Digital DSO" – A Vision and the Regulatory Environment Needed to Enable It.

13 Constellation Research by Ray Wang: Monday's Musings: Understand the Spectrum of Seven Artificial Intelligence Outcomes (2016).

14 ETP SmartGrids WG3 "Demand and Metering/Task Force Retail" (2011). Energy Retailers' perspective on the deployment of SmartGrids in Europe.

15 European Commission Joint Research Center (JRC) (2021). Smart Metering Deployment in Europe. updated report.

16 European Commission (2019). Benchmarking smart metering deployment in EU-28. Tractebel Engie report.

17 European Commission (2021). SmartGrids and Meters.

18 WEF, World Economic Forum (2016). Electricity: generating value through digital transformation.

19 ESMIG press release (2016). Energy Union proposals to unlock the benefits of demand side flexibility and energy efficiency". M. Chebbo was a former President of ESMIG (2014–2018).

20 ESMIG press release (2016). Review of the Energy Performance of Buildings Directive". M. Chebbo was a former President of ESMIG (2014–2018).

21 ENTSOE (2019). The Cyber Physical System for the Energy Transition. Digital Challenges, Opportunities and Projects from TSOs and ENTSO-E.

22 ETIP Europe Batteries Digital Task Force chaired by M. Chebbo (2020). Digitalization position paper.

23 ETP SmartGrids WG3 (2016). "Demand & Metering" chaired by M. Chebbo: The Digital Energy System 4.0.

24 Flexiciency H2020 project (2017).

25 ETIP Europe Batteries Digital Task Force chaired by M. Chebbo (2020). Digitalization position paper.

26 ETIP SNET WG4 Digital Energy group (2020). "Big idea" proposal for a one stop shop integrated digital platform cross-stakeholders.

27 ETIP SNET WG4 chaired by M. Chebbo (2018). Digitalization of the Electricity System & customer participation (ETIP SNET WG4 (M. Chebbo), 2018).

28 ETIP Europe Batteries Digital Task Force chaired by M. Chebbo (2020). Digitalization position paper.
29 Based on the ETIP SNET WG4 chaired by M. Chebbo (2018). Digitalization of the Electricity System & customer participation.
30 ETIP SNET WG4 chaired by M. Chebbo (2018). Digitalization of the Electricity System & customer participation.

Further Reading

Accenture (2015). ICT Solution for the 21st Century. GeSI, Global e-Sustainability Initiative. SMARTer2030.
Bloomberg New Energy Finance (2017). Digitalization of Energy Systems.
ENTSOE (2019). The Cyber Physical System for the Energy Transition. Digital Challenges, Opportunities and Projects from TSOs and ENTSO-E.
ESMIG (2016). ESMIG press release: Review of the energy performance of buildings directive. https://www.esmig.eu/esmig-welcomes-the-revision-of-the-european-energy-efficiency-directive/.
ESMIG (2016). Press release : Energy Union proposals to unlock the benefits of demand side flexibility and energy efficiency. ESMIG. https://eubac.org/wp-content/uploads/2021/03/2016.06.16_IDEAS_Proposals_for_DSFEE.pdf.
ETIP Europe Batteries Digital Task Force and Chebbo, M. (2020). Digitalization Position Paper.
ETIP SNET WG4 and Chebbo, M. (2018). Digitalization of the Electricity System & Customer Participation.
ETP SmartGrids and EEGI (2010/2017). SmartGrids R&DD Agenda and Technologies Mapped.
ETP SmartGrids WG3 and Chebbo, M. (2011). Energy Retailers' perspective on the deployment of SmartGrids in Europe: Demand and Metering/Task Force Retail.
ETP SmartGrids WG3 and Chebbo, M. (2016). Value Chain of the Unbundled Market, Generation, TSOs, DSOs, Retailers: "The Digitalisation of the Energy system 4.0".
European Commission (2021). SmartGrids and Meters.
European Commission Joint Research Center (JRC) (2021). Smart Metering Deployment in Europe (updated report).
European Distribution System Operators (EDSO) for SmartGrids (2016). "Digital DSO" – A Vision and the Regulatory Environment Needed to Enable It.
GE Renewables & Power and Chebbo, M. (2018). The Digital Energy Transformation, white paper.
Times Economic (2018). How power grid digitalization is unlocking new possibilities. EnergyWorld.com. (accessed 2 November 2021).
Tractebel Engie report (2019). Benchmarking Smart Metering Deployment in EU-28. European Commission.

Wang, R. (2016). Monday's musings: understand the spectrum of seven artificial intelligence outcomes. *Constellat. Res.* https://www.constellationr.com/blog-news/monday-s-musings-understand-spectrum-seven-artificial-intelligence-outcomes.

WEF, World Economic Forum (2016). Electricity: Generating Value Through Digital Transformation.

7

Smart and Sustainable Mobility Adaptation Toward the Energy Transition

Carla Silva, Catarina Marques, Mariana Raposo, and Angelo Soares

da Universidade de Lisboa, Instituto Dom Luiz (IDL), Departamento de Engenharia Geográfica, Geofísica e Energia, Faculdade de Ciências, Campo Grande, Lisboa 1250-096, Portugal

7.1 Smart and Sustainable Mobility Definitions and Metrics

The 2030 Agenda for Sustainable Development contains the agreement on 17 Sustainable Development Goals (SDG), 169 targets (in effect since 1 January 2016), and replacing the Millennium Development Goals set in 2000. Monitoring the progress is accomplished through the use of 232 indicators [1, 2].

As mobility needs increase in a growing globalized world, more people and goods are moved across the globe; thus, it urges this mobility to be equitable, efficient, safe, and climate responsive. These mobility needs are in line with several of the key development goals: reduced inequalities (SDG 10), sustainable cities and communities (SDG 11), affordable and clean energy (SDG 7), good health and well-being (SDG 3), and climate action (SDG 13). To further boost seamless and efficient dislocations and therefore facilitate people and goods mobility, the use of information and communications technologies (ICT) is unavoidable. Smart mobility is sustainable mobility taking advantage of ICT (vehicle-to-vehicle communication, vehicle–infrastructure communication, and vehicle/network–user communication). It provides innovative services relating to different modes of transport and traffic management while enabling various users to be better informed and make safer, more coordinated, and "smarter" use of transport networks (DIRECTIVE 2010/40/EU). These principles of smart mobility are also in line with SDG 12 (responsible consumption), SDG 9 (industry, innovation, and infrastructure), and SDG 8 (decent work and economic growth). The concept of smart mobility frequently appears in scientific literature and international policies and is always linked with the concept of smart cities [3].

Figure 7.1 illustrates the relationship between smart mobility and the SDGs mentioned above.

To achieve these SDGs, it is necessary to use indicators and metrics to monitor and track progress. For example, SDG 3 "By 2020, halve the number of global deaths and

Technologies for Integrated Energy Systems and Networks, First Edition.
Edited by Giorgio Graditi and Marialaura Di Somma.

Figure 7.1 Illustrative SDG 2030 related to smart mobility (own illustration).

injuries from road traffic accidents," and "By 2030 substantially reduce the number of deaths and illnesses from hazardous chemicals and air, water, and soil pollution and contamination," requires using the indicators "death rate due to road traffic injuries" (number of deaths per capita) and "mortality rate attributed to household and ambient air pollution" (number of deaths per capita) [4]. The SDG 7 states that "By 2030, increase substantially the share of renewable energy in the global energy mix" have related indicators such as "renewable energy share in the total final energy consumption" (percentage of renewables) and "energy intensity" (primary energy and gross domestic product [GDP] ratio). Another example would be SDG 11 that states "By 2030, provide access to safe, affordable, accessible and sustainable transport systems for all, improving road safety, notably by expanding public transport, with special attention to the needs of those in vulnerable situations, women, children, persons with disabilities and older persons", which is translated in the indicator "Proportion of population that has convenient access to public transport, by age, sex and persons with disabilities". Besides the highlighted SDG aspects, the mobility/transport sector also has significant intersections with other SDGs. For example, without providing reliable, safe, and sustainable transport systems, people cannot attend schools (SDG 4), women cannot be assured opportunities for employment and empowerment (SDG 5), and the needs of the population cannot be met and/or countries cannot provide food security (SDG 2).

The main problem with sustainable mobility indicators and metrics is that they are yet to be universally agreed on [5], but it is clear they are essential for monitoring progress as previously mentioned. Therefore, reaching a consensus on these indicators and metrics becomes increasingly relevant as the global demand for passenger mobility and freight demand is expected to increase threefold from 2015 until 2050, as well as the growing importance of e-commerce and the accompanying boom in last-mile delivery [6]. In 2017, in Europe, the transport sector accounted for almost a quarter of all greenhouse gas (GHG) emissions, with road transport alone representing 20% and being the only economic sector where GHG was higher than in 1990 with emissions growing despite the mitigation efforts undertaken. The European Green Deal has set the key objective to deliver a 90% reduction in transport-related GHG emissions by 2050.

7.1.1 Sustainable Mobility KPI (Key Performance Indicators)

There are at least 18 different key performance indicator (KPI) systems to characterize and track the level of sustainability in mobility [7], from which a sample of four can be selected: the International urban mobility index 3.0 developed by Artur D Little [8], the international sustainable urban mobility project 2.0 developed by the World Business Council for Sustainable Development (WBCSD) [9], the European Indicators to assess the sustainability of transport activities [10], and the European sustainable urban mobility indicators (SUMI) [11]. All these KPI systems have mobility-related indicators divided into certain categories or dimensions. Table 7.1 describes the transparency of the metrics, availability of metrics, and each approach complexity, in other words, if they are fully and freely available, fully described, and how many indicators and metrics there are. "Dimensions" indicates how many aspects are considered in each approach. "Complexity" refers to the quantity of core indicators that each approach considers.

Data availability, which assumes open data periodically updated, is highly dependent on the region and country policies. The ease of access to open data and the ability to compute each indicator metric are very important. Likewise, a lower complexity brings simplicity to the approach. In this way, to demonstrate how these indicators can be applied to any city, two Portuguese cities, Lisbon and Porto, will be characterized as a practical example guide. Open metrics, open data, and data availability will be further explored using the sustainable mobility project (SMP) 2.0 and European SUMI.

In this manner and for the sake of simplicity, we will employ the following common and easy-to-apply indicators from the SUMI and SMP2.0 that give us great insights into mobility trends and if the SDG targets are being met for these two Portuguese cities. In the following sections, we will introduce and explore these indicators by explaining what they measure and why they are important. Ultimately, this macroanalysis will lead us to an easy-to-read circular graph that characterizes the cities in analysis.

Table 7.1 Comparison of a sample of people sustainable mobility indicator systems.

KPI system	Urban Mobility Index 3.0	SMP2.0	Dobranskyte-Niskota et al.	SUMI
Dimensions	3 (maturity, innovation, and performance)	4 (global environment, quality of life in the city, economic success, and mobility system performance)	5 (environmental, social, economic, institutional, technical, and operational)	NA but based on [9]
Complexity (# core indicators)	27 (9 for each dimension with different weights)	19 (5–20 in each dimension equal weights)	55 (8–19 in each dimension equal weights)	13 (based on [9])
Indicators examples	Share of public transportation in modal split; share of zero-emission modes; road density; cycle-path network density; urban agglomeration density; bicycle-sharing performance; car-sharing performance; transport-related CO_2 emissions; NO_2, PM_{10}, and $PM_{2.5}$ concentration; mean travel time to work; and traffic-related fatalities	Affordability of public transport for the poorest group; accessibility for mobility-impaired groups; air polluting emissions; noise hindrance; fatalities; access to mobility services; quality of public area; urban functional diversity; commuting travel time; economic opportunity; net public finance; mobility space usage; emissions of greenhouse gases (GHG); congestion and delays; energy efficiency; opportunity for active mobility; intermodal integration; comfort and pleasure; and security	Volume of transport relative to GDP (tonne km; passenger km); motor vehicle fuel prices and taxes (for gasoline and gas/diesel); private car ownership; cases of chronic respiratory diseases, cancer, headaches, respiratory restricted activity days, and premature deaths because of motor vehicle pollution; persons killed in traffic accidents (number of fatalities – 1000 vehicle km; per million inhabit ants); average passenger journey time; and NO_x, VOCs, PM_{10}, $PM_{2.5}$, sulfur oxides (SO_x), O_3, CO_2, N_2O, and CH_4 emissions (per capita)	Total energy use by urban transport per passenger km and tonne km (annual average over all modes); well-to-wheels GHG emissions; road deaths by all transport accidents in the urban area on a yearly basis; fatalities of active mode users in traffic accidents in the city in relation to their exposure to traffic; and infrastructure for active mobility, namely, walking and cycling
Transparency (Are metrics extensively described?)	No	Yes	Yes	Yes
References	[8]	[9]	[10]	[11]

PT, public transportation; VOC, volatile organic compound; GDP, gross domestic product; NA, not applicable.

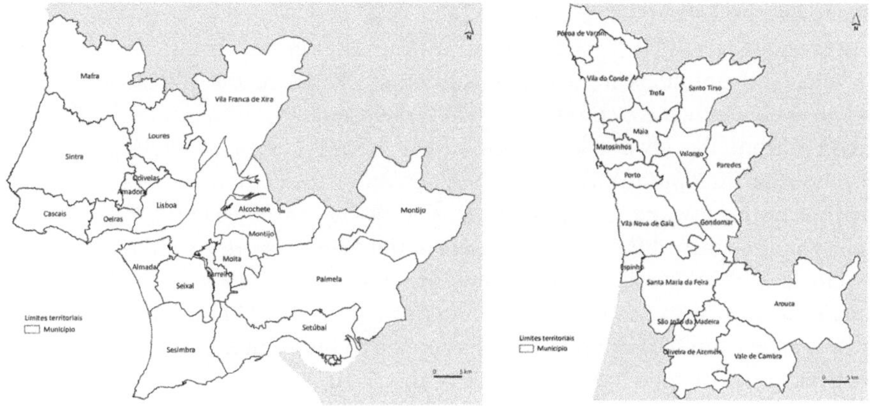

Figure 7.2 (a) AML and (b) AMP, regions of Portugal (own illustration).

7.1.2 KPI of Urban Mobility in Two European Cities

Figure 7.2 shows the Lisbon metropolitan area (AML) that encompasses Lisbon city and the revolving 17 municipalities (2 827 514 inhabitants in 2017) as well as the Porto metropolitan area (AMP) that includes Porto city and the revolving 16 municipalities (1 719 362 inhabitants in 2017).

Fatalities, commuting travel times, air polluting emissions, emission of GHG, and access to mobility services were analyzed for Lisbon AML and Porto AMP for 2017. This year was selected because of the wide open data availability, including travel times and passenger per kilometer (pkm) from a mobility survey at the Instituto Portugues de Estatistica (INE) [12].

Fatalities indicator, population per area, and fuel sales are available through statistical databases, in this case, the Portuguese database PORDATA and INE. Bus, train, and metro station stops' geographic distribution is available through geographic information systems (GIS). The information used for each fossil fuel (liquefied petroleum gas [LPG], diesel, gasoline) is observable in Table 7.2 and is based on international sources [13, 14]. In this Portuguese case study, the processes of producing, transporting, manufacturing, and distributing are translated into indicators that cover all steps from extracting, capturing, or growing the

Table 7.2 Fossil fuel properties and selected emissions [13, 14].

		TTW			WTT	WTW
	Low heaving value (LHV) (MJ/kg) [13]	CO_2 (g/MJ) [13]	Tier 1 NO_x (g/kg) [14]	Tier 1 $PM_{2.5}$ (g/kg) [14]	CO_2 (g/MJ) [13]	CO_2 (g/MJ) [13]
Diesel (B7)	43.1	73.2	8.73	0.03	18.9	92.1
Gasoline (E5)	42.3	73.3	12.96	1.1	17	90.3
LPG	46	65.7	15.2	0	7.8	73.5

primary energy carrier used to then create and fuel vehicles. These indicators are tank-to-wheels (TTW), well-to-tank (WTT), and well-to-wheels (WTW). TTW represents the emissions from the consumption of fuel in the vehicle engine to move the vehicle through combustion. WTT represents the average of all the GHG emissions from the production process (e.g. refinery) until the delivery of the fuel produced (fuel pump). Lastly, WTW constitutes the combination of energy or emissions from TTW and WTT [15]. When electric cars are considered, the meaning of the term is slightly different: TTW refers to the battery energy depletion to move the vehicle and WTT refers to electricity generation, transmission, and charging losses. To restate, WTW combines the energy or emissions of WTT and TTW. These concepts arise from the life cycle analysis applied to the vehicle energy sources and can be better understood in Silva and coworkers 2010 [16].

Fossil fuel consumption (referring to the TTW part) because of transport activity (LPG, diesel, and gasoline fuels) was nearly 5.5 million tonnes in Portugal, with AML and AMP covering 30% of that value.

Electricity CO_{2eq} intensity has been decreasing, and in 2017, it was 353 g/kWh for the generation and, considering low voltage losses of 6.7%, the WTT value was 380 g/kWh.

For the indicator commuting travel time, the duration of commute to and from work or an educational establishment, T_{comav}, the following formula applies and the data must come from a survey:

$$T_{\text{comav}} = \frac{\sum_{i=1}^{n}(T_{\text{outi}} + T_{\text{returni}})}{n} \tag{7.1}$$

where

T_{comav} is the average commuting time (minutes/day);
T_{comi} is the averaged commuting time surveyed person i;
T_{outi} is the commuting time from home to work/school by person i (minutes/day);
T_{returni} is the commuting time to home by person i (minutes/day);
n is the number of persons in the survey.

Taking the survey [12] data, it was estimated that AML and AMP have a T_{comav} of 48.6 minutes and 43.6 minutes, respectively.

For the indicator fatality rate (FR), the following formula applies:

$$\text{FR} = \frac{\sum_{i=1}^{n} K_i * 100\,000}{\text{Cap}} \tag{7.2}$$

where

FR is the fatality rate (#/100 000 population/year);
K_i is the number of persons killed in the transport mode i (#/year);
Cap is the capita or number of inhabitants in the city (#); and
i is the transport mode (passenger car, freight traffic, tram, bus, train, motorcycle, river transport, etc.) (type).

According to PORDATA yearly statistics of deaths per municipality, there are 2.8 and 3.8 deaths per 100 000 population in the AML and AMP, respectively.

For the indicator energy efficiency, the following formula applies:

$$E = \frac{\sum_i \sum_j A_{ij} * \left(\sum_k S_{jk} * I_{jk} * \mathrm{EC}_k \right)}{\mathrm{TV}_{\mathrm{pass}} + \mathrm{TV}_{\mathrm{fre}} * 1/8} \tag{7.3}$$

where

E is the energy consumption rate (MJ/pkm);
$\mathrm{TV}_{\mathrm{pass}}$ is the transport volume passenger transport (million passenger km);
$\mathrm{TV}_{\mathrm{fre}}$ is the transport volume freight transport (million tonne km);
S_{jk} is the share of fuel type k per vehicle type j (fraction);
I_{jk} is the energy intensity per distance driven for vehicle type j and fuel type k (l/km or MJ/km or kWh/km);
A_{ij} is the activity volume (distance driven by transport mode i and vehicle type j) (million km/year);
EC_k is the fuel energy content for fuel k (l/km or MJ/km or kWh/km);
k is the fuel type (type).
i is the transport mode (passenger car, tram, bus, train, motorcycle, inland vessel, freight train, truck, etc.) (type);
j is the vehicle class (if available, specified by model, e.g. sport utility vehicles [SUV], etc.) (type).

Knowing the specific fleet composition circulating in the AML and AMP is not an easy task. By taking a macro approach, the energy consumption can be used to estimate the fleet composition using the yearly fuel sales statistics and electricity use for transport, as can be seen in Figure 7.3 and Figure 7.4. Thus, LPG and gasoline are car and motorcycle related. Diesel has contributions from car, bus, and freight. For the people mobility volumes, the INE mobility survey can be used to acknowledge the average distance per day and per person by mode. Assuming 48 weeks in a year,

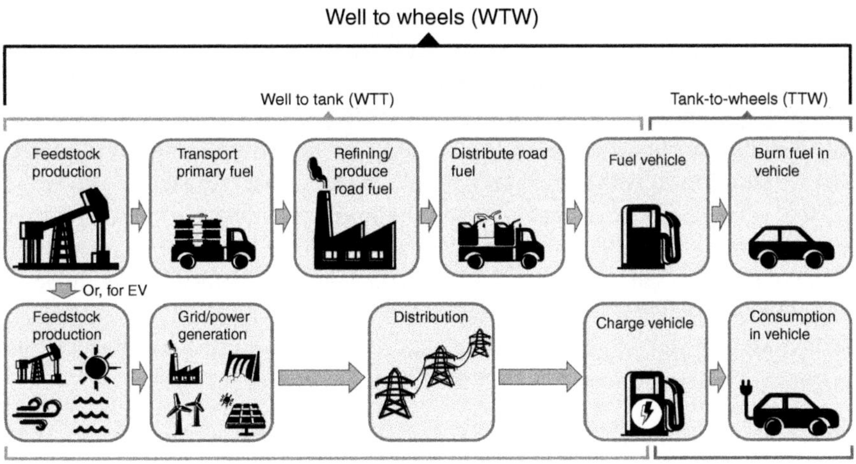

Figure 7.3 Diagram of WTW, WTT, and TTW (own illustration).

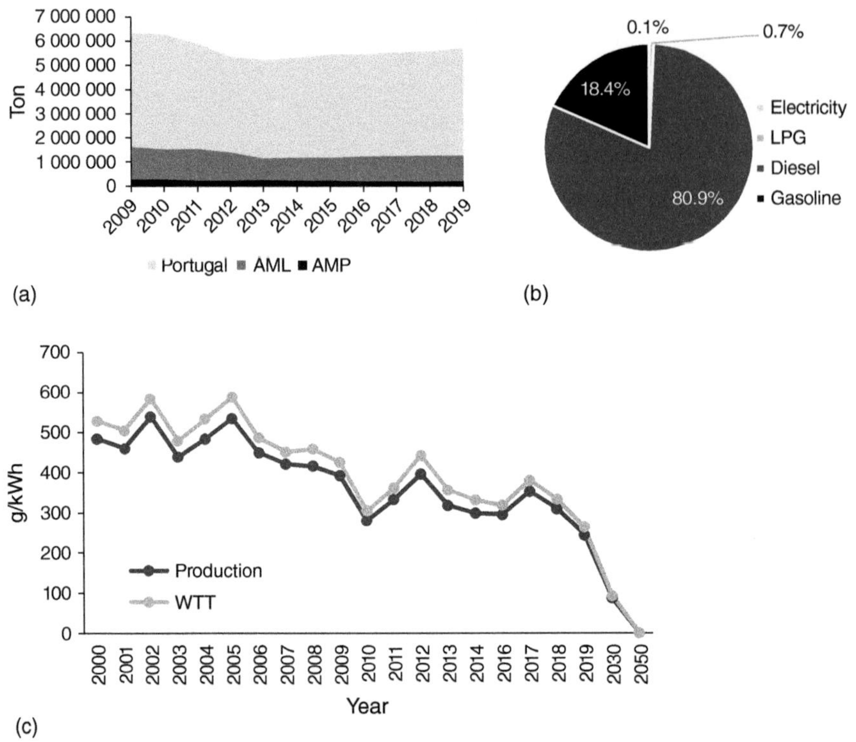

Figure 7.4 (a) Liquid fossil fuel consumption for transport; (b) electricity for transport share in final energy 2017 (metro, train); and (c) electricity carbon intensity evolution with distribution losses (WTT).

7 d/wk, the passenger mobility levels by mode can now be calculated. It should be noted that often there are data inefficiencies that need to be overcome. Freight data in Portugal is only available at a national level, making freight fuel sales per municipality hard to estimate. Still, the information available should be used if possible, even if it represents a rough estimate. At a national level in 2017, transport volume in the freight sector amounted to 34.1 thousand million tonnes-kilometer (tkm). For passenger volume, using statistical data from INE, 12 927 000 000 pkm was registered in public transport and 116 343 000 000 pkm in private transport. Overall, 129 270 000 000 pkm in both public and private transport was registered. However, freight fuel sales occur more likely outside the municipalities, and therefore, this approach has huge uncertainty. For the sake of simplicity, the survey approach of mobility volumes (people mobility) and its use to weight fuel efficiency by mode is used, see Table 7.3 and Table 7.4 information.

For the indicator emission harm index (Table 7.5), air polluting emissions of all passenger, and freight transport modes, the following formula applies:

$$\mathrm{EHI} = \frac{\sum_{s} E_{\mathrm{eqs}} * \left(\sum_{j} A_{ij}\right) * \left(\sum_{\mathrm{ck}} S_{ijk} * I_k * E_{\mathrm{ijkcs}}\right)}{\mathrm{Cap}} \tag{7.4}$$

Table 7.3 People mobility volumes and energy efficiency (TTW) by mode

Transport mode	AML (pkm)	%	AMP (pkm)	%	MJ/pkm[a)]
Car	10 647 916 842	76	6 702 046 543	84	2.17
Bus	1 716 754 149	12	782 712 423	10	1.47
Metro	480 022 090	3	213 966 412	3	0.45
Train	1 113 885 228	8	282 654 922	3	0.27
Boat	108 301 716	1	0	0	2.70
Total billion	14.1	100	NA	100	NA
Average MJ/pkm (E Eq. (7.3))	1.88	NA	1.98	NA	NA

a) Average values taken from the sustainability reports of the public transport companies and top brand and car models combined sales, including urban and extra-urban driving.

Table 7.4 Final energy consumption for transport- and emission-related indicators.

	AML	AMP
LPG	376 536 000	266 521 600
Gasoline	11 818 884 000	7 902 356 000
Diesel	41 060 088 000	31 485 369 600
Electricity	53 793 442	40 054 795
Final energy consumption (MJ)	53 793 442 424	40 054 795 152
NO_x	11 902 106.01	8 777 580.28
$PM_{2.5}$	323 986.05	219 423.74
EHI	4.34	5.24
TTW GHG	3.85 602E+12	2.87 362E+12
WTT GHG	9.70 309E+11	7.24 907E+11
G	1.87	2.25

Table 7.5 Conversion of indicator values in a 0–10 scale [9].

Indicator/scale	0	10
Commuting travel time	≥90 min/d	≤10 min/d
Fatality rate	≥35/100 000 inhabitants	0
Energy efficiency (TTW)	≥3.5 MJ/pkm	≤0.5 MJ/pkm
GHG emissions (WTW)	≥2.75 tonCO_{2eq}/capita/yr	0 tonCO_{2eq}/capita/yr
Emission harm index (TTW)	≥55 kg NO_{xeq}/capita/yr	0 kg NO_{xeq}/capita/yr

Source: Based on WBCSD Mobility [9].

where

EHI is the emission harm equivalent index (kg NO_{xeq}/cap/year);
E_{eqs} is the emission substance-type equivalent health impact value (factor);
E_{ijkcs} is the emission of pollutants per unit of energy consumed for fuel type k, emission class c of vehicle type j of transport mode i (g/l, g/kg);
A_{ij} is the activity volume (distance driven by transport mode i and vehicle type j) (million km/year);
S_{ijk} is the share of fuel type k per vehicle type j and per transport mode i (fraction);
I_k is the energy intensity per distance driven per fuel type k (l/km or kWh/km or kg/km);
Cap is the capita or number of inhabitants in the city (#);
k is the energy type (petrol, diesel, bio-fuel, electricity, hydrogen, etc.) (type);
i is the vehicle-type transport mode (passenger car, tram, bus, train, motorcycle, inland vessel, freight train, truck, etc.) (type);
j is the vehicle class (if available specified by model, e.g. SUV) (type);
s is the type of substance (type) limited to NO_x and particles with 10 μm diameter or less (PM_{10}); and
c is the emission class (euro norm) (type).

Again, it is easier to take a macro approach and look at fuel sales by using fuel specifications in Table 7.2 and then deduct NO_x and particles with 2.5 μm diameter or less ($PM_{2.5}$) emissions after converting them to NO_x equivalent by multiplying by the factor 1.06 so that they are in the same units of measurement.

For the GHG emissions (Table 7.4), these are in tonne CO_2 equivalent (WTW) per annum and per capita:

$$G = \frac{\sum_{i}\sum_{j} A_{ij} * \left(\sum_{k} S_{jk} * I_{jk} * (C_k(1 + F_{ijk}) + W_k) \right)}{\text{Cap}} \tag{7.5}$$

where

G is the GHG emission (tonne CO_2(eq)/cap/year);
C_k is the TTW CO_2 emission per energy type unit considered (kg/l or kg/kWh);
W_k is the WTT CO_2 equivalent emission per energy type unit considered (factor);
A_{ij} is the activity volume (distance driven by transport mode i and vehicle type j) (million km/year);
S_{jk} is the share of fuel type k per vehicle type j (fraction);
I_{jk} is the energy intensity per distance driven for vehicle type j and fuel type k (l/km or MJ/km or kWh/km);
Cap is the capita or number of inhabitants in the city (#);
F_{ijk} is the non-CO_2 GHG correction (CO_2 equivalent) (factor);
k is the energy type (petrol, diesel, bio-fuel, electricity, hydrogen, etc.) (type);
i is the transport mode (passenger car, tram, bus, train, motorcycle, inland vessel, freight train, truck, etc.) (type); and
j is the vehicle class (if available specified by model, e.g. SUV) (type).

To reiterate, it is easier to take a macro approach and look into fuel sales by analyzing fossil fuels, electricity use, and WTT factor and then deduct WTW CO_{2eq} emissions using Table 7.2 information.

Following the proposed SMP2.0 example analysis for AML and AMP, each indicator must be converted into a scale of 0 to 10 points and depicted in a radar graph. The conversion follows the nomenclature depicted in Table 7.5.

An analytical and visual representation of data is fundamental to understand the mobility patterns and manage spatial (social and geographical) data. General transit feed specification (GTFS) files collected from several Portuguese open data sources (Transporlis, European data portal, STEPP, and dados.gov) contain the geographical positions of the main transport operators and include stops and trajectories for bus, train, and metro transport. Using this data in conjunction with population data by municipality, available through the latest Census data at INE portal, it is possible to identify the accessibility of the population to public transport, as can be seen in Figure 7.5. Furthermore, each municipality's minimum and maximum value of opportunities per 1000 inhabitants to transport was calculated as 0 and 48, respectively. In Figure 7.6, these values were translated as access to shared transport on a scale of 1 to 10.

Affordability (AF) belongs to the social dimension and can be represented by

$$AF = \frac{\text{PT monthly fare}}{\text{Income25\%}} * 100\% \quad (7.6)$$

where PT stands for public transport and Income25% represents the percentage of money spent in transport by each working person belonging to the lowest 25% income (usually population gross income by tier is available through annual statistics). Therefore, affordability captures the ability of transport system users to pay for transport based on how big of a percentage this expenditure is compared to their income. An affordable system is one that consumes a smaller share of users' incomes. If this expenditure is higher than 35% of a user income, it would be represented as a 0 in Figure 7.6, and if is less than 3.5%, it represents the 10 in the 0–10 scale. According to the INE statistics, the 25% income, per worker, in AML was €7532/yr and AMP was €6192/yr. Public transportation (PT) fare without social discount was, in the same year, €90 AML (AF = 14%) and €96.6/mo AMP (AF = 18%), representing a six and a five affordable system, respectively.

7.2 Smart Mobility Applied to Bicycle Sharing in Urban Context and Impacts on Sustainability

The Lisbon bicycle-sharing system "GIRA" depicted in Figure 7.7 started in mid-2017 with a pilot program of 100 bicycles and 10 stations. As of August 2021, and through its continued expansion, it is composed of 1410 bicycles (940 electric and 470 conventional) and 140 stations. Consider the following two scenarios that possibly explain the GIRA network growth: It could indicate that people abandoned the car and shifted to the bicycle-sharing system, decreasing fossil fuel usage (optimistic scenario), or the people who used to walk started using the bicycle-sharing

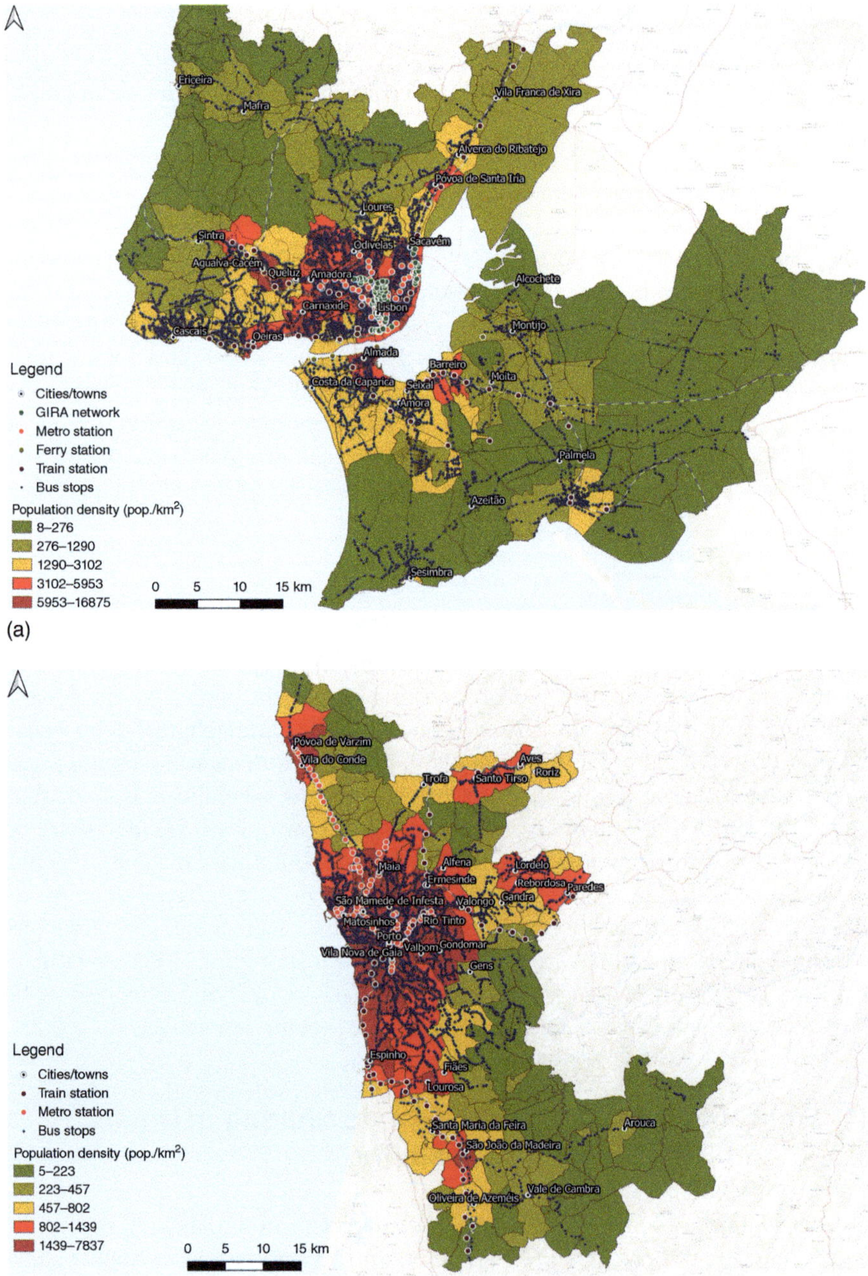

Figure 7.5 Metropolitan areas of (a) Lisbon and (b) Porto with population density and all public transport stops, and stations depicted using GIS (own illustration).

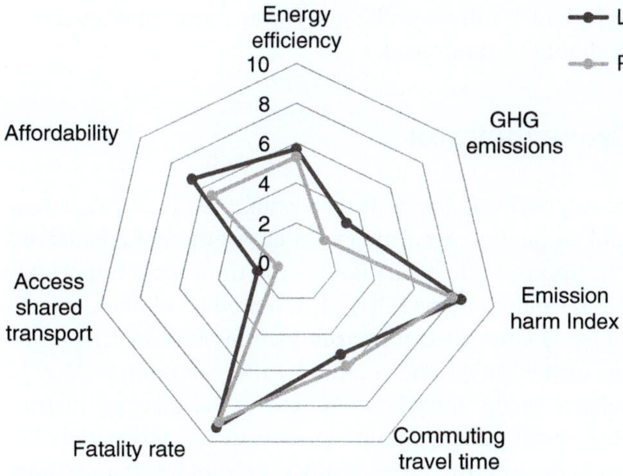

Figure 7.6 Radar for seven sustainable mobility indicators, in AML and AMP, two regions within the same European country (Portugal) year 2017(own illustration).

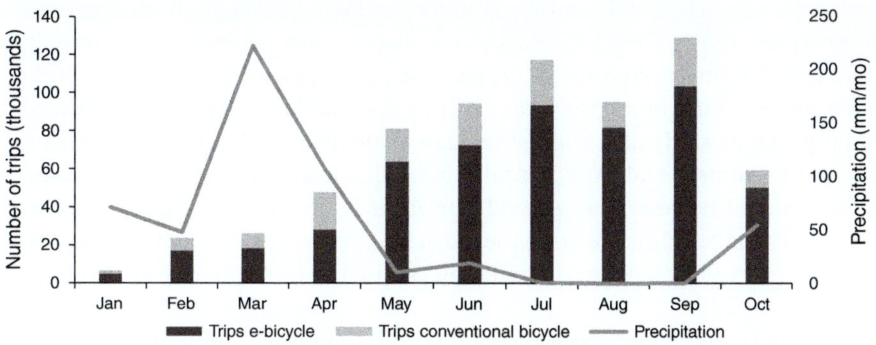

Figure 7.7 Influence of precipitation on bike-sharing transport use (own illustration).

system, increasing electricity consumption in transport (pessimistic scenario). This last scenario would indicate that the increase in electricity use for transport would cause an increase in the WTW GHG emissions. On this account, through mobility indicators, better data gathering, and open data, it would be possible to get a clearer picture about the reason for the network adherence and evolution. Furthermore, other important dynamics of this system are related to climate variables. Portugal is a relatively sunny country with a Mediterranean climate. Culturally speaking, people are not used to cycle in the rain, and in this case, precipitation can heavily influence the activity of the network as illustrated in Figure 7.9. Going back to the possible scenarios, these adverse meteorological events can negatively impact the optimistic scenario because in rainy days, people will go back to using private transportation increasing WTW GHG emissions. Data is an important construct to truly understand the mobility changes in our society and understand where the focus resides and possible solutions for these problems. For instance, in rainy days, use of public transportation should be reinforced.

7.3 Ground-Level Ozone Indicator

Ground-level ozone (O_3) or tropospheric ozone is a secondary pollutant that has effects on global warming and air quality. According to Intergovernmental Panel on Climate Change (IPCC) [17], tropospheric ozone (O_3) is the third most important GHG after carbon dioxide (CO_2) and methane (CH_4). It is a product of photochemistry, and its future abundance is controlled primarily by emissions of CH_4, carbon monoxide (CO), nitrogen oxides (NO_x), and volatile organic compounds (VOC). There is now greater confidence in the model assessment of the increase in tropospheric O_3 since the preindustrial period, which amounts to 30% when globally averaged. Ozone abundances in the troposphere typically vary from less than 10 ppb over remote tropical oceans up to about 100 ppb in the upper troposphere and often exceed 100 ppb downwind of polluted metropolitan regions. Ozone is an irritant gas that causes oxidative stress and inflammatory reactions in the lungs. Epidemiological studies have shown that short-term exposure to O_3 results in reduced respiratory function, increased hospital admissions, and even death. Although short-term exposure to O_3 has shown robust time series correlation with morbidity and mortality factors [18], long-term exposure to O_3 has had mixed results, often because of the limited nature of the monitoring data [18] or methodology. Nonetheless, the current large cohort study data suggest enough evidence to link long-term exposure to O_3 and respiratory and cardiovascular mortality. Regarding long-term O_3 exposure and morbidity, there is no general consensus on the matter [19], but there are some studies suggesting correlations between long-term exposure to O_3 and asthma admissions, especially among children [20], and increased respiratory symptoms in asthmatics [21, 22].

The World Health Organization (WHO) guideline value of 100 $\mu g/m^3$ [23] is still in the range, which has been found to increase the risk of mortality by about 1–2%. This residual risk was accepted as the ozone concentrations at the set guideline level may be due occasionally to natural phenomena, such as intrusion of stratospheric ozone into the troposphere.

In summary, O_3 has been associated with both acute and chronic health impacts ranging from asthma episodes, chronic obstructive pulmonary disease (COPD), increased hospital admissions, and overall cardiovascular and cardiopulmonary morbidity and mortality. Respiratory system diseases are the fifth leading cause of internment and the first cause of death in Portuguese hospitals [24]. A study conducted in Porto reported that an increase of 10 $\mu g/m^3$ in the daily O_3 eight hours maximum moving average corresponds to an increase of 0.95% and 1.58% in non-accidental mortality and cardiovascular mortality, respectively [25]. In Lisbon, there are similar results with an increase of 1.11% for all-cause mortality in the population group $\geq$65 years and an increase of 0.96% for the general population [26]. The societal impacts and continuous strain caused to the health system by air pollution, especially in big cities, is non-negligible [27].

The number of exceedances a year could be an attempt indicator to compare environmental-related sustainability (additional indicator in Figure 7.6). Based on the recent findings [28, 29], weekends and summer school holidays, with less traffic levels, are associated with an increase of tropospheric O_3, so it is expectable that the electrification of the fleet would be detrimental to this emission rather than beneficial in the short-term.

7.4 Energy Transition

Europe aims to achieve full decarbonization by 2050 and claims that electrification is the best way to decarbonize all sectors. Electrification through battery technology is seen as the best way to decarbonize the transport sector. However, long-range transport such as road goods transport, maritime bunkers, and aviation are harder to depend on a pure battery powertrain, making biofuels and green hydrogen better contenders. Several scenarios for 2050 were benchmarked [30], with eight scenarios expected to achieve more than 50% reduction of GHG emissions by 2030 compared to 1990 and 16 scenarios aiming for climate neutrality by 2050. Renewable energy (including bioenergy) share in final energy consumption was at 5% in 2017 but is expected to reach almost 20% in the near future. The scenarios consistently project similar developments for transport in 2030 in terms of final energy demand, reduction of fossil fuels, and penetration shares of new technologies as seen in one of the example scenarios depicted in Figure 7.8.

Battery electric vehicles, plug-in hybrids, and fuel cell vehicles are technological solutions that can reduce the environmental burden of road transport. Conventional and advanced biofuels are also considered technologically feasible options, although there are uncertainties regarding their environmental impact, especially because of the indirect land-use change effects. Other scenarios explore social innovation to alleviate the demand for transport (e.g. modal shift, teleconferencing, and carpooling). It is widely agreed that at least 50% emission reduction by 2030 with electric vehicles penetration is possible, even when adjusting for the potential increase in

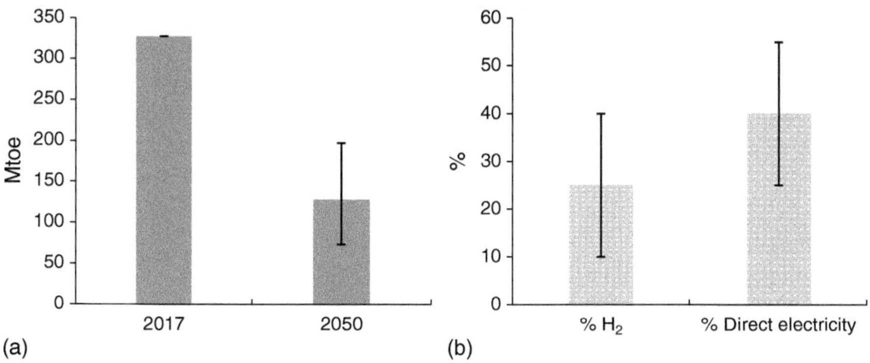

Figure 7.8 Illustration of 2050 benchmark visions for transport with energy transition [30]: (a) final energy consumption and (b) hydrogen and direct electricity participation. Source: Based on Ioannis et al. [30].

total electricity consumption in the transport sector by a factor of 3 to 7 compared to today. This expected growth is mainly attributed to road transport but is counteracted by energy scenarios, which predict 7 to 90 million battery electric vehicles on European Union (EU) roads by 2030.

In 2050, the transport sector is expected to consume lower electricity amounts than other sectors (up to 780 terawatt hour [TWh]). Most scenarios for transport predict that electricity will cover between 30% and 45% of the final demand. It will be the most transformed sector considering that in 2017, roughly 65 TWh was consumed, covering less than 2% of the demand. Most of this consumption was in the rail sector without multipliers. For instance, a multiplier of 5 indicates that for every 2 electric cars, 10 will be counted in the final analysis. Moreover, transport is the only sector that increases its demand for electricity across all scenarios. The share of renewable energy and bioenergy in scenarios that achieve at least 90% emission reduction in 2050 is 60–100%. The electric vehicle fleet reaches 100 to 220 million battery electric vehicles in most scenarios. Hydrogen and e-fuels are fully deployed and become key elements in decarbonizing transport; they supply 15% to 50% of the sector's energy needs as demonstrated in Figure 7.8. Oil is expected to still be used in transport in most scenarios (2–50 Mtoe, 1% to 16% of today's consumption; only 6 out of 17 scenarios phase out oil completely). The share of electricity (battery electric vehicles) and hydrogen use (fuel cell vehicles) in road transport ranges from 32% to 69% in 2050.

Regarding the seven indicators seen in Figure 7.6, the impact of transport electrification and renewables will only change the GHG WTW emissions and the emission harm index (two out of the seven indicators) if the same mobility volumes by mode and PT prices are maintained. Thus, the electrification of the fleet would be detrimental to ground-level ozone emission rather than beneficial (see Section 7.3).

7.5 Resilience of the Mobility System

Day-to-day living in both personal and business settings was drastically affected by COVID-19 pandemic. The vision scenarios in Figure 7.8, Section 7.4, do not entail such disruptive events. Teleconferencing, home school, and deeply reduced traffic volumes [31, 32] all had consequences in energy consumption and air quality. Near-traffic locations registered an overall decrease of air pollutants, but ground-level ozone increased, giving rise to the suggestion that there is considerable pressure to add the ozone indicator as an additional indicator of sustainable mobility in Figure 7.6. Lisbon near-roundabout monitoring data seen in Figure 7.9 shows this effect. Although NO_2 shows a reduction because of the lockdown and high traffic circulation restrictions from mid to end march in Portugal, the O_3 level increased, in relation with the homologous period in 2016–2019.

On the other hand, the extensive future transport system reliance on renewables and energy storage such as batteries, electrolysis, and fuel cells may pose another challenge as rare earth materials shortage, including increased demand for copper,

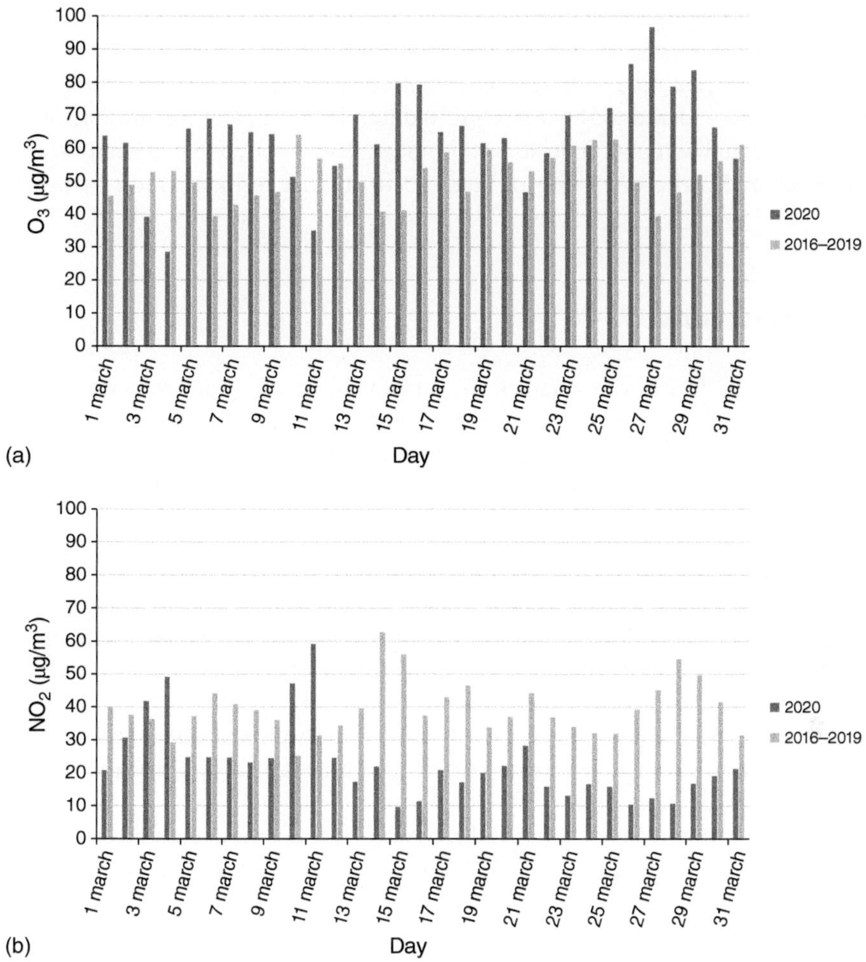

Figure 7.9 COVID-19 impact on a near-traffic location in Lisbon city, NO_x and ground-level O_3 during pre-COVID adaptation, and full lockdown 11–31 March.

silver, nickel, lithium, and cobalt, as well as steel, can take place [33]. In general, mining and processing rare metals such as indium, cobalt, and platinum generate larger environmental impacts to obtain a given amount of minerals than bulk minerals such as steel and aluminum [34]. In this context, resource depletion could serve as an indicator of sustainable mobility in Figure 7.6.

Last but not least, climate change adaptation strategies for precipitation and temperature extremes should also be considered [35]. For example, extreme precipitation causes the reduction in traction, leading to increases in vehicle collisions and casualties, thus causing a higher fatality rate and/or road and lane closures affecting commuting times. Such extreme events, if frequent, would impact active and shared mobility (e.g. bicycle-sharing systems) whose utilization is largely decreased on rainy days (see Section 7.2 and [35]). It would also be useful to add an indicator of system resilience in Figure 7.6. High temperatures would require more ventilation needs in both above ground and underground rail systems. Underground railway systems heat up over time. If inadequate ventilation occurs, the temperature can be 11 °C when compared to the above ground ambient temperature and, in some instances, be equal or higher than 40 °C promoting discomfort, one of the indicators not explored in this chapter. There are attempts to produce resilience indicators, for example Ref. [36].

7.6 Conclusions

This chapter describes how to characterize sustainable mobility and focuses on the most internationally recognized indicators exploring affordability, climate change, air pollution, energy efficiency, commuting time, and fatality rate. Possible extensions include ground-level ozone, resource depletion, and adaptability to extreme events indicators. The key aspects found in this chapter are as follows:

- Monitoring and tracking smart and sustainable mobility effects on economic, environmental, and social aspects must use a set of indicators;
- Using sustainable mobility indicators based on surveys and yearly statistics, avoiding time-consuming and expensive measuring campaigns or complex simulations, easy to calculate, is the main advantage;
- Energy transition in transport seems focused on battery electrification and renewables except for freight, aviation, and maritime bunkers that will probably move to green hydrogen or biofuels;
- Impact of transport electrification and renewables, if the same mobility volumes by mode, will only improve GHG WTW emissions and the emission harm index, this is, two out of seven indicators; and
- Ground-level ozone as a short-lived GHG and its exacerbating effect on respiratory problems deserve more attention and possible incorporation in environmental and health indicators.

COVID-19 impact on transport is yet to be fully evaluated, but preliminary results reveal its adverse impact on ground-level ozone near traffic. Global energy transitions could fundamentally change the demand for both minerals and energy resources over time. Extreme weather events occurring more often also lead to poorer performance of transport systems.

Acknowledgments

The authors would like to acknowledge the financial support FCT through project UIDB/50019/2021–IDL, and Angelo Soares would like to acknowledge the support through FCT PhD grant 2021.06878.BD.

List of Abbreviations

AF	affordability
AML	Lisbon metropolitan area
AMP	Porto metropolitan area
CH_4	methane
CO	carbon monoxide
CO_2	carbon dioxide
COPD	chronic obstructive pulmonary disease
E	energy consumption rate
EU	European Union
FR	fatality rate
g	grams
GDP	gross domestic product
GHG	greenhouse gas
GIS	geographic information system
GTFS	general transit feed specification
ICT	information and communications technologies
INE	Portuguese Institute of Statistics
IoT	Internet of Things
kg	kilograms
KPI	key performance indicators
kWh	kilowatt hour
LHV	low heaving value
LPG	liquified petroleum gas
MJ	megajoules
NA	not available
N_2O	nitrous oxide
NO_2	nitrogen dioxide
O_3	ozone
Pkm	passenger per kilometer
PM_{10}	particles with 10 μm diameter or less
$PM_{2.5}$	particles with 2.5 μm diameter or less
PT	public transportation
SDG	sustainable development goals
SMP	sustainable mobility project
SO_x	sulfur oxides
SUMI	sustainable urban mobility indicators
Tkm	tonne per kilometer
TTW	tank-to-wheels
TWh	terawatt hour
VOCs	volatile organic compounds
WBCSD	World Business Council for Sustainable Development
WTT	well-to-tank
WTW	well-to-wheels

References

1 UN (2015). Transforming our World: The 2030 Agenda for Sustainable Development. A/RES/70/1.

2 Moyer, J.D. and Hedden, S. (2020). Are we on the right path to achieve the sustainable development goals? *World Dev.* 127: 104749. https://doi.org/10.1016/j.worlddev.2019.104749.

3 Šurdonja, S., Giuffrè, T., and Deluka-Tibljaš, A. (2020). Smart mobility solutions-necessary precondition for a well-functioning smart city. In: *Proceedings of the Transportation Research Procedia*, vol. 45, 604–611. Elsevier B.V.

4 UN (2017). Work of the Statistical Commission pertaining to the 2030 Agenda for Sustainable Development. A/RES/71/313.

5 Sustainable Mobility for All - SuM4All (2017). Global Mobility Report 2017: Tracking Sector Performance, Washington, DC. https://www.sum4all.org/publications/global-mobility-report-2017.

6 Mangiaracina, R., Perego, A., Seghezzi, A., and Tumino, A. (2019). Innovative solutions to increase last-mile delivery efficiency in B2C e-commerce: a literature review. *Int. J. Phys. Distrib. Logist. Manag.* 49: 901–920. https://doi.org/10.1108/IJPDLM-02-2019-0048.

7 Nemoto, E.H., Issaoui, R., Korbee, D. et al. (2021). How to measure the impacts of shared automated electric vehicles on urban mobility. *Transp. Res. Part D Transp. Environ.* 93: 102766. https://doi.org/10.1016/j.trd.2021.102766.

8 Audenhove, F., Smith, A., Rominger, G., and Bettati, A. (2018). The Future of Mobility 3.0, Reinventing mobility in the era of disruption and creativity.

9 WBCSD Mobility (2011). *Sustainable Mobility Project 2.0 (SMP2.0)*, Vol. 0.

10 Dobranskyte-Niskota, A., Perujo, A., and Pregl, M. (2007). Indicators to assess sustainability of transport activities.

11 EC (2020). Technical support related to sustainable urban mobility indicators (SUMI). MOVE/B4/2017-358.

12 INE (2018). Inquérito à Mobilidade nas Áreas Metropolitanas do Porto e de Lisboa 2017.

13 Prussi, M., Yugo, M., De Prada, L. et al. (2020). JEC Well-To-Wheels report v5, Luxembourg.

14 Albrektsen, R., Hutchings, N., Hjorth, M. et al. (2019). EMEP/EEA emission inventory guidebook, Luxembourg, Vol. 11.

15 Edwards, R., Hass, H., Larivé, J.-F. et al. (2013). WELL-TO-TANK (WTT) Report. Appendix 2 - Version 4a - Summary of energy and GHG balance of individual pathways. ISBN 9789279338885.

16 Baptista, P., Tomás, M., and Silva, C. (2010). Plug-in hybrid fuel cell vehicles market penetration scenarios. *Int. J. Hydrogen Energy* https://doi.org/10.1016/j.ijhydene.2010.01.086.

17 IPCC (2001). IPCC Climate Change 2001. Synthesis Report. IPCC Third Assessment Report (TAR).

18 World Health Organization (2013). Review of evidence on health aspects of air pollution – REVIHAAP Project: Technical Report.
19 U.S. Environ. Prot. Agency (2020). US EPA Integrated Science Assessment (ISA) for ozone and related photochemical oxidants (Final Report).
20 Tzivian, L. (2011). Outdoor air pollution and asthma in children. *J. Asthma* 48: 470–481. https://doi.org/10.3109/02770903.2011.570407.
21 McConnell, R., Islam, T., Shankardass, K. et al. (2010). Childhood incident asthma and traffic-related air pollution at home and school. *Environ. Health Perspect.* 118: 1021–1026. https://doi.org/10.1289/ehp.0901232.
22 Salam, M.T., Islam, T., and Gilliland, F.D. (2008). Recent evidence for adverse effects of residential proximity to traffic sources on asthma. *Curr. Opin. Pulm. Med.* 14: 3–8. https://doi.org/10.1097/MCP.0b013e3282f1987a.
23 Krzyzanowski, M. and Cohen, A. (2008). Update of WHO air quality guidelines. *Air Qual. Atmos. Heal.* 1: 7–13. https://doi.org/10.1007/s11869-008-0008-9.
24 ONDR (2017). DGS Programa nacional para as doenças respiratórias - 2017.
25 de Almeida, S.P., Casimiro, E., and Calheiros, J. (2011). Short-term association between exposure to ozone and mortality in Oporto, Portugal. *Environ. Res.* https://doi.org/10.1016/j.envres.2011.01.024.
26 Garrett, P. and Casimiro, E. (2011). Short-term effect of fine particulate matter (PM2.5) and ozone on daily mortality in Lisbon, Portugal. *Environ. Sci. Pollut. Res.* https://doi.org/10.1007/s11356-011-0519-z.
27 de Bruyn, S. and de Vries, J. (2020). *Health Costs of Air Pollution in European Cities and the Linkage with Transport*. CE Delft.
28 Soares, A.R., Deus, R., Barroso, C., and Silva, C. (2021). Urban ground-level O_3 trends: lessons from portuguese cities, 2010–2018. *Atmosphere (Basel)* 12: https://doi.org/10.3390/atmos12020183.
29 Soares, A.R., Neto, D., Avelino, T., and Silva, C. (2020). Ground level ozone formation near a traffic intersection: Lisbon "rotunda de Entrecampos" case study. *Energies* https://doi.org/10.3390/en13071562.
30 Ioannis, T., Wouter, N., Dalius, T., and Pablo, R.C. (2020). Towards net-zero emissions in the EU energy system by 2050.
31 Tian, X., An, C., Chen, Z., and Tian, Z. (2021). Assessing the impact of COVID-19 pandemic on urban transportation and air quality in Canada. *Sci. Total Environ.* 765: 144270. https://doi.org/10.1016/j.scitotenv.2020.144270.
32 Okazaki, Y., Ito, L., and Tokai, A. (2021). Health risk of increased O_3 concentration based on regional emission characteristics under the Unusual State of the COVID-19 Pandemic. *Atmosphere (Basel)* 12: 335. https://doi.org/10.3390/atmos12030335.
33 Watari, T., McLellan, B.C., Giurco, D. et al. (2019). Total material requirement for the global energy transition to 2050: a focus on transport and electricity. *Resour. Conserv. Recycl.* 148: 91–103. https://doi.org/10.1016/j.resconrec.2019.05.015.

34 Nuss, P. and Eckelman, M.J. (2014). Life cycle assessment of metals: a scientific synthesis. *PLoS One* 9: e101298. https://doi.org/10.1371/journal.pone.0101298.

35 Love, G., Soares, A., and Püempel, H. (2010). Climate change, climate variability and transportation. *Procedia Environ. Sci.* 1: 130–145. https://doi.org/10.1016/j.proenv.2010.09.010.

36 Das, R. (2020). Approach for measuring transportation network resiliency: a case study on Dhaka, Bangladesh. *Case Stud. Transp. Policy* 8: 586–592. https://doi.org/10.1016/j.cstp.2020.04.001.

8

Evolution of Electrical Distribution Grids Toward the Smart Grid Concept

Lucía Suárez-Ramón[1], *Pablo Arboleya*[2], *José Lorenzo-Álvarez*[1], *and José M. Carou-Álvarez*[1]

[1] *E-Redes (EDP Group), Plaza del Fresno, 2, Oviedo 33011, Asturias Oviedo, Spain*
[2] *University of Oviedo, LEMUR Research Group, Campus de Gijón, s/n, Gijón, 33204 Asturias, Spain*

8.1 Smart Grid Concept

The smart grid concept is being introduced in the management of electrical networks thanks to the digitization of the low-voltage (LV) network. Until a few years ago, LV networks were planned, managed, and operated by feel based on the knowledge of the field personnel who had accumulated experience over years of doing this work by hand and taking measurements using clamp meters.

The deployment of smart meters (SMs) and the other behind-the-meter (BTM) digital devices and other sensors that are being installed in the LV network gives rise to start talking about the smart grid concept.

The LV network does not change, but the perception of the people who are working in the field on the network is greatly expanded. The digital equipment provides abundant and detailed information that makes it possible to understand the situations that are occurring on the grid, to get closer to real time, and even to predict what is going to happen in the near future.

In addition to increasing the observability of the LV network, it is also possible to gain a better understanding of the interaction between the medium-voltage (MV) and LV networks since transformer stations are also digitized by sensors and smart meters that make it possible to identify in time space the path of energy flows.

In the new smart grid concept, end users take on a leading role, from being simple domestic consumers to start playing other roles at the grid end points, such as generators thanks to the installation of photovoltaic panels in isolated or shared self-consumption mode, storage managers in battery mode, or consumers with assets such as electric vehicle (EV) chargers or heat pumps. In other words, the end user becomes a prosumer with a constant bidirectional interaction with the LV distribution grid.

Smart grids are the engines that have enabled the accelerated expansion of distributed energy resources. To control this entire digital ecosystem, the development

Technologies for Integrated Energy Systems and Networks, First Edition.
Edited by Giorgio Graditi and Marialaura Di Somma.

of the Advanced Metering Infrastructure has been promoted because of the imperative need to be able to intelligently manage the MV/LV distribution grid and the loads or generations connected to it. Advanced Metering Infrastructure (AMI) is the name given to the set of devices and systems that have been developed over the past few years to enable intelligent management of the terminal distribution network. The following sections of this chapter detail the parts that integrate an AMI system and its operating rules in order to manage the new smart grid concept.

8.2 Advanced Metering Infrastructure (AMI) General Description

LV network monitoring is the biggest new challenge facing energy distribution companies. Since the implementation of smart meters and remote management systems, the observability of the LV network has increased enormously. In addition, over time, the different variables, information, and measurements that can be obtained from the set of digital devices installed in the network are increasing. New equipment and new applications are also emerging that allow us to go deeper and deeper into the operating details of the LV power system.

All this is making it possible to advance in the understanding of the key operating parameters of the whole system, i.e. electrical grid system and remote management system, the latter being understood as the set of digital devices and systems that provide information on the connection points where they are installed and other sensors and BTM devices, such as batteries or EV charger, that control the performance parameters of any prosumer with loads or generations connected to the grid.

Taking into account all these new variables, grid management requires the implementation of a model that reliably replicates the joint and individual behavior of all the elements involved in the LV power grid or digital twin grid. This will make possible to manage the network in real time, and it will be able to simulate the behavior of the electrical grid with the introduction of new loads or generations, i.e. it will be possible to increase the penetration of distributed resources by containing the investment in network extensions or reinforcements, making them only where they are really necessary.

In order to be able to interpret the information in its entirety and move toward a digital network twin that reliably replicates the operation of the whole system, it is necessary to deepen the understanding of the remote management system and its interaction with the power grid.

Utilities are currently deploying different remote management infrastructure technologies. The most common is currently power line communication (PLC) based because it does not require a third-party company for installation and subsequent operation. Its main drawback, however, is its limited bandwidth. On the other hand, its main advantage is that it provides fundamental topological information to define more precisely the digital twin of the network. Because of the economic and technical advantages, this type of technology is expected to continue to be used in the medium term or even hybrid communication scenarios are being

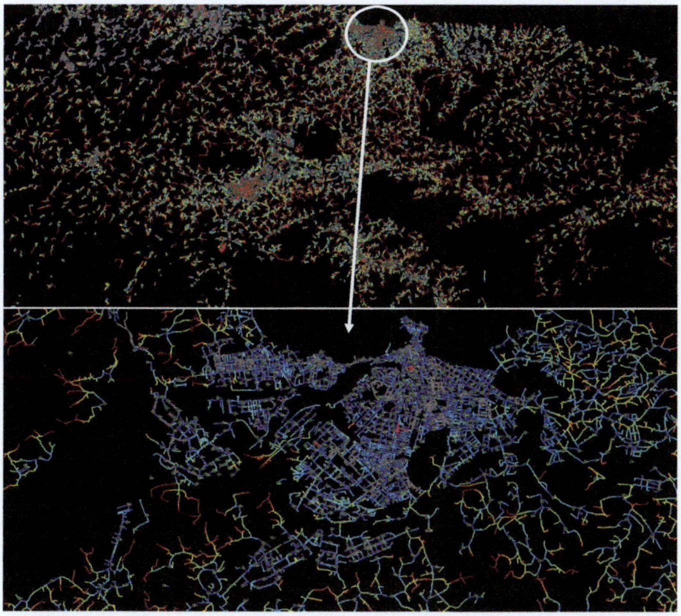

Figure 8.1 Geographic representation of the actual network taken as an example. Zoom of a compact urban area.

considered to make up for the disadvantages of PLC. Because of this, this chapter will focus on the in-depth analysis of an LV power grid management based on PLC technology, as this aspect is important to discover hidden information in the data due to the fact that the PLC communication is supported by the LV power cable as the physical communication medium.

This section will provide an in-depth explanation of the current data acquisition systems using as an example a real distribution network of an electric power distribution company that manages 646.000 LV supply points and for this purpose has 6.080 station transformers (STs) with their corresponding associated LV network. The demographic distribution of these supply points is approximately one-third in rural areas and two-thirds in urban and industrial areas.

Figure 8.1 shows a geographic view of the network taken as an example, with a zoomed-in view of a compact urban area.

The network contains 646.000 loads grouped into 186.000 electrical supply points in a general protection box (GPB) that are connected by 18.400 feeders to 6.080 transformation centers. This network serves approximately 900.000 inhabitants. A summary of the characteristics of the network used as an example through the whole chapter can be found in Tables 8.1 and 8.2.

Table 8.1 shows the detailed distribution of feeders, GPB, transformers, and supply points according to the type of area, urban and rural.

This indicates that, in rural area, the transformer stations, which will be outdoors, feed an average of 35 supplies, and in urban area, where the population is more

Table 8.1 Total number of feeders, GPBs, supply points, and transformer stations in the described network.

	Number of feeders	Number of GPB	Number of supply points	Station transformer
Rural area	6.800	108.700	123.000	3.540
Urban+industrial area	11.600	77.800	523.000	2.540
Total area	18.400	186.500	646.000	6.080

Table 8.2 Average number of supplies per feeder, GPB, and transformer station.

	Average (supplies/feeder)	Average (supplies/GPB)	Average (supplies/transformer)
Rural area	18	1	35
Urban+industrial area	45	7	206
Total area	35	3	106

compact, the average number of supplies per transformer rises to 206. In the case of urban areas, transformer stations can have up to two LV transformers per station.

Table 8.2 shows the average distribution of supply points for each feeder, GPB, and transformer according to the type of area, urban and rural.

This section will provide an overview of how this real LV distribution network is structured based on concrete examples of the network mentioned above. The installation and operation scheme of the digital equipment that monitors the LV network will also be described. This data and its structuring will help to understand the importance of having all this information conveniently ordered, classified, and validated in order to be able to make a reliable interpretation of what is happening in the LV network and in the devices that monitors it.

The LV distribution network used as the basis for the development of this section is a European-type network. In this type of network, the topology downstream of the transformer station is relatively complex, and although it operates radially, there are multiple and different possible configurations that are obtained by changing the position of the different switching elements present in the network.

The different switches that allow changing the configuration of the LV feeders and the switch boxes installed at different points of the LV lines are not remote-controlled, they are manually operated, because they are traditionally made up of fuse boxes that do not allow remote control. There is also currently no economical solution on the market to monitor their open/closed position. The nominal voltage of the LV network described is 416 V (line voltage), the distribution system is a four-wire system with distributed neutral isolated from ground at the consumers and grounded only at the secondary side of the delta-star transformer stations.

The ecosystem described here consists of two major systems: the electrical system and the remote management (tele-management) system, the latter considering the set of digital devices, communication media, and central system that monitor the measurement points of the electrical system.

Breaking this ecosystem down into parts, each of them individually is easily understandable and manageable.

The complexity of managing this ecosystem comes from the joint treatment of the two systems, electrical and remote management systems. It is therefore necessary to understand each of the systems separately, both the individual functioning of each system and the inter-relationships between them.

Needless to describe in deep the electrical system and its physical and mathematical formulation as they are well known and there is extensive literature to explain the working principles of the electrical network as, for example, described in Ref. [1], the focus of this section will be on the operating parameters of the different parts and mechanisms of the remote management system.

Regarding the electrical system, it is necessary to mention that it is essential to know the hierarchy of the electrical network that is monitored with the remote management devices and its trace to the consumption and generation points, as well as the constructive characteristics and technical parameters of the parts that make it up: MV/LV transformer and its tap changer, switches, fuses, disconnectors, LV switchboards, LV feeders, line feeder trace, section, length and impedance of conductors, configuration of switch boxes and GPBs, and connection of smart meters (SMs).

Having a complete and accurate inventory of the LV network is a major challenge for distribution companies because of its size, geographical dispersion, and the large number of elements that make it up. See Figure 8.1 where a typical geographical layout of a European-type LV system is represented.

Overall, electric utilities have made great efforts in inventorying the asset of their networks for various reasons, and their remuneration depends largely on their investments in this type of assets, and also without a reliable asset inventory, it is impossible to address with guarantee a digitization process that brings added value to the company. An asset inventory without any error is practically impossible to achieve at least in a first iteration; however, the digitization process provides tools capable of correcting minor inconsistencies in it automatically. On the other hand, the digitization of the asset inventory is an essential step in order to have reliable digital twins that can be used in both operational and network planning processes.

The remote management system emerged from the need to digitize metering at consumption points in order to reduce the tasks of local reading of traditional meters and to be able to execute remote actions for switching off or connecting supplies and modifications derived from changes in electricity supply contracts. These were the first applications to result from the digitalization of the LV network.

To this end, distribution companies have implemented plans to replace traditional meters with smart meters. Now, distribution system operators (DSOs) are fully aware of the great advantages that smart meters bring to the operation of the LV network, and they are working on designing new applications to optimize network planning and operation based on the information provided by smart meters and complementing this information and new functionalities with new sensors and additional digital devices. This situation will lead to a deeper digitization that will allow the management of the network through digital twins thanks to the incorporation of big data and analytics techniques to obtain detailed information about the network such as customer segmentation, loss reduction, fraud detection, and so on.

There is a wide variety of smart meters on the market, both in terms of the parameters they measure and the type of communication by which they transmit these values. The most widely adopted solution at the European level is the remote management system with smart meters communicated by PLC technology. In Spain, the PRIME Alliance [2] solution based on the open standard PLC PRIME (PoweRline Intelligent Metering Evolution) protocol has been mostly adopted. The main motivation of the DSOs for a technological deployment based on PLC is the great savings in the deployment of the communications infrastructure as the LV power conductor where the smart meters are connected is used as the physical path of communication.

The usual configuration in this type of remote management systems based on PLC PRIME communication is the installation of a data concentrator unit (DCU) for each transformer, which communicates with the meters installed downstream of the transformer in the LV network. More details on this type of communication will be given in a later section of this chapter.

This indicates that it is possible to consider the set composed of the transformer, its associated network, and the equipment installed in it as a simple entity (SEN) whose configuration will be repeated as many times as transformers make up the complete LV network of the distribution company.

This simplification mechanism makes it possible to reduce the initial complex ecosystem into SENs that will have similar operating rules, where only the inventory data and physical and geographic parameters change. It must be taken into account that these SENs can vary dynamically when switching the network configuration, so that elements that, at a given time, are associated with a simple entity (SEN_1) can be associated with a different simple entity (SEN_2) at another time.

For rural networks, the use of SENs will significantly help to reduce the complexity of the ecosystem because the LV network of a rural transformer does not usually have interconnections with other nearby transformer networks and will therefore be isolated networks. In these cases, it will be enough to consider a single SEN, or at most if there are interconnections of feeders from different transformers, it will be enough to consider interactions between two or three SENs.

However, in urban areas, it is necessary to take into account the complex interactions between multiple SENs because their topological composition may change dynamically over time as network sections that are fed by one transformer (T_1) at a given instant may become fed by another transformer (T_2) because of a change of position in the switches that determine the network configuration.

In industrial areas, in addition to these interactions between LV SENs, the MV network must also be taken into account because in these distribution areas, there is naturally a high penetration of consumption and generation prosumers connected

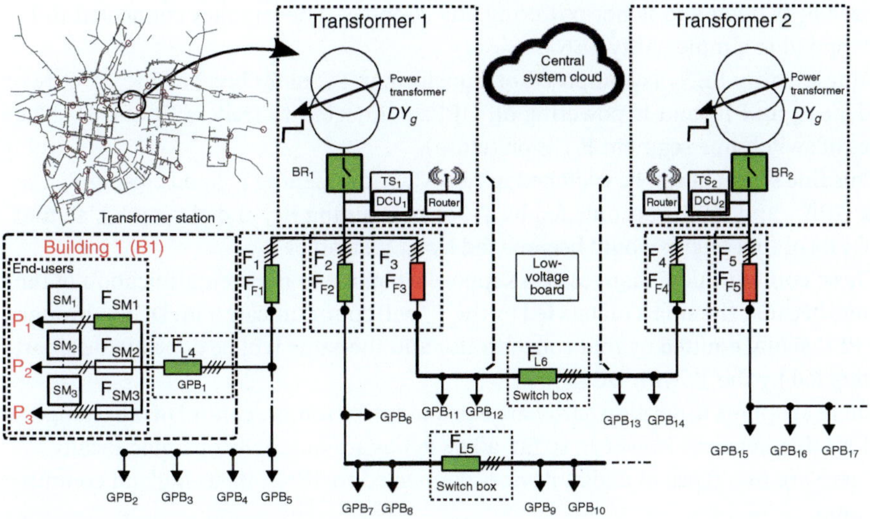

Figure 8.2 Schematic of two simple entities (SEN_1 and SEN_2) that are interconnected through one of their LV feeders.

to the MV network, which can have a more direct impact on the management of the LV network.

From now on, the focus will be on defining the operating rules and configuration of these LV SENs that are composed of a transformer, its associated LV network, the data concentrator that centralizes the information and the metering equipment, and other elements connected downstream in the distribution network.

Figure 8.2 shows a schematic of two simple entities interconnected with each other.

In Figure 8.2, it can be seen that each SEN is composed of the following elements:

- **On the power grid side**: One MV/LV transformer and one or more associated LV feeders ($F_1 \ldots F_n$), on which there are between 1 and 50 connection points ($GPB_1 \ldots GPB_n$) each connected to the grid via a GPB. Each connection point (GPB) can supply one or more prosumers.
- **On the remote management system side**: One DCU that will collect the information provided by the smart meters and will send action commands to them thanks to bidirectional communication, one transformer supervisor (TS) that will measure the energy delivered to the LV network (TS), and one smart meter for each prosumer ($SM_1 \ldots SM_n$).

This configuration has electrical switching elements, but none of them are remotely controlled: the main switch of the transformer (BR_1 and BR_2), the tap changer transformer, the fuses of the line headers (F_{F1}, F_{F2}, F_{F3}, F_{F4}, and F_{F5}), and the interconnection switches of the line segments that allow changing the topological configuration of the network (F_{L5} and F_{L6}).

Simple entity SEN_1 is composed of transformer T_1, main breaker BR1, and three feeders F_1, F_2, and F_3 and is powering all GPBs connected to feeders F_1 and F_2.

Feeder F_3 is open and is not powering any supplies. The supplies connected to F_3 correspond to simple entity SEN_2.

Simple entity SEN_2 is composed of transformer T_2, main breaker BR_2, and two feeders F_4 and F_5 and is powering all GPBs connected to feeders F_4 and F_5. The fuses of switch line segment F_{L6} is on (close).

This line segment can be switched in two ways: by opening F_{L6} and closing F_{F3}, so that GPB_{11} and GPB_{12} become fed by T_1, or by opening F_{F4} and closing F_{F3} so that all GPBs of that feeder would become fed by T_1 instead of T_2.

These configuration changes also suppose a change in the communication of the meters because the smart meters fed by the T_1 will communicate with DCU_1 through the PLC signal emitted by that concentrator and the same will be done by the smart meters fed by the T_2 with DCU_2.

There are pilots to develop hardware equipment for remote control of tap changers and LV line headers. However, so far, all switches are manually operated locally.

There are two types of digital devices connected to SEN_1: metering and communication.

The metering equipment is the smart meter ($SM_1 \ldots SM_n$) at the consumption points of the prosumers ($P_1 \ldots P_n$) and the transformer supervisor (TS_1) that measures all the energy delivered by the T_1 transformer to the LV network.

The communication devices are two, DCU and 4G router. A 4G router is needed only if there is no other way of communication in the station transformer, like optical fiber.

The configuration of each of these SENs can vary both in the physical part of the network and in the diversity of digital devices connected and their connection points.

In the SENs, there are only two types of network monitoring devices (a prosumer smart meter and a transformer supervisor) and one communication device (a DCU installed in the station transformer) as can be seen in the photo in Figure 8.3.

A 4G router is needed only if there is no other way of communication in the station transformer (ST), like optical fiber.

As all of them, monitoring and communication devices, are digital devices, they work as small computers that are programmed to follow specific roles configured in the firmware installed in them. For smart meters and transformer supervisors, it is necessary to tell the difference between the metrological firmware and the communications firmware. The communications firmware in the smart meters has been developed in such a way that the measured data can be sent via PLC PRIME technology to the data concentrators.

The transformer supervisors installed in the transformer stations are devices that are embedded in the housing of the DCU itself and are connected to the network in such a way that they measure all the energy that flows into the LV network. In this case, the communication with the DCU is not PLC, but the DCU directly receives the measurement data from the TS and can directly send the measurement data to the central system through the router. This is an advantage in terms of the possibility of increasing the sampling frequency of the measurements in these CT supervisors, something that is not possible in PLC communications.

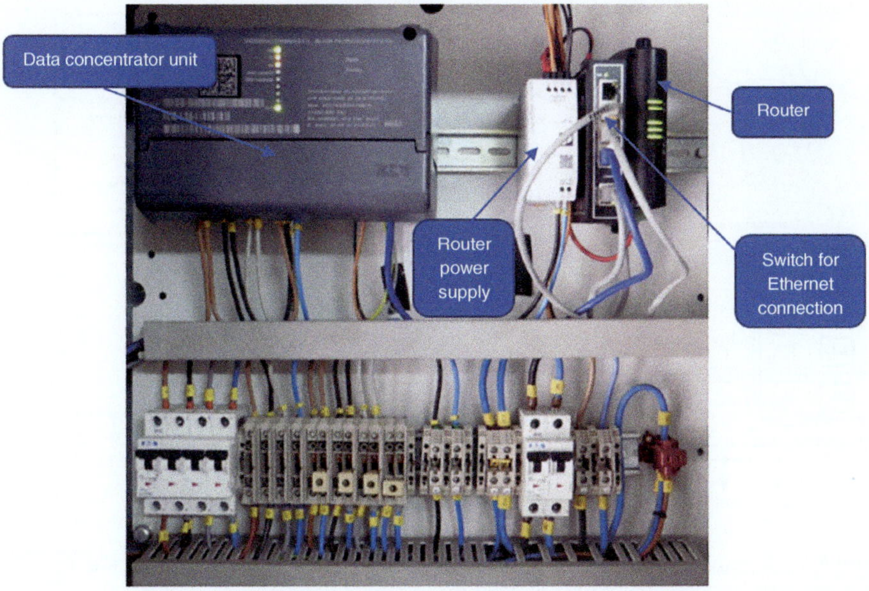

Figure 8.3 Photos with the layout of the digital communication devices in an electrical cabinet of a simple entity.

As shown in Figure 8.3, the DCU has the transformer supervisor (TS) embedded in it, and therefore, it is necessary to connect the current measurement of the three phases to this device as well. Depending on the DCU model, the power supply of the concentrator may or may not be independent of the supervisor voltage measurement. In the case of independent power supply, the concentrator will be supplied from one of the phases and will have a single-phase power supply. If the concentrator is supplied from the same measurement connection of the TS, the DCU power supply will be three phase.

The advantage of a three-phase supply will allow the concentrator to keep communicating in the event of failure of one or two phases. The 4G router required in most cases to communicate with the central system is also installed in the electrical cabinet next to the DCU and is supplied by a single-phase power supply. Figure 8.2 also shows the communication connections between the router and the concentrator by a dashed line.

Communication in these cases is done by transmission control protocol/internet protocol (TCP/IP) through the 4G router or by optical fiber in the stations' transformers where this solution is available because it is a more reliable and secure communication solution.

It is quite often the case that the router becomes unavailable and stops communicating, either because of coverage problems or failure of the router itself. In these cases, the concentrator will continue to receive data from the smart meters and manage the PLC communications network, but the router will not be able to transmit this data to the central system. When communication to the central system is done through a router, an element is introduced that needs power supply to work and

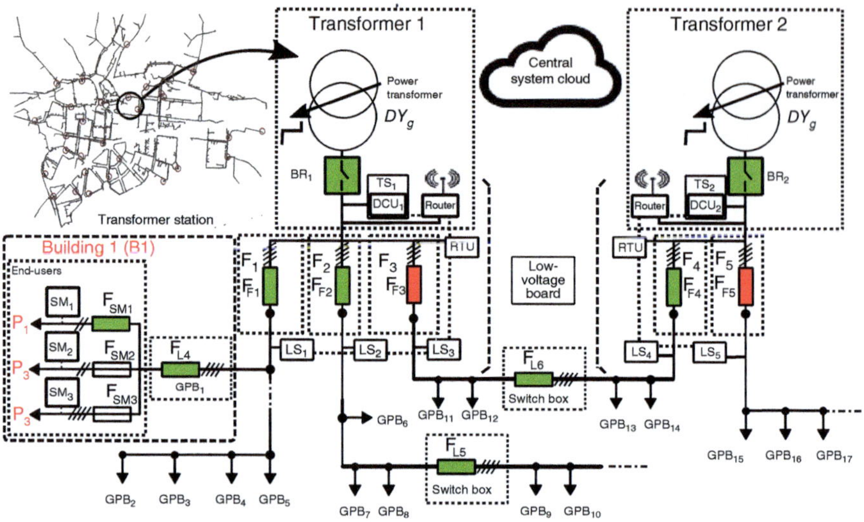

Figure 8.4 Schematic of two advanced simple entities ($ASEN_1$ and $ASEN_2$) that are interconnected through one of their LV feeders.

therefore will be affected by power failures when there are failures in the MV electrical network.

It is necessary to have a holistic view of the whole process and of the elements involved in the advanced metering infrastructure in order to have a reliable and resilient measurement system that allows us to know in real time or near-real time what is happening in the LV network.

In transformer stations with an MV remote control, there are batteries that allow the operation of digital devices even in the case of LV power supply failure. However, this configuration is not common in transformer stations without a MV remote control, and depending on the strategy of the distribution company, it is not common for the DCU electrical cabinet to be supplied with power from the MV remote control batteries even in the first scenario described earlier.

There are also other types of configurations, such as those shown in Figure 8.4, which represents an advanced simple entity (ASEN), which is a less common configuration and differs from the SEN in terms of the metering equipment installed which, in this case, allows each LV feeder to be individually supervised by means of line supervisors (LSs) at each of the LV line headers. This configuration is less widespread because of the additional investment involved in its adoption.

As shown in Figure 8.4, an advanced simple entity $ASEN_1$ is composed of transformer T_1, main breaker BR_1, and three feeders F_1, F_2, and F_3 and is powering all GPBs connected to feeders F_1 and F_2. Feeder F_3 is open and is not powering any supplies. The supplies connected to F_3 correspond to an advanced simple entity $ASEN_2$.

The advanced simple entity $ASEN_2$ is composed of transformer T_2, main breaker BR_2, and two feeders F_4 and F_5 and is powering all GPBs connected to feeders F_4 and F_5. The fuses of switch line segment F_{L6} is on (close).

Figure 8.5 Photos of an electrical cabinet of an LV transformer station where line supervisors have been installed as a part of an advance simple entity (ASEN).

As in the case of the SEN explained in Figure 8.2, the network configuration can be changed with the switches, and load can be moved from one transformer to another if necessary.

For ASENs, there are also two types of digital devices connected: metering and communication. The metering equipment for each ASEN are the smart meters ($SM_1 \ldots SM_n$) at the consumption points of the prosumers ($P_1 \ldots P_n$), the transformer supervisor (TS_X) that measures all the energy delivered by the T_X transformer to the LV network and the line supervisors ($LS_1 \ldots LS_n$) that measure the whole energy that flows into each feeder (see Figure 8.5).

In the ASENs, there are three types of network monitoring devices (prosumer smart meter, transformer supervisor, and line supervisor) and two communication devices (DCU and RTU) as can be seen in the photo in Figure 8.6.

A 4G router is needed only if there is no other way of communication in the station transformer (ST), like optical fiber.

In other words, an ASEN has two more devices than an SEN, which increase the monitoring of the LV network. The additional monitoring device incorporated in an ASEN compared to an SEN is the line supervisor. The incorporation of this device also requires the addition of the communication device that transmits the data from the line supervisors, and which is called RTU.

The advanced RTU is a device with a similar performance to a DCU. The protocol used by the RTUs is similar to that of the PRIME PLC DCU and therefore also an open protocol.

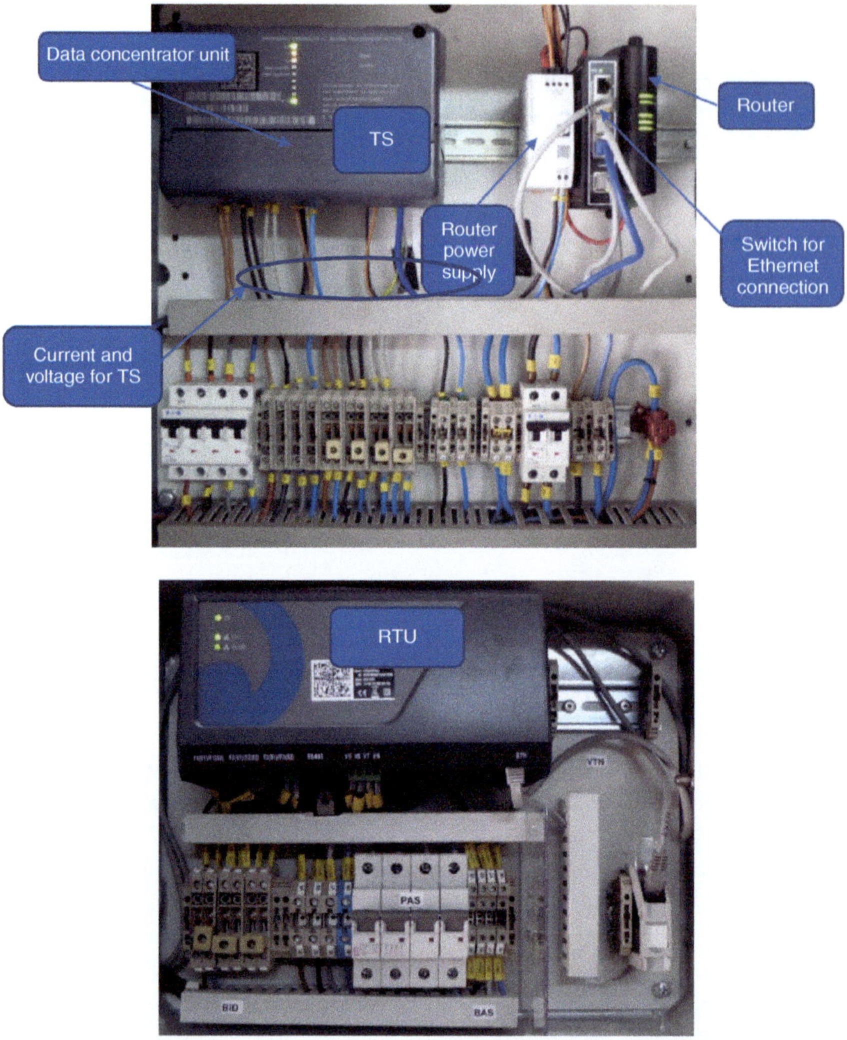

Figure 8.6 Photos with the layout of the digital communication devices in an electrical cabinet of an advanced simple entity.

The RTU only manages the line supervisors (LS*n*), and the communication is low latency because both the RTU communication devices and the line supervisors are installed in the same physical location and the communication can be established directly through an RS485 cable. In this case, the communication between the RTU and the line supervisors can be simultaneous and not sequential as it happens with the SM and the DCU because of the intrinsic characteristics of PLC communication.

Despite the investment cost of installing an ASEN over the installation of an SEN, the advantages of adopting the ASEN-type configuration are unquestionable because it allows simplifying the reference entity (ASEN) in as many SENs as feeders has each station transformer. In this way, each of the feeders of the station transformer can be considered as an SEN whose operating rules will be like those of other feeders with the same configuration of metering equipment.

This configuration has been spreading and will increase its penetration rapidly as distribution companies carry out a technological watch on the possibilities of network digitalization and improvement of their AMI) and are adopting solutions that increase the capabilities of LV network management.

In the network taken as an example, this configuration already exists in 600 MV/LV transformers and the plan is to increase the number of transformers equipped with this solution by 10% each year.

It is important to test technological advances by implementing small-scale pilots that allow reliable conclusions to be drawn in terms of the functionalities provided and the cost of implementing the solution. An example of these new technologies is the PLC portable supervisor that can be installed at any point downstream of an LV feeder and that allows the energy of a network section to be measured in a simple and continuous way. This type of connection configuration, being portable, allows great flexibility when it is necessary to increase the observability of the LV network in a specific area of the same, where the SEN does not allow enough sampling frequency of monitoring of electrical variables to obtain conclusions in the analysis of the data obtained on the standard measurement points.

It is necessary once again to highlight the importance of providing LV networks with the capacity to change the topological configuration by linking some LV feeders to others. Normally, the operating configuration of the LV feeders will be radial, but their layout is meshed, and it will be possible to change the power supply of each line segment by transferring the load from one of the line segment to another and therefore from one of the transformers to another.

Some areas of the network will be highly meshed and others less so. Some zones will have only one switch possibility and others will be isolated, and it will not be possible to make any changes in their LV topological configuration. Redundancy is usually higher in urban networks with a high concentration of users and in industrial networks, while in rural networks with dispersed users, configurations without any redundancy are more common, with a completely static topology.

Regarding the physical layout of the digital devices, the DCU is installed in an electrical cabinet with the necessary protections next to the LV switchboard of the transformer station and is connected in such a way that it takes the power supply from the LV switchboard and injects the PRIME signal through the phases of the LV conductor. This connection is made downstream of the main breaker of the transformer (BRx).

8.3 Communications and Impact on Remote Management

AMI in its current stage of development can greatly help the management of the LV network. However, there are many functionalities that could be developed if the equipment installed in the network had more monitoring capacity. Advances in communications will be a key factor in the development of new applications.

In addition, digital devices are managed and controlled by firmware that is also subject to developments. On the other hand, the combined management of the different devices monitoring the network using different communication channels and the use of different computing layers open the door to other dimensions of management not yet explored.

8.3.1 PLC PRIME Communication

The PRIME standard for smart meters was created to meet a need of DSOs to digitize meter reading tasks. In 2009, the PRIME Alliance [2] was created to bring together and guide all the stakeholders necessary for the development and use of this technology. The PRIME Alliance includes equipment manufacturers and distribution companies that have been using this technology for years to manage the smart grids. The most widely used PLC PRIME communication standard is version V1.3.6 [3]. Currently, a more advanced version V1.4 [4, 5] is available, but the meters are not yet mass produced and only a few pilots are deployed. PRIME PLC communication technology is used only for meters installed downstream of transformers, when the consumption to be measured is outside the transformer station and distributed anywhere in the LV network. The main advantage of the PRIME PLC is the cost reduction at the time of deployment of the advanced measurement system because the physical communication layer is already available and there is no need to invest additional costs to set it up.

However, communication via PRIME PLC is narrowband, which severely limits the amount and speed of data to be transmitted. How to make the best use of the communication capacity of the PRIME PLC meters will be explained later on. This limitation, which is implicit in PLC communication, does not apply to a transformer supervisor (TS) and line supervisors (LS_X), which can communicate directly by the Ethernet cable or via 4G with the central system, making it possible to obtain measurement data much faster and with greater sampling frequency.

Smart meters and portable supervisors communicating via the PRIME PLC technology are limited in their capability to send data to the central system via the data concentrator (DCU) that manages their communication. The normal configuration of the PLC PRIME communication of a concentrator installed in a transformer station involves the creation of a PLC network managed by the DCU.

The basis of PLC PRIME V1.3.6 is the injection of a signal by the DCU in the Comité Européen de Normalisation Électrotechnique (CENELEC) A frequency band using the orthogonal frequency division multiplexing (OFDM) modulation scheme. The meters receive the signal sent by the DCU through the electric cable and communication is established between smart meters that work as service nodes (SNs) and DCU that acts as the base node (BN).

The PLC PRIME protocol is organized in three communication layers: physical (PHY layer), media access control (MAC layer), and convergence layer (CL).

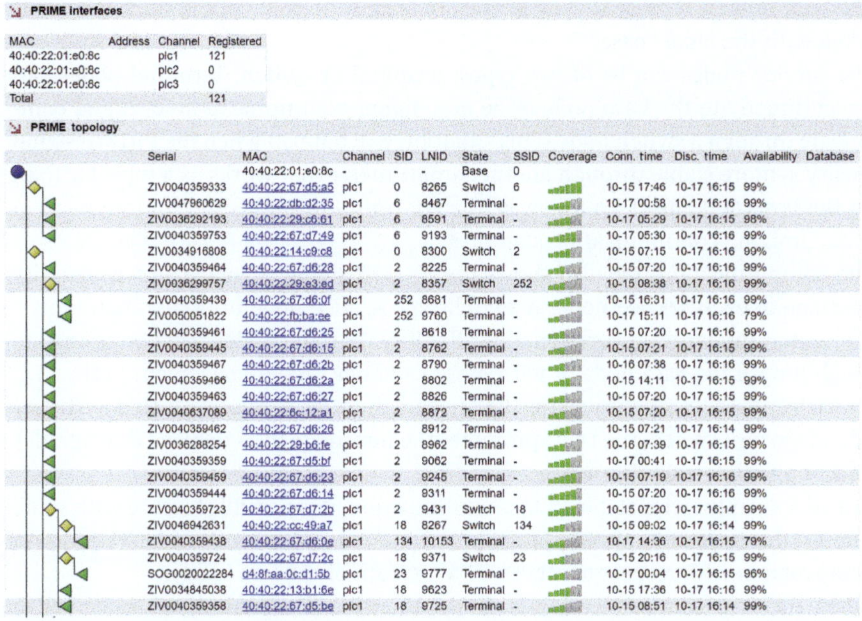

PRIME interfaces

MAC	Address	Channel	Registered
40:40:22:01:e0:8c		plc1	121
40:40:22:01:e0:8c		plc2	0
40:40:22:01:e0:8c		plc3	0
Total			121

PRIME topology

Serial	MAC	Channel	SID	LNID	State	SSID	Coverage	Conn. time	Disc. time	Availability	Database
	40:40:22:01:e0:8c	-	-	-	Base	0					
ZIV0040359333	40:40:22:67:d5:a5	plc1	0	8265	Switch	6		10-15 17:46	10-17 16:15	99%	
ZIV0047960629	40:40:22:db:d2:35	plc1	6	8467	Terminal	-		10-17 00:58	10-17 16:16	99%	
ZIV0036292193	40:40:22:29:c6:61	plc1	6	8591	Terminal	-		10-17 05:29	10-17 16:15	98%	
ZIV0040359753	40:40:22:67:d7:49	plc1	6	9193	Terminal	-		10-17 05:30	10-17 16:16	99%	
ZIV0034916808	40:40:22:14:c9:c8	plc1	0	8300	Switch	2		10-15 07:15	10-17 16:16	99%	
ZIV0040359464	40:40:22:67:d6:28	plc1	2	8225	Terminal	-		10-15 07:20	10-17 16:16	99%	
ZIV0036299757	40:40:22:29:e3:ed	plc1	2	8357	Switch	252		10-15 08:38	10-17 16:16	99%	
ZIV0040359439	40:40:22:67:d6:0f	plc1	252	8681	Terminal	-		10-15 16:31	10-17 16:16	99%	
ZIV0050051822	40:40:22:fb:ba:ee	plc1	252	9760	Terminal	-		10-17 15:11	10-17 16:16	79%	
ZIV0040359461	40:40:22:67:d6:25	plc1	2	8618	Terminal	-		10-15 07:20	10-17 16:16	99%	
ZIV0040359445	40:40:22:67:d6:15	plc1	2	8762	Terminal	-		10-15 07:21	10-17 16:15	99%	
ZIV0040359467	40:40:22:67:d6:2b	plc1	2	8790	Terminal	-		10-16 07:38	10-17 16:16	99%	
ZIV0040359466	40:40:22:67:d6:2a	plc1	2	8802	Terminal	-		10-15 14:11	10-17 16:15	99%	
ZIV0040359463	40:40:22:67:d6:27	plc1	2	8826	Terminal	-		10-15 07:20	10-17 16:15	99%	
ZIV0040637089	40:40:22:6c:12:a1	plc1	2	8872	Terminal	-		10-15 07:20	10-17 16:15	99%	
ZIV0040359462	40:40:22:67:d6:26	plc1	2	8912	Terminal	-		10-15 07:21	10-17 16:14	99%	
ZIV0036288254	40:40:22:29:b6:fe	plc1	2	8962	Terminal	-		10-16 07:39	10-17 16:15	99%	
ZIV0040359359	40:40:22:67:d5:bf	plc1	2	9062	Terminal	-		10-17 00:41	10-17 16:15	99%	
ZIV0040359459	40:40:22:67:d6:23	plc1	2	9245	Terminal	-		10-15 07:19	10-17 16:16	99%	
ZIV0040359444	40:40:22:67:d6:14	plc1	2	9311	Terminal	-		10-15 07:20	10-17 16:16	99%	
ZIV0040359723	40:40:22:67:d7:2b	plc1	2	9431	Switch	18		10-15 07:20	10-17 16:14	99%	
ZIV0046942631	40:40:22:cc:49:a7	plc1	18	8267	Switch	134		10-15 09:02	10-17 16:14	99%	
ZIV0040359438	40:40:22:67:d6:0e	plc1	134	10153	Terminal	-		10-17 14:36	10-17 16:15	79%	
ZIV0040359724	40:40:22:67:d7:2c	plc1	18	9371	Switch	23		10-15 20:16	10-17 16:15	99%	
SOG002022284	d4:8f:aa:0c:d1:5b	plc1	23	9777	Terminal	-		10-17 00:04	10-17 16:15	96%	
ZIV0034845038	40:40:22:13:b1:6e	plc1	18	9623	Terminal	-		10-15 17:36	10-17 16:16	99%	
ZIV0040359358	40:40:22:67:d5:be	plc1	18	9725	Terminal	-		10-15 08:51	10-17 16:14	99%	

Figure 8.7 PRIME communication tree of a DCU communicating with 121 smart meters. Source: PRIME Alliance AISBL.

This section will only explain the use of the MAC layer information in terms of the applicability of this layer information in the performance of the electrical network.

It is essential to understand the rules of performance of the MAC layer and the PRIME communication tree because the establishment of hierarchical relationships between nodes is based on the PRIME signal intensity that runs through the LV cable and reaches each service node, allowing the establishment of communication connections with the base node.

Figure 8.7 shows the connection structure of the service nodes and the base node, i.e. the PRIME communication tree, in the form of a tree.

The DCU concentrator of Figure 8.7 manages the PRIME communication network of 121 smart meters. The selected example corresponds to a case where communication is stable, i.e. with few state changes at the MAC layer.

The upper side shows the base node and the three possible PRIME signal injection channels (pl_{c1}, pl_{c2}, and pl_{c3}). In this case, the signal is being injected only through the pl_{c1} channel representing the R phase and the 121 smart meters are connected as service nodes to the base node in this pl_{c1} or R phase channel.

It is possible to configure the concentrator parameters to change the injection to any of the other two channels or even to activate one of the three-phase injection or intelligent injection functions. In the three-phase injection function, the concentrator will inject the PRIME signal through the three phases simultaneously, splitting the total signal intensity between the three phases, and in the intelligent injection function, it will automatically choose the injection function that maximizes communication with all the service nodes it manages.

The base node is represented by a blue circle and the value of the state column is marked with the label "base."

The service nodes can be of two types: terminal or switch. Terminal nodes can connect directly to the base node or be at a higher communication level and communicate through a switch node. This situation will happen when the PRIME signal intensity is more stable through another smart meter that works as a repeater than from the base node itself. In this way, the concentrator builds the communication tree maximizing communication stability.

The communication tree also shows the PRIME signal coverage and availability percentage of each service node. A switch-type service node normally maintains a higher communication coverage and availability than the nodes that communicate through it with the base node or with other switches. The management of the PLC network by the DCU is optimized so that, in case there is a service node that is a better candidate to be a switch, the DCU promotes the new node to switch and downgrades the other one to terminal.

In a concentrator with stable PLC communication, it is possible to trace with some reliability the connection phase of the smart meters and even the relative position of some meters with respect to others, downstream of the line header.

When the PLC communication is stable, the PRIME signal injection is kept in phase R, which is normally selected by default. In this case, the smart meters that are directly connected to the base node will be in phase R and normally the smart meters that depend on a switch node will be electrically close to that smart meter that has the role of switch. In any case, it is normal to have to wait a few hours after restarting a DCU for the tree to become stable.

This phase detection method, by itself, is not applicable because it has not yet been possible to determine the variables to be measured, nor has enough work been done on the statistical data to determine the stability limit from which the conclusions obtained are valid; however, the existence of indications that allow us to think of correlations between network stability, communication speeds of the devices, and the connection phase is clear.

In addition, this information about the communications system can be used to identify errors, i.e. to ensure when the network inventory is not correct. Instead of specifying which situations are correct among all the possible ones, what can be determined are the situations that are not possible. For example, when a smart meter is located in our inventory in an LV network that is several kilometers away from the DCU with which it communicates. If, in addition, this erroneously geolocated meter is communicating through a switch node, checks have been made to determine that the correct geographic location is indeed electrically close to the switch meter.

In Figure 8.8, there is an example of a smart meter that is incorrectly positioned in the network of the transformer I004222. This smart meter communicates with the DCU installed in the transformer C004470. Both transformers are located several kilometers apart. By consulting the communication tree of the transformer C004470, it is found that the meter is incorrectly positioned. In addition, the communication tree shows that the smart meter is a terminal node that depends on the switch with the service set identifier (SSID) label "1." The switch with SSID "1" is geographically

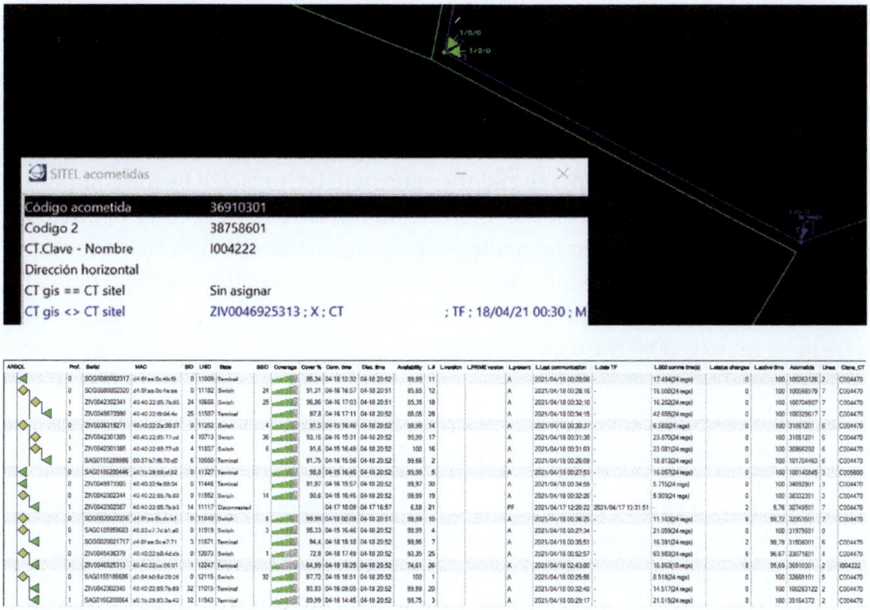

Figure 8.8 Smart meter in blue is incorrectly positioned in the network of transformer I004222. The communication tree corresponds to the transformer C004470. Source: PRIME Alliance AISBL.

located in another area several kilometers away and belongs to the LV network of the transformer C004470.

In order to try to make this information more reliable, new parameters and statistics to be measured are being tested, such as the average communication speed (S02 Communication time [seconds]) of each node or the variation of daily communication availability together with the number of status changes.

The situation of stable communication is the optimum for PLC PRIME management, but it is not always the one that occurs. There are several cases that can temporarily or continuously prevent this stability. The most common cases are as follows:

- Interference in the communication band reserved for the PLC PRIME signal, such as different types of domestic and industrial electronic equipment (inverters, rectifiers, mobile phone antenna, etc.).
- Significant attenuation of the PRIME signal due to the length of the electrical cable combined with the absence of meters in a long section of conductor or any other situation that involves a significant increase in impedance, such as deteriorated connections or conductors with poor insulation, network failures, etc.
- Failure or malfunction of the concentrator or of any service node that may be key as a switch.
- Very large networks with a high number of service nodes to be managed (more than 500) and which require the dedication of many computing resources to manage the PRIME communications tree.

It is therefore important to classify the PRIME communication network of each concentrator in terms of stability. In stable networks, it will be possible to obtain useful information for network management and will facilitate near real-time monitoring of the network.

The monitoring of electrical values measured by the meters is totally dependent on the PRIME communication tree and therefore will be impacted by the communication stability. Therefore, the decision of the monitoring strategy to be applied for each case to be analyzed must be made considering factors such as the communication speed of each device, which depends on its position in the electrical network and therefore on its position or level in the communication tree and the number of meters to be managed by the concentrator.

8.3.2 Data Concentrator Unit (DCU) Description

Based on the SEN formed by a transformer and its associated network, together with the digital devices installed in the station transformer and downstream of it, the data DCU is the device that takes the responsibility of centralizing the events and coordinating the management of all the elements. A DCU is like an industrial processor unit with a firmware that has been custom developed in order to optimize the management of PRIME communication, the collection of smart meter readings for billing, and the execution of the connection and disconnection orders of the metering devices, as this was the main motivation for the deployment of remote management.

Based on this scenario, the new functionalities required for the intelligent management of the LV network are also incorporated into the DCU already installed by modifying the firmware or, if this is not possible because of the limited memory capacity of this device, new control units also installed in the station transformers are incorporated, such as the RTUs that manage the line supervisors (LS_X).

It is quite possible that in the medium term, the functionalities of the data concentrator and the RTUs and others that may appear will be merged into a single modular equipment or industrial PC in which SEN control software packages can be loaded. In this way, in addition to the current manufacturers of hardware and software necessary for the control and monitoring of smart meters, there may also be software developers of functionalities for network management who, using the data obtained by the modular equipment, may develop functionalities at the transformer computing level (edge computing level).

The concentrator is the only element of the station transformer that communicates with the smart meters, and therefore, the monitoring of the electrical values and other parameters to be controlled, as well as the transmission to the central systems of the events generated by the meters, are actions that are performed through the data concentrator.

The concentrators are prepared to program repeated tasks over time, to collect and send daily, monthly, and hourly readings, such as collecting the quarter-hourly load profile at the end of the day, and can also receive requests via Web services from the remote management system to obtain other types of information or modify operating parameters, such as obtaining instantaneous electrical values from a smart meter.

In any case, any information to be obtained from the meters is collected through the PLC PRIME channel, and therefore, it must be sequential because it is a single channel communication, i.e. the information is obtained from one meter after another, it is not possible to obtain simultaneous information from two or more meters.

Therefore, in order to use the data concentrator as an LV network monitoring device, the data concentrator time will be available once its regular tasks have been performed. However, if necessary, it is possible to interrupt the tasks being performed by the concentrator for actions that are considered a priority for the management of the meters or the network: switching off or reconnection of meters or requests for electrical values in order to resolve a fault in the LV network.

The data concentrator, in addition to managing communication with the meters, must also manage communication with the central system. Metering data and information requests from the remote management system are sent via File Transfer Protocol (FTP) or Web services, but it is also possible to consult certain data published in Simple Network Management Protocol (SNMP). Therefore, the ways for the concentrator to send data to the central systems are varied, and the most favorable option will have to be chosen in each case and according to the type of information.

In general, periodic information on readings, configuration parameters, hourly load profile, and event history is provided via FTP; real-time events, specific requests for action, and extraction of meter values are provided via Web services; and the most volatile and dynamic information related to PRIME PLC communication is obtained via the SNMP protocol.

8.3.3 Smart Meter Description

Smart meters communicating via the PLC PRIME technology are also industrial PCs of smaller size and lower computing capacity than DCUs. In smart meters, a distinction must be made between two types of control firmware: metrological firmware and communication firmware.

The metrological firmware is designed for taking measurements and related parameters in such a way that these meters replicate the operation of traditional mechanical meters and expand their data storage capacity, for example, in event logging and hourly load profile recording.

The communication firmware allows the meters to communicate with the data concentrator through the PLC channel. This firmware has evolved both in improving the quality of communication and in incorporating different levels of security, encryption, and authentication of the communication to prevent data loss or attacks to manipulate the devices.

Manufacturers are currently working on a new generation of smart meters to cover short-, medium-, and long-term needs, improving the capabilities of current meters and providing them with new data and measurement recording functionalities and expanding communication possibilities for use at points in the network where PLC PRIME communication is not feasible.

The new-generation smart meters are designed to overcome the limitations of the current meters in terms of connection phase identification, recording of electrical values, communication latencies, and obtaining real-time snapshot values, among others.

The advantages of PLC technology are greater than the disadvantages, so the medium-term scenario will be a mixed use of meters with PLC PRIME communication in more advanced protocols and meters with wireless communication. Smart meters that include both communication possibilities in the same device are even being tested.

8.3.4 Future Scenario: Evolution of Communications Toward Hybrid Systems

The continuous development of new communication technologies such as Internet of Things (IoT) and 5G will allow evolving toward hybrid scenarios where solutions are adopted to overcome the limitations of PLC [2]. PLC communication equipment manufacturers, driven by distribution companies, are working on new solutions, such as the development of a new version of the PRIME protocol (version 1.4) and smart meters that, in addition to the prime communication module, have other alternative communication modules based on radio frequency (RF) technologies.

The introduction of this type of smart meters with hybrid communication will make it possible to upgrade the observability of the LV network in places where it is now unfeasible because of the limitations of the PLC PRIME protocol in version 1.3.6.

On the one hand, DCUs and smart meters working with the PLC PRIME version 1.4 protocol will be able to find the optimum PRIME signal injection configuration for each case, avoiding the problems caused by interference in the CENELEC A band that hinder or even prevent communication with the smart meters of the current version 1.3.6.

On the other hand, it will be possible to switch the communication to the other module available in the smart meter as an alternative to PRIME and based on RF technology. In this way, all the advantages of PLC communication will be maintained, and at the same time, it will be possible to increase the monitoring frequency where necessary without having the current limitations of the PLC channel.

8.4 Central System for Data Reception and Analysis

The AMI system for digital device management is composed of several modules that optimize data collection and remote device operation. It is not easy to define the scope of an AMI system, as they are systems in constant evolution. New modules are constantly emerging around the management of advanced metering devices, which provide the system with new monitoring and action functionalities on the devices and the associated LV network. It is difficult to set a boundary between the AMI system and the LV SCADA system that allows operating the LV network. Ideally, both systems should be integrated in such a way that they evolve as a single system.

The main function of the AMI system is to communicate with the smart meters through the DCUs to obtain the daily and hourly measurements and to perform the necessary actions on the meters for the management of the supply contracts.

The following can also be considered as functions of the AMI system:

- Inventory control of digital devices and management of its configuration parameters (DCUs, RTUs, smart meters, transformer supervisor, line supervisors, portable supervisors, routers, webcams, and other peripherals installed in transformer stations or in the LV network).
- Monitoring of communications at all levels and protocols: TCP-IP, SNMP, and PLC PRIME.
- Monitoring of the LV network snapshot or average electrical values and smart meter events.
- Storage and processing of metering values and other parameters.
- Remote operations on digital devices.
- Combined analysis of all data obtained, whether measurement or monitoring data.
- Identification, diagnosis, and management of the failure resolution process on the remote management devices and the LV network.

The development and evolution of these multiple modules that can be considered within an AMI system require hardware and software capable of supporting this enormous amount of data, communications, processes, and therefore must be designed taking advantage of edge technologies that allow a high computing capacity and handling of a high volume of data. To cite specifically some of the new technologies that are taken into account to design and evolve this type of systems, it is necessary to talk about artificial intelligence, machine learning, IoT, big data, edge computing, advanced data analytics, and real-time data processing.

In the coming years, an evolution of this type of systems will be necessary to cover all the challenges posed by the management of LV power grids and the remote management systems that monitor them, and these emerging technologies will help to configure solutions to overcome them.

The basic functionalities for the LV network management through the AMI system are the creation of strategies for monitoring electrical values of metering equipment (transformer and line supervisors and smart meters), automatic diagnosis of incidents for ecosystem maintenance, real-time event management of digital devices, and real-time management of transformer and line supervisors.

8.4.1 Real-Time Event Management

Remote management devices with the PLC PRIME technology allow the collection of events generated by the devices connected to the LV network.

In addition, the firmware of the counters and concentrators is prepared to configure any of these events as spontaneous, in real time, so that they are received as they occur and thus be able to react immediately to certain situations that arise.

The classification of all the alerts and other warnings that come to us as events from the remote management devices or monitoring systems will allow us to increase the number of cases analyzed and classified in such a way that it is possible to apply artificial intelligence and supervised automatic learning algorithms.

The arrival of a real-time event showing that there is a loss of supply quality will allow monitoring to be started at those points in the network where this defect is being observed. Because of the monitoring limitations of PRIME technology that require sequential communication with the smart meters, it will be necessary to design different monitoring strategies when it is necessary to obtain information from several smart meters located in LV networks in the same transformer station where only one concentrator will be installed.

8.4.2 LV Network Monitoring

The monitoring capability of electrical values is the most important functionality for obtaining near-real-time information about what is happening in the electrical network. Although this functionality is limited by the intrinsic characteristics of PLC communication, it is possible to design strategies of exploration of electrical values that allow obtaining very valuable data with the appropriate sampling frequency to develop algorithms for calculation and analysis of what is happening in the network such as those based on power flows or state estimation.

The interpretation of information in near-real time also requires understanding the historical data captured, calculating trends, and classifying behaviors as normal or anomalous, so that it is possible to compare these historical trends with the data obtained in near-real time, and this allows us to diagnose current events, classifying them as situations that are within normality or as anomalous situations.

When anomalous situations are detected, it is necessary to increase the sampling frequency of the monitoring or start-specific monitoring strategies adapted to the case to be analyzed, if necessary, reducing the number of devices to be scanned and ignoring those that seem to be not affected by the detected situation.

In remote management systems with smart meters communicating via the PLC technology, the data concentrator is responsible for orchestrating communications with the meters, and communication management should be oriented to maximize the obtaining of snapshot values, taking advantage of the concentrator's free time or even freeing the concentrator from routine tasks to dedicate it exclusively to obtaining information to clarify the anomalous behavior detected.

8.4.3 Automatic Diagnostic

From the different sets of monitoring data of the ecosystem, both those obtained from the monitoring of the communications of the remote management devices and those specific to the electrical network, it is possible to detect anomalous situations that will be necessary to classify according to the type of failure, such as network failure, device failure, or fraud.

By analyzing the communications monitoring data from the concentrators and routers and by aggregating them at the higher hierarchical level of their connection point, it will be possible to determine whether the failure is due to a trip of the MV line or is due to an isolated failure of the communication device.

Another automatic diagnostic functionality based on monitoring the communication of meters with data concentrators will make it possible to determine whether a massive change of meter communication from one LV line to another concentrator is due to a network switching or whether it may be due to a fault in the LV cable or even interference in the PLC communication channel that prevents the meters from communicating but allows the continuity of the power supply.

8.5 DSO Challenge: AMI for LV Network Management

As can be concluded from the detailed description of each of the elements, mechanisms, and functionalities that make up the so-called SEN and ASEN, the main challenge faced by distribution companies is to try to manage all the possibilities of communications, configuration, monitoring, and information gathering systems in such a way as to achieve proper management of the LV network, both by identifying possible errors in the network topology and inventory of devices and by diagnosing possible faults in the electrical network or in the metering devices, or even reducing non-technical losses.

To do this, it will be necessary to develop a control application that takes advantage of the right information at the right time to identify what is happening in the network and decide on the right actions to be taken either remotely or in the field directly on the devices or on the network.

Moving forward in the development of new functionalities for ecosystem management, it will be seen that certain cases could be more easily solved by identifying some additional key variables, which are not currently contemplated among the data being captured. In order to have these new key variables, it will be necessary to agree with the manufacturers of the devices to incorporate them via software or hardware or both.

It may also be necessary to bring the advanced data analysis algorithms closer to the device that obtains the data in order to avoid massive data processing and storage and reduce it to alerts or alarms generated by the devices from data processing. This mechanism is known as "edge computing" and consists of decentralizing the execution of algorithms on site, providing the digital device that obtains the data with information storage, computing, and decision-making capacity.

The following is a list of the current set of variables that can help to define a series of functionalities that will allow progress in the development of control software for remote management systems and LV network management and that will help to decide on the actions to be taken for the management of the ecosystem:

- LV network model, having a digital replica of the transformer station, digital devices, and network configuration in terms of LV connections between networks or working as an isolated network.
- Classifying transformer stations according to the inventory of devices and network (data capture errors, phase identification, ordering of supplies in relation to the distance to the transformer, and inventory validation).
- Classifying transformer stations in terms of technical and non-technical losses (suspicion of fraud).
- Communication configuration of the transformer station and digital devices.
- Supply of the DCU and TS of the transformer station.
- Classifying transformer stations in terms of TCP-IP communication to central systems (failures in communication devices or failures in the power grid).
- Classifying transformer stations to the stability of the PRIME network of the DCU.
- Classifying transformer stations according to the communication speed of the DCU (depending on the number of meters and the configuration of the LV network of the transformation center).
- Configuration parameters of the devices:
  - Smart meters and portable supervisors (PLC communication).
  - Transformer supervisor and line supervisors
  - DCUs
  - Routers
- Measurement information of the devices: smart meters and supervisors.
- Monitoring information on communications and AMI system operation.
- Communication statistics of devices.
- Monitoring information of the LV network through the remote management devices (variation of the sampling frequency of the information depending on the use case to be analyzed).
- Events generated by the system.
- Programmed operations in the smart meters and remote management devices.
- Alarms, alerts, and warnings due to anomalies in the operation of remote management devices.
- Network failures and scheduled outages.

By capturing, properly classifying, and analyzing this data set (some are discrete data, and some are time series or calculated data), algorithms can be created to identify situations that are happening in the grid or that are going to happen soon. To go deeper into these functionalities, the next sections will describe some smart grid operation scenarios that can be identified through algorithms to interpret this set of data from the AMI system and the network system.

8.6 Digital Twin of the LV Network

The basis for all management of the LV network is to have a reliable LV network model that functions as a digital twin of the LV network. In order to create a real replica of the LV network, it is necessary to define the basic data model, consisting of the network elements and the connections between these elements. Referring to Figure 8.4, where the ASEN is described, the elements that make up the LV network model are represented as follows:

- Transformer station consisting of the transformer with its tap changer, transformer main breaker, LV switchboard with the fuses of the line headers, and the feeders tracing to the consumption and generation points.
- LV network consisting of conductors that make up the line segments, the switch boxes with the fuses for changing the topology configuration, and the GPBs that reach the smart meter centralizations from where the distribution lines trace to the customers.
- Digital metering devices: transformer supervisor, line supervisors, portable supervisors, and smart meters at consumption and generation points.
- Digital communication devices: DCU, advanced supervision RTU, and router where necessary.

In addition to having the technical and configuration characteristics of each of the elements, it is necessary to model the relationships and connections between them and to handle the switchable elements in a differentiated way since they will change their position over time, and it is necessary to store a history in order to be able to replicate past network configuration situations reliably.

The switchable elements are the main circuit breakers of the transformer, the fuses located at the line headers, and the switches that allow interconnection between networks of two or more LV feeders.

For digital devices, it will also be necessary to keep a history of their connection points with start and end dates. Measuring devices are quite static and normally remain installed during their entire life cycle in their original location, but, like any other type of device, they occasionally malfunction or fail and must be removed and replaced by new ones, and therefore, it will be necessary to keep a historical inventory of the different measuring devices that are installed at each measuring point.

Similar is the case with the communication devices, which also fail or malfunction in their configuration and need to be removed and repaired before being re-installed elsewhere in the network. As the transformer supervisors (TSs) are embedded in the housing of the DCU, when a DCU is removed or relocated, this change also impacts the TS. Therefore, it is also necessary to keep a historical inventory of the different supervisors that have been installed in each transformer in order to be able to compose the measurements by consulting the data of the different devices according to this date range.

To collect all this information and to be able to reproduce past situations, human intervention is essential because some elements, such as fuses and switches, are not devices that communicate their position, nor can they be remotely operated. In addition, the digital devices that are reachable remotely do not provide their geographical position and require manual intervention in the compilation of the inventory and history.

However, it is possible to detect the switching positions of fuses and switches in the network using data analytics. It is a current challenge to develop these functionalities to obtain information about the network configuration without the need to send workforce to the field to report feedback.

8.7 Evolution of the Functionalities for LV Network Management

As a next step to having a digital twin of the LV network, it will be possible to develop software functionalities based on it that will allow us to identify anomalous situations that are occurring in the network or to make decisions that allow an optimized exploitation of the LV network while minimizing losses. In addition, it will be possible to simulate future scenarios both for the connection of new loads such as new supplies or EVs and to simulate the connection of other types of distributed energy resources such as photovoltaic or wind power plants or energy storage batteries.

The basis for developing these functionalities is all that has been described in the previous sections of this chapter. On the one hand, it will be essential to have a validated model that represents the digital twin of each of the simple entities of the network, and on the other hand, it is necessary to have the validated data from the monitoring of snapshot values as well as the load profile and events. These two groups of validated information will allow us to make decisions on network management with guarantees.

The representation on the digital twin of the real-time events and the monitoring directed toward the points of the network that are returning values indicating a loss of quality of supply in the network or even high energy losses allows the development of automatic anomaly detection functionalities.

The loss optimization functionality based on the dynamic change of the network configuration will be possible thanks to the analysis of losses by simulating different LV network operation scenarios. In the case of isolated networks, the most practical option is to look for the most balanced possible phase distribution of loads or generations, as well as the possibility to remotely operate the transformer tap changer and keep the voltage stable.

In meshed grids, in addition to the two options described for isolated grids, it will also be possible to transfer load or generation from one grid to the other, looking for the most optimized operating scenario for both grids.

In the design of new networks or new network segments needed to connect photovoltaic plants or centralized batteries, it will be possible to simulate the advantages of connecting these plants or batteries to two nearby networks, creating a connection between them and therefore the possibility of generating new network operation scenarios.

The functionality of integration of new supplies and distributed resources can be developed as a simulation, so in this case, it will be enough to have the historical hourly or quarter-hourly consumption data aggregated by connection point. The hourly or quarter-hourly profiles of the new supplies or generators to be connected will be simulated, and by using a power flow, it will be possible to determine whether or not the capacities of each zone of the network are overcome.

With the appropriate processing of this historical data on the digital twin of the network, it will be possible to represent a capacity map of the LV network, differentiating the network segments according to the percentage of free capacity.

8.8 Conclusions

This chapter has tried to explain in detail the new management capacity of smart grids, breaking down the management system of a smart grid into two main blocks, the distribution grid system, and the AMI system or remote management system and has gone deeper into the fundamental parts to be taken into account in both systems, paying special attention to the AMI system because it is the most innovative and is in continuous development thanks to the incorporation of digitization in the distribution grid.

In addition, the parts that make up an AMI system, the digital equipment, and applications that are part of the AMI and a multitude of configuration parameters of this ecosystem have been described.

An attempt has been made to explain all the factors that influence the operation of an AMI system and how to use these variables in the interpretation and management of smart grids. However, the possibilities are endless, and as the AMI system is in continuous evolution, the use and the new smart grid management functionalities will increase more and more from different points of view, being the end user or prosumer and the DSOs the ones that will benefit the most from the development of these functionalities. The prosumer will gain advantages in the management of their energy packages and DSOs will be able to improve efficiency in the operation, maintenance, and planning of distribution networks.

Currently, the distribution network is no longer managed "by feel" as it was done until now by the distribution company's field workforce. The new smart grid concept, using AMI system data, has made it possible to move to much faster, safer, and more efficient data-driven decision making.

In the literature, solutions to problems in the LV distribution network are often found that make unrealistic assumptions about the operation of the network and are therefore not applicable in practice. This chapter provides a description of a real management system for a European-type distribution network that is common to most networks of this type. This description will allow researchers to know the scope and real applicability of the proposed solutions.

List of Abbreviations

AMI	advanced metering infrastructure
ASEN	advanced simple entity
BN	base node

BTM	behind-the-meter
CENELEC	Comité Européen de Normalisation Électrotechnique
CL	convergence layer
DSO	distribution system operator
DCU	data concentrator unit
EV	electric vehicle
FTP	file transfer protocol
GPB	general protection box
LV	low voltage
MAC	media access control
OFDM	orthogonal frequency division multiplexing
PHY	physical
PLC	power line communication
PRIME	PoweRline Intelligent Metering Evolution
RF	radio frequency
RTU	remote terminal unit
SEN	simple entity
SM	smart meter
SN	service node
SNMP	Simple Network Management Protocol
ST	station transformer
TCP/IP	Transmission Control Protocol/Internet Protocol

References

1 Kersting, W.H. (2012). *Distribution System Modeling and Analysis*, 3e. Abingdon, VA: CRC Press.

2 https://www.prime-alliance.org/ (accessed 04 2021).

3 P. A. T. W. Group. https://www.prime-alliance.org/wp-content/uploads/2020/04/PRIME-Spec_v1.3.6.pdf (accessed April 2021).

4 P. A. T. W. Group (2014). https://www.prime-alliance.org/wp-content/uploads/2020/04/PRIME-Spec_v1.4-20141031.pdf [En línea] (Último acceso April 2021).

5 https://www.smart-energy.com/industry-sectors/smart-meters/prime-hybrid-solution-combines-plc-and-rf/ (accessed April 2021).

9

Smart Grids for the Efficient Management of Distributed Energy Resources

Roberto Ciavarella, Marialaura Di Somma, Giorgio Graditi, and Maria Valenti

Italian National Agency for New Technologies, Energy and Sustainable Economic Development, ENEA, Department of Energy Technologies and Renewable Sources, Rome, Italy

9.1 Electrical System Toward the Smart Grid Concept

The energy transition, an enabling element for the European decarbonization strategy, requires the migration from an energy system based on fossil sources to a new one based on renewable energy. A massive integration of distributed generation (DG) in the power system infrastructures substantially changes the layout of the electricity grids [1–4]. Current electrical systems have largely been conceived as unidirectional networks designed to manage power flows from large conventional generators connected to the transmission network level to distribution networks until the end users. In the presence of DG, numerous generation plants from non-programmable renewable sources of small size, located along the distribution network, become nodes for the generation of non-programmable power flows, and therefore, the electricity system is called upon to manage the bidirectional power flows [5–10].

In such a context, distribution networks, originally designed as passive terminations of the transmission network, play an active role because of the presence of non-programmable renewable sources which, however, are generally not manageable or centrally coordinated by the distribution system operator (DSO). The presence of DG therefore represents a concrete element of criticality for the distribution network and for the electricity system in general; network operators, in fact, are called upon to ensure the power supply to end users in conditions of safety, reliability, and continuity even in the presence of resources that they cannot actually program or control [11].

Guaranteeing a safe management of the electricity system, even in the face of significant incremental shares of non-programmable renewable sources, requires restructuring interventions that foresee reasonably short times and sustainable costs. Therefore, actions are necessary that aim at the adoption of technological solutions for the reconversion – rather than the replacement – of the current network assets and, more specifically, which allow the introduction of new management

Technologies for Integrated Energy Systems and Networks, First Edition.
Edited by Giorgio Graditi and Marialaura Di Somma.

logics of the current infrastructures through projects of modernization and digitization of networks.

The concept of smart grid originates from these premises. The smart grid can be defined as an electricity grid capable of efficiently integrating the behavior and actions of all users connected to it, namely, units of generations, transmission and distribution network operators, market operators, consumers, and prosumers, in order to ensure the achievement of an electricity system that, in addition to being economically efficient and sustainable, is characterized by low losses and an adequate level of quality and safety of the electricity service [12–14].

Compared to the traditional electricity grid, which can only transmit and distribute electricity, the smart grid is able to activate an integrated and optimized management paradigm of the constituent technologies through the active involvement of all users connected to it.

The main factors favoring the evolution of the electricity system toward the smart grid are outlined below [15]:

- High DG penetration rates, especially with reference to non-programmable renewable sources.
- Adoption of an appropriate communication infrastructure (Information and Communication Technologies – ICT) to connect the various "active" elements of the electricity grid.
- Decreasing separation between the planning and operation phases of the network.
- Implementation of solutions that favor a more efficient use of the existing transmission and distribution systems, considering, at the same time, the limits of the current infrastructures and the problems associated with the creation of new systems.
- Use of energy storage equipment, both in combination with generators from renewable sources and as a tool for optimizing the network parameters.
- Gradual replacement of the supply of ancillary services, carried out until now by fossil fuel generation, with renewable sources, storage systems, and consumption units in the aggregated form.

The infrastructure of a smart grid essentially consists of the assets described in Table 9.1.

The smart grid is therefore configured as an innovative concept of network infrastructure, which, with the limits imposed by the complexity and extension of the existing infrastructure, has the primary objective of supporting a reliable, safe, and competitive energy supply system.

Globally, smart grids are considered an essential aspect for the evolution of the electricity system, essential for coping with the changing energy needs. In fact, smart grids will guide the energy transition toward an optimized management approach of an infrastructure capable of exchanging information, in a bidirectional way, with all its nodes and adapting to different conditions with a high degree of flexibility. Based on the smart grid concept, the electricity grid and communication technologies are connected in an integrated infrastructure, where producers, consumers, prosumers,

Table 9.1 Assets of the smart grid infrastructure.

Physical infrastructure	Smart grid network backbone that includes all the equipment for transporting the electricity carrier from generation systems to the end user.
Communication infrastructure	Communication backbone of the smart grid that includes all ICT devices that allow the communication and transmission of data related to energy flows, enabling their management according to the logic integrated into the digital infrastructure. The communication infrastructure is a fundamental element for the purpose of connecting the elements of the electricity system and the transmission of information to the electricity market.
Digital infrastructure	Set of all platforms, technologies, and digital logics that implement advanced control functions aimed at automation and intelligent management of the network.

and storage systems interact in a free market in order to efficiently manage the entire system.

The enabling factors for smart grids are all those technologies, in particular ICT, that favor the integration of renewable energy and DG, both in terms of energy and power, as well as the related business models, which provide for relationships, transactions, and remuneration that are as appropriate as possible to the real services offered/requested to/by the different actors and the system as a whole.

An essential feature of smart grids is the ability to manage, through protocols and information flows, the generators and the active loads available in the network by coordinating them in order to perform certain functions in real time, such as coping with peak loads. A protocol similar to peer-to-peer (P2P) for information management used in computer networks is then applied to the electricity network. In such a network, the hierarchical relationship between the nodes is practically canceled, thus obtaining all equal nodes, including the nodes of the end users of the distribution network, which, from this point of view, do not act as simple users of the network, but they constitute nodes themselves by sharing and exchanging information with the rest of the network. In this way, the devices of the electrical network become the active parts of an extended control system, which goes from the generation units up to the individual users.

In order to describe the potential benefits deriving from this approach, the attention can be focused on the "smart" management of one of the major critical issues typical of today's electricity system, namely, the presence of peak energy demands. During the occurrence of such peaks, in order to avoid network contingencies, auxiliary standby generators are generally used, which are activated to remedy the voltage drop due to the increase in the load on the network. However, this solution is inefficient both from an economic and an energetic point of view. A "smart" solution is instead characterized by the management of these peaks by means of the regulation, itself "smart," of the end-user loads, with the joint use of smart meters (digital meters capable of communicating with the rest of the network) and automated load management systems that act at the end-user level. By following this approach, it is possible to flatten the peak loads by also obtaining a considerable economic/energy saving because of the non-use of the auxiliary generators in standby. In addition, the end user becomes an active part in the management of network contingencies through appropriate demand response (DR) services. Moreover, in this context, storage also plays a fundamental role, thanks to the ability to provide flexibility to the network and to contribute to the supply of ancillary and dispatching services, thereby favoring the implementation of this new energy model.

9.1.1 Technology Areas of Smart Grids

The technological areas of smart grids, each in turn made up of a series of technologies, affect the entire electrical system from generation to transmission and distribution, up to the various types of end users [12].

The wide-area monitoring and control technological area, through real-time monitoring and display over large areas of the components and performance of the power supply system, supports system operators in optimizing their performance. Advanced system operating tools avoid critical situations such as blackouts and facilitate the integration of renewable energy and DG. Advanced system monitoring and control technologies and analytics – including wide-area situation consciousness (WASA), wide-area monitoring system (WAMS), and wide-area adaptive protection, control and automation (WAAPCA) – generate data to aid the decision-making process by the system operator, to mitigate disturbances over large areas, and to improve the transmission capacity and reliability of the entire system.

With reference to the integration of ICT technologies into the network, the related communication devices, system control software, and resource planning software support the bidirectional exchange of information between the various stakeholders involved, allowing a more efficient management of the electricity grid.

The integration of renewable sources and DG in the grid, which can affect both the transmission and the distribution grid up to end users, as already discussed above, can present considerable challenges because of the problems related to dispatching and control of these resources with a direct impact on the functioning of the entire electricity system. Smart grids can help ensuring the balance between supply and demand, not only through the integration of energy storage, which allow for the decoupling of production and consumption, but also through the automation of the control logics of generation and loads, the latter obtainable by defining appropriate DR programs.

With reference to the transmission network, there are various technologies that allow improving its management and related performance. Flexible alternating current transmission systems (FACTS) are mainly used to improve the controllability of transmission networks and maximize the power transfer capacity. The implementation of these technologies on existing lines can improve their efficiency and postpone the need for investments in the network [16, 17]. The high-voltage direct current

(HVDC) technology, on the other hand, represents a competitive solution both for long-distance power transmission and for the interconnection of different electrical systems. In fact, it allows reducing losses by also guaranteeing greater controllability of the system, thereby fostering an efficient use for those renewable sources located at a certain distance from the load [18].

As regards, the management of the distribution network, in a "smart" perspective, through the implementation of advanced automation systems, it is possible to process in real time the information coming from the sensors and meters distributed on the network for a series of purposes, such as localization of faults, automatic reconfiguration of power supplies, optimization of voltage and reactive power, as well as control of DG units and loads. Therefore, sensor technologies can help optimizing equipment performance by ensuring the efficient use of resources.

The advanced metering infrastructure (AMI) provides for the implementation of a series of technologies such as advanced and smart meters that allow the bidirectional flow of information between utilities and end users, providing data relating to prices and consumption that are essential for the correct management of the electricity distribution process. Such systems offer a wide range of features, such as

- ability to collect, store, and process data on users' energy consumption for any required time interval, even for real-time applications;
- improvement of energy diagnostics thanks to the analysis of more detailed load profiles;
- ability to remotely identify the location and extent of interruptions through a measurement function that sends a signal when the instrument turns off and when power is restored; and
- detection of network losses.

The electric vehicle charging infrastructure is also another key element in smart grid context, able to manage, during periods of low loads, the billing and programming phase for "smart" charging of electric vehicles for grid-to-vehicle applications. A smart management of this type of infrastructure also allows providing ancillary services such as reserve capacity and peak shaving through vehicle-to-grid applications.

Finally, the systems related to the end user, used for the "smart" management of consumption at industrial, tertiary, and residential level, include energy management systems, storage units, and applications for building automation. In these applications, the DR services are based on the automatic response of the user through the use of automatic devices and thermostats sensitive to price changes, which are all connected to an energy management system and controlled downstream of the receipt of a signal by the utility or system operator.

9.1.2 Services and Functionalities of the Smart Grids

In order to characterize the smart grids, it is essential to analyze the services that they are able to offer to the various actors of the electricity system that are described in the following sections.

9.1.2.1 Needs to Integrate New Emerging Technologies

Smart grids represent a valid solution to guarantee the integration of renewable energy and DG and ensure the energy necessary for new final electrical uses in an electrified energy scenario, such as heat pumps for space heating/cooling and electric mobility. This service is carried out by the DSO, while the advantages are at the level of the entire system. The purpose of this action is to ensure the integration of renewable energy and DG into the network while maintaining an adequate level of service safety.

9.1.2.2 Improve the Operation of the Network

Smart grids help reducing the out-of-service periods in the case of breakdowns or anomalies by helping to improve the continuity of the electricity service. Taking advantage of the technological areas described above, smart grids are in fact able to implement automatic and optimal network reconfiguration functions that dynamically adapt to the network topology. In addition, through the use of "smart" monitoring systems of network components, smart grids allow applying advanced maintenance techniques to reduce disruptions and optimize network asset management. This service is carried out by the DSO, while the advantages affect the entire electricity system.

9.1.2.3 New Investment Planning Criteria

Thanks to a greater degree of knowledge of the network, it is possible to integrate optimized network development methodologies by taking into account the needs of the various players connected to the network. In this context, the use of methodologies for an active operation of the distribution network, which allows DSOs to intervene on the input profiles of the generators connected to the network, as well as on the load profiles of the end users, allows a better exploitation of the network in its current configuration by postponing the need for further investments. This service is carried out by the DSO, while the main beneficiaries are producers, end users, and balancing operators (including the owners of the storage systems).

9.1.2.4 Improve the Functionality of the Market and Services to End Users

Through the smart grid and its intrinsic feature that provides information on the status of the network to be used by end users, it is possible to increase their awareness, favoring, in fact, a more efficient energy use and a more informed access to the market. This type of approach can be adopted through the use of smart meters, which allow putting DR mechanisms into practice, which, in turn, allow applying a process of flexibilization of energy demand also through the direct control of loads. A flexible demand also determines a reduction in energy prices during peak hours, which correspond to the hours in which the peak load usually occurs. Finally, this framework favors the emergence of new market players, such as aggregators, which offer new services to network operators by aggregating a series of users, who otherwise could not access the market.

9.1.2.5 Active Involvement of the End User

Through the smart meters, the data regarding withdrawals, feed-in, and price signals represent information at the service of the end user. The end-user involvement is, in fact, a key element in the smart grid, especially because of the need for modulation and flexibility required for the management of network contingencies. The interface between the end user and the electricity network is given by the digital meter managed remotely, with the ability to record the feed-in and withdrawal of active power with the proper granularity. The advantages of smart meters are manifold. From the point of view of network operators, the collection of consumption data is made more efficient, replacing manual reading with automatic acquisition. In addition, diagnostics and immediate detection of faults allow a more efficient use of the network and increase its reliability thanks to timely repairs. The operator is also able to provide additional services to the user, for example, by providing, in real time, a pricing system based on time slots at differentiated costs, which can bring economic benefits to the user. Finally, this application allows the end user to be more responsible, increasing awareness of their consumption. This, in fact, involves a saving of energy resources, as well as a reduction in network losses.

9.1.2.6 Increased Energy Efficiency and Reduced Environmental Impact

According to the International Energy Agency (IEA), smart grids, with the key objective of promoting the integration of renewable energy and increasing energy efficiency in the transmission and distribution of electricity, are able to provide an essential contribution to the creation of a low-carbon and high-efficiency energy system. In a study conducted by the US Department of Energy, various mechanisms were identified by which smart grids can contribute to the reduction of primary energy use through the increase of energy efficiency and to the reduction of the environmental impact in terms of CO_2 emissions, as shown in Table 9.2 [19]. In detail, two types of impact were analyzed, namely:

- the direct reduction of the primary energy use and the environmental impact, in which the functions of the smart grid allow increasing energy savings and a reduction of CO_2 emissions;
- the indirect reduction in which the functions of the smart grid allow generating economic savings to be reinvested later in terms of energy efficiency and/or use of RES.

9.2 Need of a Multi-Domain Optimization in Smart Grids

As discussed in previous sections, large-scale integration of RES needs to be carefully monitored in power system operation across the whole energy chain. Bidirectional flows of energy make the grid management more complex than a monodirectional production–transmission–distribution energy scheme. On the

Table 9.2 Mechanisms of smart grid and related impacts.

Category of functionality	Mechanism	Impact: Direct reduction of primary energy consumption and CO_2 emissions	Impact: Indirect reduction of primary energy consumption and CO_2 emissions
Energy efficiency	Use of smart meters to encourage end-user involvement	Energy savings achieved through the active role of the end users in managing network contingencies and making themselves responsible for their consumption	–
Energy efficiency	Use of energy diagnostic systems in residential and tertiary buildings	Energy savings achieved by improving the performance of heating, ventilation, and air conditioning (HVAC) and lighting systems	–
Energy efficiency	Use of measurement and verification systems for energy efficiency programs	Energy savings achieved through the use of measurement and verification systems for energy efficiency programs, through which end users are made aware in real time of the energy savings obtained and the reduction of CO_2	Economic savings obtained through the use of measurement and verification systems
Energy efficiency	Load shifting in order to favor the use of more efficient generation systems	Reduction in the use of fuel and related emissions resulting from load shifting mechanisms that allow using more efficient generators for satisfying loads, through DR mechanisms and distributed storage	–
Energy efficiency	Support for the integration of electric vehicles	Reduction of fuel consumption and emissions of electric vehicles enabled for connection to smart charging systems	–
Energy efficiency	Advanced control of the network voltage	Reduction of distribution losses and final energy consumption resulting from the optimization of the network voltage	–
RES integration	Support for the penetration of solar generation into the grid	Greater penetration of solar generation in the distribution network thanks to the implementation of advanced grid voltage control logics and automatic systems to manage energy flows	–
RES integration	Support for the penetration of wind generation into the grid	Greater penetration of wind generation through the use of DR mechanisms and distributed storage to provide regulation services to the grid	Cost reduction for additional generation capacity by using DR and distributed storage mechanisms instead of fossil fueled power plants to meet reserve requirements

Source: Based on Pratt et al. [19].

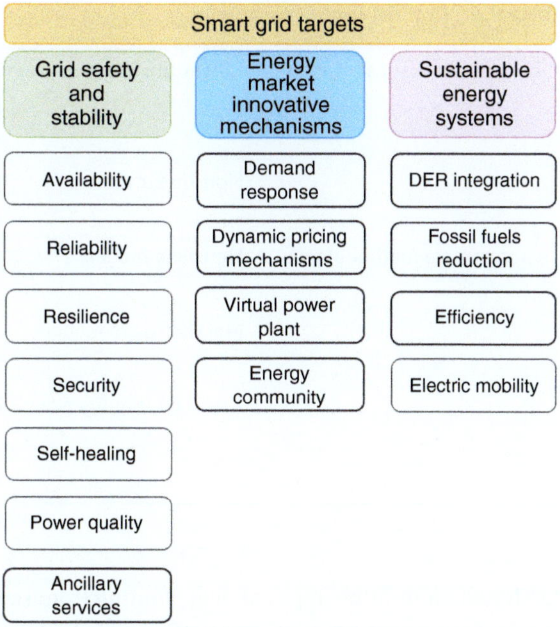

Figure 9.1 Smart grid targets.

other hand, a proper energy management should properly consider intercommunication among the several domains of the power system (i.e. generation, transmission, distribution, DER, and end user [20]). In new energy scenarios, the optimal energy management is not connected to a single domain target anymore, but it should be based on a multi-objective approach according to a multi-domain perspective. Indeed, smart grids perform many functionalities to meet several requirements for actors operating in different domains and zones (i.e. process, field, station, operation, enterprise, and market [20]) of the smart grid reference architecture (SGAM) in order to develop the best balance among the different targets shown in Figure 9.1.

As shown in Figure 9.1, three main categories of targets can be identified to be realized at the several domains of SGAM architecture. The category "Grid Safety and Stability" includes the targets aiming at guaranteeing the safety and stability of the grid and the power quality, also in the presence of a large amount of DER. The category "Energy Market Mechanisms" includes the targets aiming at involving flexible resources in the smart grids, also through an active participation of consumers/prosumers individually or in aggregation. The category "Sustainable Energy Systems" considers the targets aiming at realizing sustainable energy systems through actions for reducing the environmental impacts, such as DER integration, electric mobility, energy efficiency measures, etc.

Optimization techniques have been always utilized in electric power industry to allocate power generation to traditional generators in order to meet power requirements and technical constraints (e.g. voltage control, frequency control, etc.) [21]. Optimization models for conventional power systems are based on the estimate of the network's state. In smart grid context, optimization problems (OPs)

Table 9.3 Optimal control techniques.

Exact mathematical models	Linear models	Linear programming
		Integer linear programming
		Mixed integer linear programming
	Non-linear models	Quadratic programming
		Convex programming
Approximate methodologies	Heuristic models	Constructive algorithms
		Local search algorithms
	Metaheuristic models	Trajectory models
		Population-based methods
	Artificial intelligence	Ruled based
		Agent based
		Expert systems

are based on multiple types of data, coming from several network components, and updated from real-time information provided by devices as sensors, meters. In this case, the OP consists of finding the best solution by maximizing or minimizing an objective function (OF), while satisfying all considered constraints related to the integrated smart grid system. The OF represents the mathematical formulation of the smart grid targets (e.g. increasing reliability). The OP can be solved by using several methodologies described as follows:

- **Optimization-based methodologies**: these techniques can be based on exact mathematical models or approximate methodologies as shown in Table 9.3. Unlike the rule-based methods, exact mathematical models can provide the optimal solution, but they require more work to "translate" the problem into a mathematical formulation by properly simplifying the key phenomena and considering suitable preliminary assumptions. On the other hand, approximate methods are simpler than the exact methods, but they do not guarantee the same quality of the results because they, in general, use random research-based methodologies [22].
- **Rule-based methods**: techniques that search regularities by using data mining or machine learning analysis. These regularities are expressed in the form of an IF-THEN rule [23]. Rule-based methodologies provide results that are strictly linked to the specific scenarios, not ever corresponding to the best possible solution.

In this chapter, a set of advanced control logics, based on optimization-based techniques, are proposed for the smart management of DER with the aim to provide ancillary services in the context of smart grids by involving the resources belonging to the five domains of the SGAM. Several use cases are also presented to show the effectiveness of the control logics.

9.3 Advanced Control Mechanisms for Smart Grid

9.3.1 Architecture and Grid Model

New energy scenarios enable the possibility to use resources connected at different levels of the grid (e.g. resources connected to the distribution system) to provide ancillary services as congestion management, frequency control, and voltage control. A coordinated flow of information and commands among the different operators is required to ensure a reliable and optimal use of these resources.

This section discusses three coordination schemes and the related advanced control mechanisms to solve specific network issues by using resources connected at different voltage levels. Each scheme is described in terms of tasks, roles, responsibilities, and technical theory and assumptions.

To introduce the proposed multi-domain approach, all schemes consider the architecture below shown in Figure 9.2 as the conceptual structure of the grid.

This architecture reflects innovative business models introduced by current trends in regulations [24]. In detail, according to this approach, the following functionalities are investigated:

- A continuous cooperation and interoperability is proposed between TSO and DSO levels [25–29];

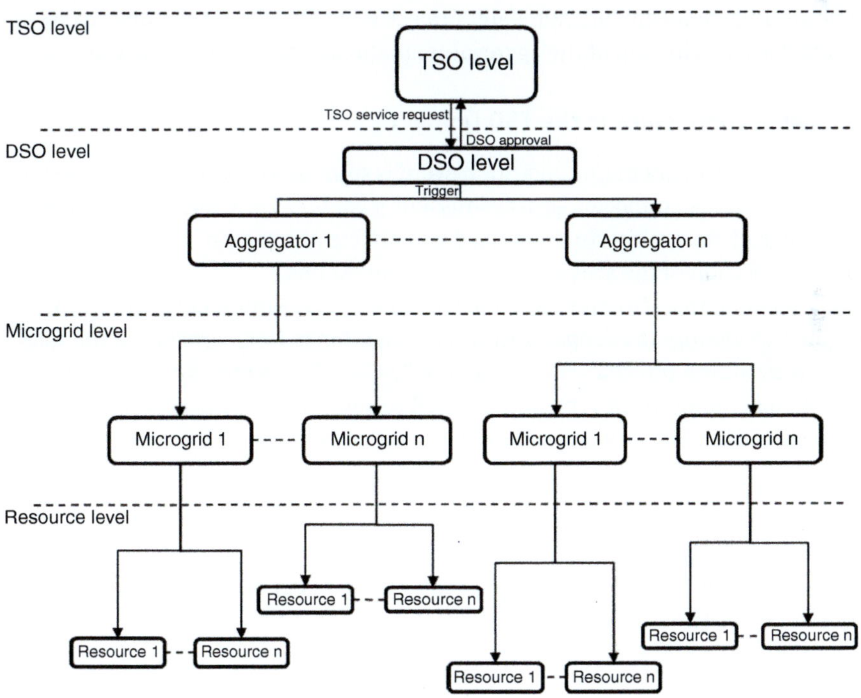

Figure 9.2 Conceptual grid structure for the proposed multi-domain approach.

- The DERs are managed by aggregators and integrated into microgrids. In particular, aggregators are operators that control DER, making a sizable capacity available to DSOs and TSOs, similarly to conventional generators [30]. In the proposed architecture, aggregators control several microgrids, each one integrating many DERs. Integration of DER in microgrids allows enabling different schemes and improve the possibility for DER to participate to the energy market [31, 32].

All the proposed control mechanisms have been developed by using Python and then tested in a Powerfactory DigSilent environment. A SimBench-based network model has been used to model the grid architecture according to the conceptual scheme in Figure 9.2. The SimBench data set is a benchmark data set – developed under the German R&D project "SimBench"[1] – for electric power system analysis, planning, and operation management. Thirteen basic network models, including electrical parameters and data at all voltage levels, are publicly available in the "Download" area of the website https://simbench.de/en/. The network model proposed in this chapter is obtained by customizing a SimBench basic network. In particular, the DERs integrated into the model are as follows:

- photovoltaic generation (PV): 23 MVA;
- wind power generation: 28 MVA;
- storage units: 76 MVA.

Each microgrid includes at least one PV system, one storage system, and controllable loads.

All control mechanisms provide daily forecasted values (24 hours time scale) to allow a better participation of the aggregators to the day-ahead energy market.

9.3.2 Congestion Issues in the TSO Domain

In the first proposed scheme, a congestion event is supposed to occur at transmission level, and resources connected at transmission level and DER integrated at lower voltage levels of the architecture are used to mitigate congestion. In practice, TSO can use the solution of the scheme to prevent congestions.

In this scheme, the TSO checks – at each time step (15 minutes) – the status of power lines to identify the proper operation of the transmission system. If the forecasted values show a possible congestion, the TSO evaluates the required resources to solve it and communicates the needed DER settings to the DSO, according to the flow represented in Figure 9.3 and the steps detailed below.

1) **Congestion Detection Function**: This function evaluates the loading level of the lines at the transmission level. If one or more congested lines are identified, it activates the optimization function. In the first
2) **Optimization Function (OF)**: This function aims to calculate the minimum active power variation to apply at each busbar [33] at both transmission and distribution levels to mitigate the detected congestion. The optimal solution

1 SimBench is a German R&D project which provides benchmark data sets for electric network analysis, planning, and operation, https://simbench.de.

calculated is also communicated at the distribution level [34]. Indeed, using the assets at the distribution level requires the DSO authorization. If the DSO authorizes the resources' use to provide the required ancillary services, the dispatching function is activated.

In detail, the OF is a multi-objective function aiming to calculate the minimum active power variation to apply at each busbar while maximizing the use of DG/RES in ancillary service provision. The use of DG/RES to provide ancillary services – considered in the OF – is formulated as follows:

$$\sum_{i=0}^{N_{\text{busbar}}} \frac{P_{i_{\text{optimal\_solution}}}}{P_{i_{\text{gen}}}} \times 100 \tag{9.1}$$

where N_{busbar} is the total number of buses, $P_{i_{\text{optimal\_solution}}}$ is the optimal active power variation at busbar i, and $P_{i_{\text{gen}}}$ is the total active power provided by generation resources at busbar i. The optimal active power variation at each busbar is obtained by minimizing the OF formulated below:

$$f = \min\left(\sum_{i=0}^{N_{\text{busbar}}} \left(\Delta P^{+}_{\text{busbar}_i} + \Delta P^{-}_{\text{busbar}_i}\right)\right) \tag{9.2}$$

where

$\Delta P^{+}_{\text{busbar}_i}$ is the positive active power variation at busbar i;

$\Delta P^{-}_{\text{busbar}_i}$ is the negative active power variation at busbar i; and

N_{busbar} is the total number of busbars.

The main constraint in the OP is formulated as follows and represents the difference between the line rating current and the measured line current based on the active and reactive power variation of the resources:

$$I_j - \frac{\sqrt{\left(P^0{}_j + \frac{\partial P_j}{\partial P_i}.\Delta P_{\text{busbar}_i} + \frac{\partial P_j}{\partial Q_i}.\Delta Q_{\text{busbar}_i}\right)^2 + \left(Q^0{}_j + \frac{\partial Q_j}{\partial P_i}.\Delta P_{\text{busbar}_i} + \frac{\partial Q_j}{\partial Q_i}.\Delta Q_{\text{busbar}_i}\right)^2}}{\sqrt{3}.\left(U^0{}_j + U_{\text{Nom}_j}.\left(\frac{\partial u_j}{\partial P_i}.\Delta P_{\text{busbar}_i} + \frac{\partial u_j}{\partial Q_i}.\Delta Q_{\text{busbar}_i}\right)\right)} \geq 0 \;\; \forall i,j \tag{9.3}$$

where

I_j is the rating current value of line j;

$P^0{}_j$ is the active power initial condition that flows through line j;

$Q^0{}_j$ is the reactive power initial condition that flows through line j;

P_j is the active power that flows through line j;

Q_j is the reactive power that flows through line j;

$\frac{\partial P_j}{\partial P_i}$ is the effect on active power of the injection of ΔP at busbar i for line j;

$\frac{\partial P_j}{\partial Q_i}$ is the effect on active power of the injection of ΔQ at busbar i for line j;

$\frac{\partial Q_j}{\partial P_i}$ is the effect on reactive power of the injection of ΔP at busbar i for line j;

$\frac{\partial Q_j}{\partial Q_i}$ is the effect on reactive power of the injection of ΔQ at busbar i for line j;

$U^0{}_j$ is the voltage initial condition for line j;

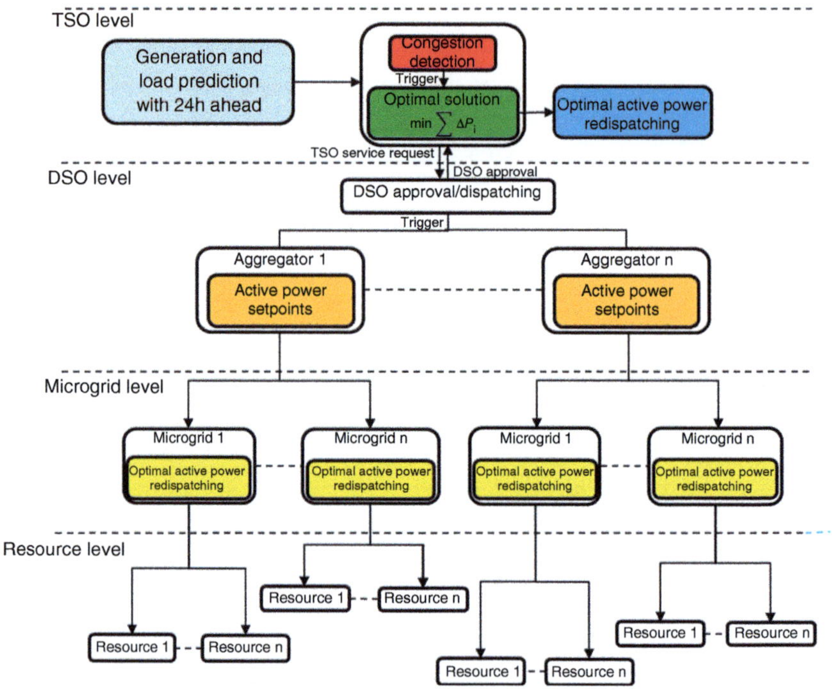

Figure 9.3 Proposed scheme to solve the congestion issue at the TSO level.

U_{Nom_j} is the nominal voltage for line j;

$u_j = \frac{U_j}{U_{N_j}}$, where U_j is the voltage of the line j.

3) **Dispatching Function**: This function sends the optimal active power set point to the aggregators. This set point is evaluated by considering the active/reactive power flexibility information exchanged among the aggregators and the DSO and the actual availability of aggregators to provide ancillary services [35].
4) **Re-dispatching Function**: Based on the information received by the dispatching function at point (3), each aggregator calculates the active power set point to assign to each microgrid. This set point is compared with the actual DER availability into the microgrids. Thus, the aggregator collects all the active/reactive power flexibility information for each controlled microgrid in order to determine the total aggregator flexibility to communicate to the DSO and – through the DSO – to the TSO [36].

9.3.3 Congestion Issues in the DSO Domain

In this scheme, the DSO checks – at each time step (15 minutes) – the status of power lines to identify the proper operation of the distribution system. If the forecasted scenario shows a possible congestion, the DSO evaluates the required resources to solve it and communicates the needed DER set points to the aggregators, according to the diagram represented in Figure 9.4 and the steps detailed below.

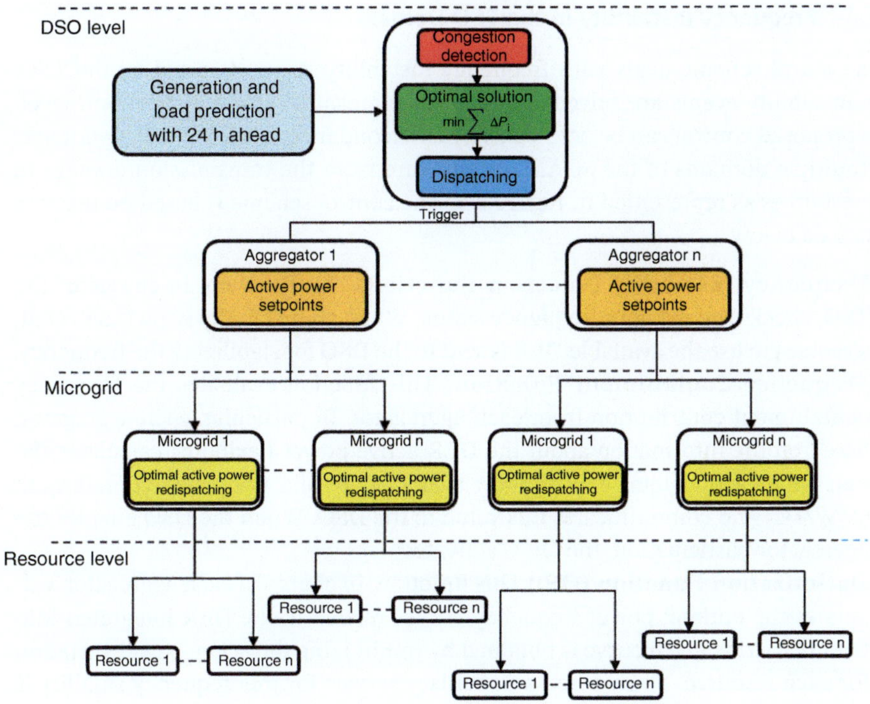

Figure 9.4 Proposed scheme to solve congestion issue at the DSO level.

1) **Congestion Detection Function**: This function checks the line loading values. When the loading values higher than the pre-set loading thresholds are identified, the OF is activated.
2) **Optimization Function (OF)**: This function aims to calculate the minimum active power variation to apply at each busbar [33] at the distribution level to mitigate the detected congestions. The dispatching function is activated. As in the previous case, the OF is a multi-objective function aiming to calculate the minimum active power variation to apply at each busbar, while maximizing the use of DG/RES in ancillary service provision.
3) **Dispatching Function**: This function evaluates how distributing the required optimal active power set point defined at point (2) among the aggregators. This set point is evaluated by considering the active/reactive power flexibility information exchanged among the aggregators and the DSO and the actual availability of aggregators to provide ancillary services.
4) **Re-dispatching Function**: Based on the information received by the dispatching function at point (3), each aggregator calculates the active power set point to assign to each microgrid. This set point is compared with the actual DER availability into the microgrid. Thus, the aggregator collects all the active/reactive power flexibility information for each controlled microgrid in order to determine the total aggregator flexibility to communicate to the DSO.

9.3.4 Frequency Instability in the TSO Domain

This control scheme deals with frequency instability issues detected by the TSO. The instability events are solved through the aggregators at the distribution level. The proposed control can be defined as a fast secondary control [37, 38], and it acts on multiple domains of the SGAM architecture from the transmission domain to the resources as represented in Figure 9.5. The control scheme is based on the step discussed below.

1) **Frequency Instability Detection Function**: This function, in charge of the TSO, checks the network frequency status. When an instability issue is detected, a request to use the available DER is sent to the DSO for stabilizing the frequency.
2) **Frequency Containment Function**: This function evaluates the frequency containment contribution from each aggregator. In particular, each aggregator, based on the information about the DER active power flexibility, calculates the maximum actual total contribution available for the frequency containment (MW/Hz) and communicates this value to the DSO. When the DSO enables the aggregator participation, the OF is activated.
3) **Optimization Function (OF)**: This function, in charge of each aggregator, calculates the optimal power frequency curves related to the DER integrated into the microgrids. Each curve is obtained by minimizing the active power variation for each resource. Thus, when an ancillary service for the frequency stability is required, each resource provides its contribution based on its own optimal power frequency curve.

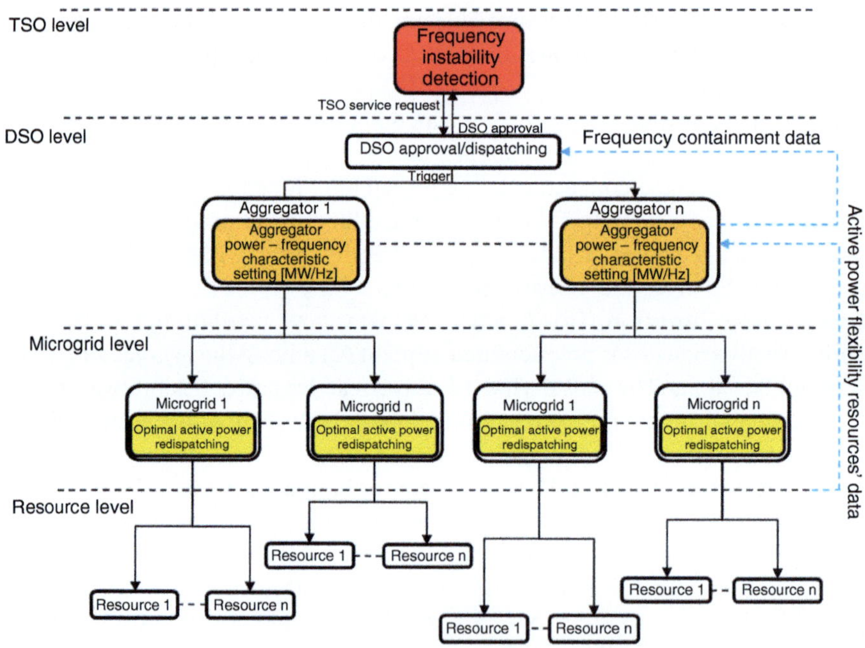

Figure 9.5 Proposed scheme to solve frequency instability issues.

9.4 Case Studies

In order to investigate the proposed control scheme effectiveness, three case studies are simulated and the corresponding scenarios are defined as listed below:

1) A congested scenario is defined at the transmission level.
2) A congested scenario is defined at the distribution level.
3) For each time step of the simulation, frequency instability events are simulated to evaluate the frequency stability ancillary service.

9.4.1 Case Study 1: Congestion Events at the Transmission Level

This case study analyses a congested scenario at the transmission level, which has been solved by using the optimization technique based on an exact mathematical model. In detail, the sequential linear-quadratic programming method is used to find the optimal solution in terms of minimum active power variation in the grid to mitigate the congestion event detected. The presented method represents a tool to prevent the grid power congested issues. For this purposes, a 24 hours ahead forecast for generation and load has been defined with a 15 minute time step. Two specific time steps (87 and 88) lead a congested issue at the transmission level. Figure 9.6 shows the "Congestion Detection" function output. The congestion analysis highlights two possible issues at time steps 87 (90.35%) and 88 (94.55%) with a maximum loading line value of over 90%.

As described in the control management scheme, when congestion criticalities are detected at the transmission level, the OF is activated to find the optimal solution. All grid power resources are involved to solve the detected congestion, but for the resources at the distribution level, the positive feedback by DSO is needed. Figure 9.7 shows the maximum line loading status before and after the control logic action.

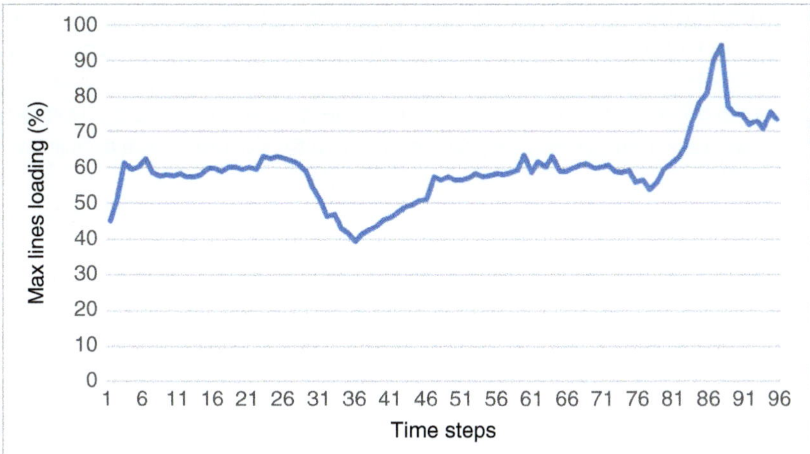

Figure 9.6 Congestion detection outcomes in case study 1.

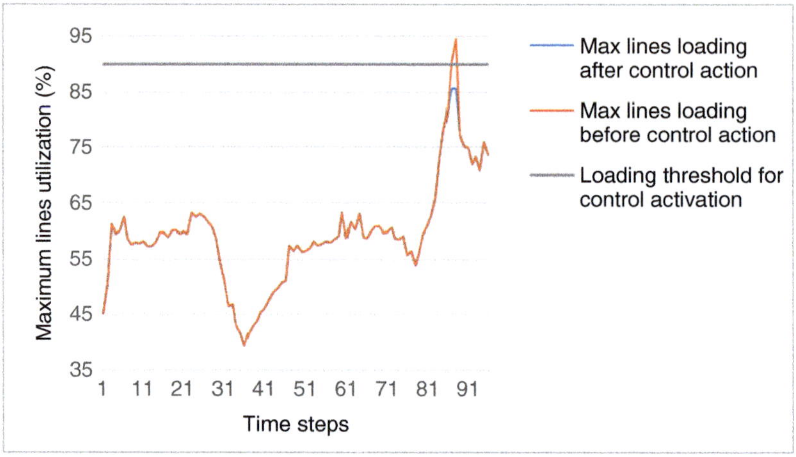

Figure 9.7 Maximum line loading status before and after the control action in case study 1.

The active power contribution from microgrids, managed by the aggregators based on the optimal solution information, leads to a maximum loading line value decreasing at the time steps with congestion problems detected by the "Congestion Detected" function. Moreover, a larger usage of DER/RES involved for the ancillary service provision is obtained. Indeed, the optimization function is developed to solve issues prioritizing the use of the DER present in the grid. This approach ensures that the contribution to ancillary services coming from local DER is maximized. In the specific case, an average increase of 3% in DER use has been gained thanks to the activation of the proposed control logic.

9.4.2 Case Study 2: Congestion Events at the Distribution Level

This use case investigates the proposed method effectiveness to solve grid congestion problems at the distribution level using only the resources available at this domain. The optimization technique used is an exact mathematical model with a sequential linear-quadratic programming method. As all the proposed methods are operational planning tools to prevent instability grid power problems, also in this case, the same 24 hours ahead forecast for generation and load defined for the previous use case has been used. In this case, two specific time steps (65 and 66) have been set to lead congested issues at the distribution level. Figure 9.8 shows the "Congestion Detection" function output.

In this case, the optimal solution is calculated at the distribution level, and the minimum variations in terms of active power to solve the detected congestion issues are communicated to aggregators based on their own active power flexibility information. Figure 9.9 shows the maximum line loading status before and after the control action.

Also in this case, the proposed methodology is able to avoid the congestion issues calculating the optimal active power set point for the resources at the distribution level to solve the detected problems 24 hours ahead.

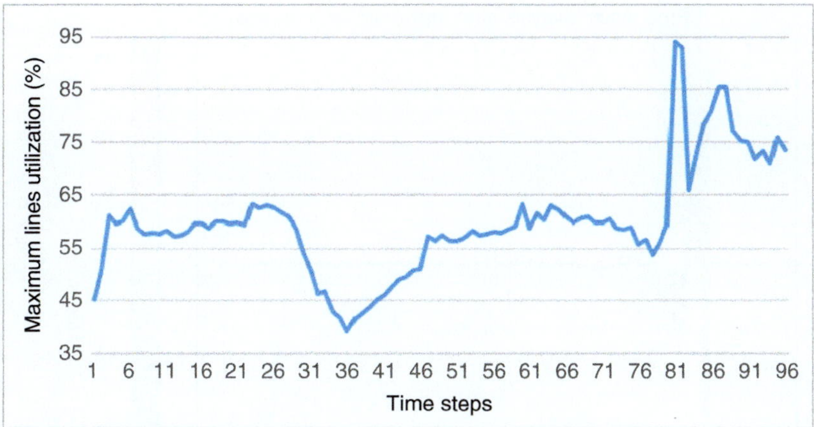

Figure 9.8 Congestion Detection outcomes in case study 2.

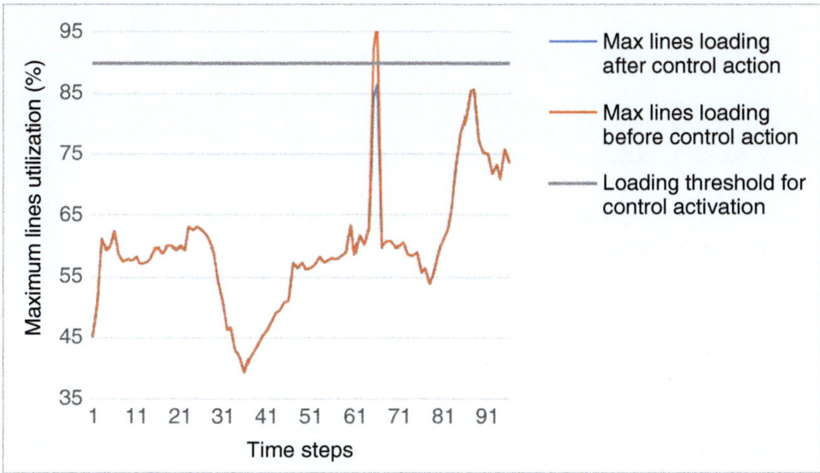

Figure 9.9 Maximum line loading status before and after the control action in case study 2.

9.4.3 Case Study 3: Frequency Instability Issues

This use case aims to show the effectiveness of the methodology proposed in Section 9.3.4 for frequency control. The 24 hours ahead generation and load forecasting employed in the previous use cases have been used by neglecting the congestion issues. Moreover, per each time step, an instability event has been defined. In particular, the maximum load per each time step has been disconnected to simulate a frequency instability event. The aggregators calculate the power-frequency curve per each controlled resource, in order to minimize, at each time step, the frequency error, such that it is lower than a set value. Figure 9.10 shows the frequency behavior after the maximum load outage.

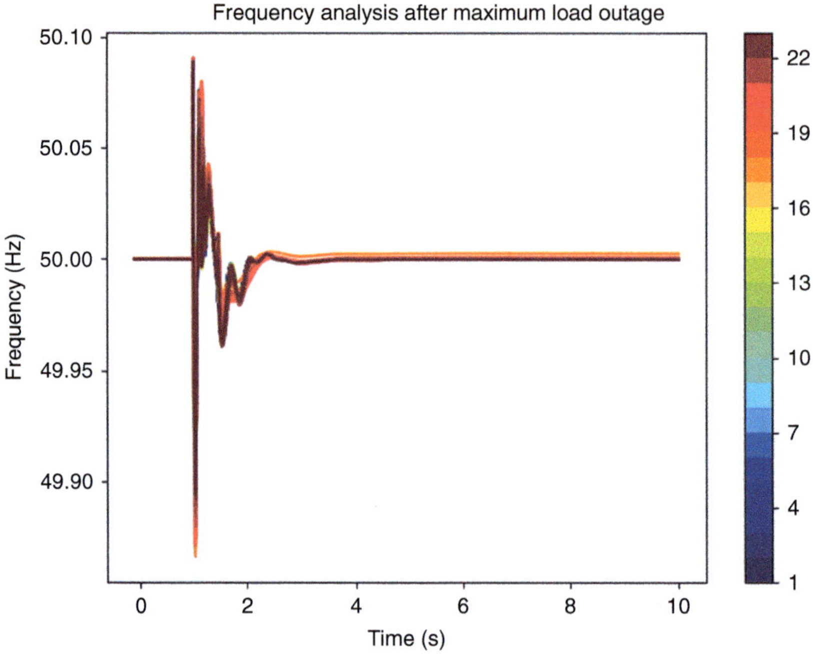

Figure 9.10 Frequency analysis after maximum load outage in case study 3.

It can be observed that, thanks to the control logic action, the frequency is well contained very quickly at each time step.

9.5 Conclusions

The "smartization" of electricity networks is a complex process both because of the difficulties associated with the problems of technical nature and "social acceptability" for the end users and the need for changes and additions to the current regulatory framework, which is not yet fully defined and consolidated. The length and complexity of the supply chain involved, made up of many subjects with different interests, also add difficulty in the identification of a single business model for all the players involved. Also in the light of these premises, the smart grids are still widespread only at the level of prototype development projects. Experiments on pilot projects are, in fact, underway in many European countries and worldwide.

The different definitions of smart grids available in the technical literature always imply, more or less explicitly, the need to migrate from the typically unidirectional and centralized management of traditional networks to the decentralized and bidirectional management of electric networks through the integration of ICT systems. In this perspective, the smart grids can be considered as electrical networks based on an IT-digital management system that implements control logics designed to coordinate, in an optimized manner, the energy flows from the high number of DERs present in the network. A set of communication equipment (Communications

Layer) enables communication between all the devices connected to the electrical system (Power System Layer) in order to send the implementation commands to the resources involved, according to the management criteria of the IT applications of the Digital Layer.

In such a context, smart grids perform many functionalities to meet several requirements from actors operating in different domains and zones of the SGAM in order to develop the best balance among the different targets. This chapter presents a set of advanced control logics for the smart management of DER with the aim to provide ancillary services in the context of smart grids by involving the resources belonging to the five domains of the SGAM. The analyzed case studies highlight the effectiveness of the proposed methods by also reflecting innovative business models introduced by the current trends in regulations through promoting a continuous cooperation between TSOs and DSOs as well as a smart management of DER through the key figure of the aggregators.

List of Abbreviations

AMI	advanced metering infrastructure
DC	direct current
DER	distributed energy resources
DG	distributed generation
DR	demand response
DSO	distribution system operator
FACTS	flexible alternating current transmission systems
HVAC	heating, ventilation, and air conditioning
HVDC	high-voltage direct current
ICT	information and communication technologies
IEA	International Energy Agency
OF	objective function
OP	optimization problem
P2P	peer-to-peer
RES	renewable energy sources
SGAM	smart grid reference architecture
TSO	transmission system operator
WAAPCA	wide-area adaptive protection, control, and automation
WAMS	wide-area monitoring system
WASA	wide-area situation consciousness

References

1 IEA (2020). World Energy Outlook, IEA, Paris. https://www.iea.org/reports/world-energy-outlook-2020.

2 IRENA, International Renewable Energy Agency (2015). *Renewable Energy, Integration in Power Grids*. Technology Brief.

3 Cleary, K. and Palmer, K. (2020). Resources for the Future, Renewables 101: Integrating Renewable Energy Resources into the Grid. https://media.rff.org/documents/Renewables_101.pdf.
4 IRENA, International Renewable Energy Agency (2020). *Global Renewables Outlook: Energy Transformation 2050*. IRENA.
5 Ullah, K., Basit, A., Ullah, Z. et al. (2021). Automatic generation control strategies in conventional and modern power systems: a comprehensive overview. *Energies* 14 (9): 2376.
6 Falvo, M.C., Mareri, F., and Cruciani, C. (2020). Active distribution grids: observability and RES-based DG forecasting. In: *2020 IEEE International Conference on Environment and Electrical Engineering and 2020 IEEE Industrial and Commercial Power Systems Europe (EEEIC/I&CPS Europe)*, 1–5. IEEE.
7 Holjevac, N., Baškarad, T., Đaković, J. et al. (2021). Challenges of high renewable energy sources integration in power systems—the case of croatia. *Energies* 14 (4): 1047.
8 Nikoobakht, A., Aghaei, J., Shafie-khah, M., and Catalão, J.P.S. (2020). Allocation of fast-acting energy storage systems in transmission grids with high renewable generation. *IEEE Trans. Sustain. Energy* 11 (3): 1728–1738.
9 Li, H., Lu, Z., Qiao, Y. et al. (2021). The flexibility test system for studies of variable renewable energy resources. *IEEE Trans. Power Syst.* 36 (2): 1526–1536.
10 Di Somma, M., Falvo, M.C. et al. Integration of renewable energy source in transmission grids: issues and perspectives. In: *2021 IEEE International Conference on Environment and Electrical Engineering and 2021 IEEE Industrial and Commercial Power Systems Europe (EEEIC/I&CPS Europe)*, 1–8. IEEE.
11 Di Somma, M., Ciavarella, R. et al. Innovative integrated operation planning tool for the current and future 2030+ European power grid. In: *CIRED 2020 Berlin Workshop (CIRED 2020)*, vol. 2020, 330–333. IET.
12 International Energy Agency (2011). *Technology Roadmap: Smart Grids*. IEA.
13 Litos Strategic Communication (2008). *The Smart Grid: An Introduction*. Washington, DC: Department of Energy (DOE).
14 Tuballa, M.L. and Abundo, M.L. (2016). A review of the development of Smart Grid technologies. *Renew. Sustain. Energy Rev.* 59: 710–725.
15 Hamiltilon, B.A. and Miller, J. (2010). *Understanding the Benefits of the Smart Grid*. National Energy Technology Laboratory.
16 Alnasseir, J., Alcharea, R., and Almaghout, F. (2021). Improving the stability of smart grids by using flexible alternating current transmission systems (FACTS). In: *12th International Renewable Engineering Conference (IREC), Amman, Jordan*, 1–3. IEEE.
17 Gomis-Bellmunt, O., Sau-Bassols, J., Prieto-Araujo, E., and Cheah-Mane, M. (2020). Flexible converters for meshed HVDC grids: from flexible AC transmission systems (FACTS) to flexible DC grids. *IEEE Trans. Power Deliv.* 35 (1): 2–15.
18 (2017). IEEE guide for establishing basic requirements for high-voltage direct-current transmission protection and control equipment. In: *IEEE Std 1899–2017*, 1–47. IEEE https://doi.org/10.1109/IEEESTD.2017.7959586.

19 Pratt, R.G., Kintner-Meyer, M.C.W., et al. (2010), The Smart Grid: An Estimation of the Energy and CO_2 Benefits. http://energyenvironment.pnnl.gov/news/pdf/PNNL-19112_Revision_1_Final.pdf.

20 CEN-CENELEC-ETSI Smart Grid Coordination Group Smart Grid Reference Architecture. (2012). https://ec.europa.eu/energy/sites/ener/files/documents/xpert_group1_reference_architecture.pdf.

21 Lin, J., Magnago, F., and Alemany, J.M. (2018). Optimization methods applied to power systems: current practices and challenges. In: *Classical and Recent Aspects of Power System Optimization*, 1–18. Academic Press.

22 Chen, Y.K., Wu, Y.C., Song, C.C., and Chen, Y.S. (2013). Design and implementation of energy management system with fuzzy control for dc microgrid systems. *IEEE Trans. Power Electron.* 28 (4): 1563–1570.

23 Kanwar, A., Rodríguez, D.I.H., von Appen, J., and Braun, M. (2013). *A Comparative Study of Optimization-and Rule-Based Control for Microgrid Operation.* Universitätsbibliothek Dortmund.

24 Directive (EU) 2018/2001 of the European Parliament and of the Council of 11 December 2018 on the promotion of the use of energy from renewable sources. https://eur-lex.europa.eu/legal-content/EN/TXT/?uri=uriserv:OJ.L_.2018.328.01.0082.01.ENG.

25 ENTSO-E (2015). *General Guidelines for Reinforcing the Cooperation Between TSOs and DSOs*. ENTSO-E.

26 Zegers, A. and Brunner, H. (2014). TSO-DSO interaction: An Overview of current interaction between transmission and distribution system operators and an assessment of their cooperation in Smart Grids. ISGAN (International Smart Grid Action Network) Discussion Paper.

27 Hadush, S.Y. and Meeus, L. (2018). DSO-TSO cooperation issues and solutions for distribution grid congestion management. *Energy Policy* 120: 610–621.

28 Gerard, H., Puente, E.I.R., and Six, D. (2018). Coordination between transmission and distribution system operators in the electricity sector: a conceptual framework. *Util. Policy* 50: 40–48.

29 Patsalides, M., Papadimitriou, C.N. et al. (2020). Frequency stability evaluation in low inertia systems utilizing smart hierarchical controllers. *Energies* 13 (13): 3506.

30 IRENA (2019). *Innovation Landscape Brief: Aggregators.* Abu Dhabi: International Renewable Energy Agency.

31 Demoulias, C.S., Malamaki, K.N.D. et al. (2020). Ancillary services offered by distributed renewable energy sources at the distribution grid level: an attempt at proper definition and quantification. *Appl. Sci.* 10 (20): 7106.

32 Oureilidis, K., Malamaki, K.N. et al. (2020). Ancillary services market design in distribution networks: review and identification of barriers. *Energies* 13 (4): 917.

33 Ciavarella, R., Di Somma, M., Graditi, G., and Valenti, M. (2019). *Congestion Management in Distribution Grid Networks Through Active Power Control of Flexible Distributed Energy Resources*, 1–6. Milan, Italy: IEEE Milan PowerTech.

34 Graditi, G., Ciavarella, R., Somma, M.D., and Valenti, M. (2019). A control strategy for participation of DSO flexible resources in TSO ancillary services provision. In: *2019 International Conference on Clean Electrical Power (ICCEP), Otranto, Italy*, 586–592. IEEE.

35 Smart Grid Task Force. (2015). Regulatory Recommendations for the Deployment of Flexibility -EG3. https://ec.europa.eu/energy/sites/ener/files/documents/EG3 Final.

36 Van den Bergh, K., Couckuyt, D., Delarue, E., and D'haeseleer, W. (2015). Redispatching in an interconnected electricity system with high renewables penetration. *Electric Power Syst. Res.* 127: 64–72.

37 Ciavarella, R., Graditi, G., Valenti, M., and Strasser, T.I. (2018). Innovative frequency controls for intelligent power systems. In: *International Symposium on Power Electronics, Electrical Drives, Automation and Motion (SPEEDAM), Amalfi, Italy*, 656–660. IEEE.

38 Kosmecki, M., Rink, R. et al. (2021). A methodology for provision of frequency stability in operation planning of low inertia power systems. *Energies* 14 (3): 737.

10

Nearly Zero-Energy and Positive-Energy Buildings: Status and Trends

Denia Kolokotsa[1], Gloria Pignatta[2], and Giulia Ulpiani[3]

[1] *Chemical and Environmental Engineering School, Technical University of Crete, Kounoupidiana, GR 73100 Chania, Crete, Greece*
[2] *University of New South Wales (UNSW) Sydney, School of Built Environment, University Mall, Kensington, NSW 2033, Australia*
[3] *University of New South Wales (UNSW) Sydney, Climate Change Research Centre, Gate 11 Botany St Library Walk, Sydney, NSW 2052, Australia*

10.1 Introduction

As the demand for energy increases worldwide in proportions with the level of urbanization [1, 2], environmental pollution and sustainability issues are becoming alarming.

The conventional production of energy from non-renewable sources, such as fossil fuels, largely contributes to the release of CO_2 into the atmosphere. To limit such emissions and mitigate their effect on the environment, a huge effort is in place worldwide to replace conventional systems for energy generation with alternative renewable solutions.

Decarbonizing the built environment is one of the main goals that the recent energy policies worldwide aim to achieve to possibly invert the trend in increasing emissions. Indeed, the building sector is a major consumer of energy and a complex polluter [3]. Conventional buildings (either residential, commercial, or industrial) generally fail at meeting sustainability criteria because of the low energy performance of their envelopes.

The use of passive energy solutions for building envelope applications [4] and other conservative energy strategies [5] is frequently proposed. Among those strategies, the demand response (DR) strategy with precooling is especially noteworthy, given its efficiency in shaving the peak load. The scientific community is investigating its potential and best condition for application in different countries and climate conditions [6].

Another approach involves the use of renewable sources to produce clean energy thanks to building integrated renewable energy technologies [5, 7, 8] or district-scale energy generation systems [9, 10]. This approach allows covering the building energy consumption, which cannot be otherwise saved, with clean energy

Technologies for Integrated Energy Systems and Networks, First Edition.
Edited by Giorgio Graditi and Marialaura Di Somma.

produced with minimal emissions of greenhouse gas (GHG) and conventional air pollution.

A third approach deals with the implementation of smart home energy management systems (SHEMSs) [11, 12] or efficient demand side management (DSM) and load control [13–15] to reduce potential energy loss.

To enable a net zero-energy balance, the connection with the national electric grid is customary. Excellent results can be achieved when the excess energy generated at the building level is appropriately stored locally to cover future energy needs or sold and sent back to the national electric grid [16].

In recent years, a number of concepts have been developed around high-performance buildings primarily based on their annual energy balance and how it is obtained.

Section 10.1.1 provides a brief overview of the concepts of nearly zero-energy building (NZEB) and positive-energy building (PEB), reporting on the most common definitions, regulations, and standards. Section 10.1.2 discusses the design strategies that are typically implemented in high-performance buildings to (i) save energy, (ii) generate renewable energy at both building and district scale, and (iii) manage and control the energy loads. Solutions span from materials selection to technical systems integration with examples of recent advancements. Section 10.2 introduces some case studies and research projects conducted in Europe and dealing with NZEB and PEB by touching on challenges, barriers, drivers, and best practices. Section 10.3 draws conclusive remarks and pathways.

10.1.1 Concept of Nearly Zero- and Positive-Energy Buildings

According to Torcellini et al. [17], the concept of zero-energy building (ZEB) was born out of the idea that the annual energy requirements of a building, featuring energy-efficient solutions, can be covered with energy that has been generated locally (on-site within the building footprint), at an affordable price, with non-polluting and renewable sources. Reducing the building load is the core target of a ZEB and can be achieved with specific design strategies that should remain in place for the entire life of the building.

At the scale of the building, the ZEB concept has led to new concepts and definitions such as follows:

- NZEB, which is an individual high-performance building, whose annual energy consumption is close to zero [18].
- Net zero-energy building (nZEB), which is an individual building, whose on-site renewable energy production is comparable with the annual building energy consumption [19].
- PEB, which is an individual building, whose on-site renewable energy production covers and exceeds its energy needs over the course of a year [20].

10.1.1.1 Definitions, Regulations, and Standards

The nearly zero, net zero, and positive-energy concepts have been introduced globally in building codes and standards to provide streamlined design guidelines toward improved building energy efficiency and environmental performance [21]. The advancement and the increased affordability of renewable energy technologies in recent years have triggered the practical implementation of nZEBs and PEBs, with a variety of virtuous examples spread around the world [20].

Despite this, a clear understanding and universal definition for ZEB and PEB with associated energy and environmental targets have not been identified and recognized yet at the international level. A plethora of definitions has been proposed on account of different energy consumption and energy generation goals, boundary conditions, key performance indicators (KPIs), and country's political targets [22]. This fragmentation hampers any straightforward comparison.

The objective and concept of NZEB have been defined in 2010 in the European Energy Performance of Buildings Directive (EPBD) Article 2(2) [18] as "a building that has very high energy performance, whereby the nearly zero or very low amount of energy required should be covered to a very significant extent by energy from renewable sources, including energy from renewable sources produced on-site or nearby."

In the same year, Pless and Torcellini [23] defined the concept of nZEB as "a residential or commercial building with greatly reduced energy needs through efficiency gains such that the balance of energy needs can be supplied with renewable technologies." and provided a useful classification system based on the type of the renewable energy supply option.

The same concept has been introduced and extended by Hernandez and Kenny [24] by accounting for the building embodied energy and the annual energy used during the operation phase of the building. This resulted in a new term, the life cycle zero-energy building (LC-ZEB) (and its classification as nearly, net, and positive [25]), defined as a building "where the primary energy used in the building in operation plus the energy embodied within its constituent materials and systems, including energy generating ones, over the life of the building is equal to or less than the energy produced by its renewable energy systems within the building over their lifetime."

In 2012, Sartori et al. [26] provided the following simple equation (Eq. (10.1)) to calculate the nZEB balance that represents the core concept of a general nZEB definition:

$$\text{nZEB balance} : \text{weighted supply} - \text{weighted demand} = 0 \tag{10.1}$$

where the weighted supply and demand may be expressed with different units of measure (e.g. kWh, CO_2, etc.).

In 2015, the US Department of Energy (DOE) Building Technologies Office defined the ZEB as "an energy-efficient building where, on a source energy basis, the actual annual delivered energy is less than or equal to the on-site renewable exported energy" [27].

In Norway, a variety of ZEB definitions that range from including just the base building load to including plug loads, embodied, and end-of-life emissions have been introduced and presented by Fufa et al. in 2016 [28].

A most recent contribution toward a collection of existing NZEB definitions and essential performance indicators has been provided by Taherahmadi et al. in 2021 [25]. In this study, it is explained that each definition corresponds to a specific design approach and different calculations and methodologies. The authors, therefore, advise specifying the associated definition for each new ZEB design so as to facilitate energy performance comparisons and the measurement of the achieved targets.

At the international level, a number of energy and climate regulations have introduced new zero-energy definitions and targets in response to the global climate change agreement adopted at the Paris Climate Conference (COP21) in 2015 [29], which several countries have committed to.

According to the objective of the European Green Deal action plan, Europe should achieve climate neutrality by 2050 [30, 31]. The new European Climate Law has been introduced to transform this political commitment into a legal obligation [32], whereas the latest EPBD dated 2018 [33] was ratified to formalize ways to meet the new long-term climate target for 2050. The Directive is currently under revision and should be recast by the end of 2021. In particular, the elements that are under revision include the mandatory minimum energy performance standards, the circularity and whole life cycle carbon emission, the harmonization of the energy performance certificates (EPCs) and building renovation passports, and the definition of "deep renovation."

At present, all new buildings in Europe are NZEBs, and most of them will release the majority of their total GHG over their 50-year life span before the occupancy and use phase. This is therefore encouraging new policies and initiatives (e.g. circularity and digitalization) to take a step forward in the achievement of the long-term climate neutrality goal.

Looking at other international zero-energy targets, the US DOE Building Technologies Office has defined, as a strategic goal, the achievement of "marketable zero energy homes in 2020 and commercial zero energy buildings in 2025" [34], while the Asia-Pacific Economic Cooperation (APEC) has set a dual goal consisting of reducing energy intensity by 45% of 2005 levels by 2035 and doubling the share of renewable energy in the energy mix between 2010 and 2030 to face the continuous escalation of building energy consumption in the Asia-Pacific region [35].

The most important aspects of nZEBs, from the concept to the framework and roadmap for their implementation, are presented and discussed by Attia in 2018 in a comprehensive book that provides the reader with a multidisciplinary perspective useful to inform design decisions [36].

10.1.2 Overview of Design Strategies

Energy conscious design strategies to achieve high-performance buildings employ energy-efficient solutions and renewable energy systems (RESs) to meet the energy and environmental targets. The following sections briefly discuss the energy conservation (Section 10.1.2.1), energy generations (Section 10.1.2.2), and building automation strategies (Section 10.1.2.3).

10.1.2.1 Energy Conservation Strategies

The energy consumption is steadily increasing because of the climatic conditions changes and extreme weather variations, especially in the urban context [37–39]. This fact poses a significant challenge for the ZEB target while predictions of future energy consumption should take into account the extreme weather conditions that might apply in each area [22, 40].

Various energy technologies are mature and can be considered for the improvement of energy efficiency and indoor comfort improvement in buildings. These technologies may be distinguished in the following basic categories:

- Measures for the improvement of the building's envelope (e.g. addition or improvement of insulation, change of color, placement of heat-insulating door and window frames, increase of thermal mass, building shaping, and super-insulated building envelopes).
- Incorporation of high-efficiency heating and cooling equipment (e.g. AC equipment with higher energy efficiency ratio [EER] and high-efficiency condensing boilers).
- Use of renewables (e.g. solar thermal systems, building integrated photovoltaics [BiPV], and hybrid systems).
- Use of "intelligent" energy management, i.e. advanced sensors, energy control (zone heating and cooling), predictive control, and monitoring systems.
- Measures for the improvement of the indoor comfort conditions in parallel with minimization of the energy requirements (e.g. increase of the ventilation rate, use of mechanical ventilation with heat recovery, and improvement of boilers and air-conditioning [AC] efficiency use of multi-functional equipment, i.e. integrated water heating with space cooling).
- Use of energy-efficient appliances, compact fluorescent lighting, light-emitting diode (LED), etc.

Taking into consideration the recent developments, the present section is focusing on innovative materials for the building envelope and smart heating, ventilation, and air conditioning (HVAC) systems.

10.1.2.1.1 Innovative Materials for Buildings' Envelope

Recent developments and innovation actions in the materials for the building envelope provide a wide list of energy-saving building envelope components. Trends show that the main areas of recent developments can be distinguished in the following categories:

- Prefabricated building envelope elements.
- Building envelope components with enhanced thermal and optical properties.

Research activities for prefabricated building envelope elements include thermal/acoustic insulation, low moisture content, low air leakages, resistance to water penetration, daylight and solar shading, heat recovery ventilation, fire and burglary protection, productive outfitting for renewable energy production, and rainwater collection. These components intra-communicate via sensors (e.g. for climate

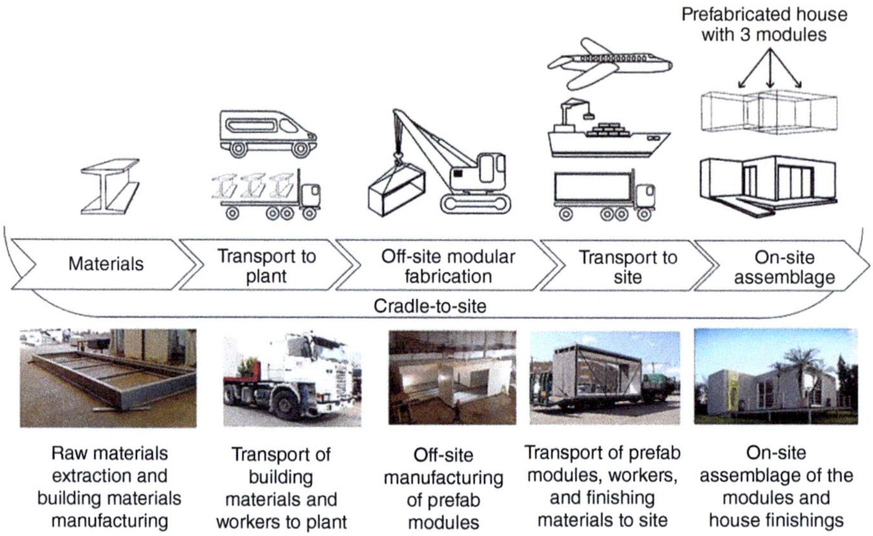

Figure 10.1 A prefabricated house with reduced carbon emissions in a cradle-to-site analysis. Source: Tavares et al. [44]/with permission of Elsevier.

control), with "plug&play" connections (mechanical, hydraulic, air, electric, and prefab airtight joints), that can be individually tailored and customized while can be mass produced [41]. In addition, 3D printing has expanded considerably the design of building envelope components with multifunctional characteristics and green infrastructure integration [42]. An example is the 3D printed vertical concrete green wall system (3D-VtGW) system proposed in Ref. [42] with a considerable reduction of wall surface combined with evapotranspiration and heat storage from the soil. Another study proposes prefabricated elements to address specific requirements with respect to their structural, hygrothermal, energy, fire, acoustical, and environmental performance incorporating phase change materials (PCMs) for increased latent thermal heat storage [43].

Efforts are performed to analyze the embodied energy and GHG emissions of fully prefabricated modular houses [44] and industrial buildings [45], showing that the selection of materials and transportation plays the most significant role. The impacts of transportation (of modules, workers, and finishing materials) vary significantly for the various building final locations and can be sizable for overseas locations, which can jeopardize the potential benefits of modular prefabrication (Figure 10.1).

New concrete materials and concrete-based building envelope components are conceived to provide better acoustic insulation/absorption, fire resistance, dimensional stability, and indoor air quality optimization, while lowering the embodied energy and improving the insulation properties. Sustainable aerogel-based insulation products and systems (thermal coating, plaster, patching filler, render, and clay bricks) are proposed by various researchers. These products improve the insulation properties at the component level by more than 25% and provide improved comfort and indoor air quality and increased fire safety, durability, and sustainability compared to the established materials [46–48]. Concerning insulation, aerogel-enhanced products are very promising materials that can be incorporated in plasters, lightweight concretes, blankets, and glazing systems. For example, the thermal conductivity of plasters with more than 80%vol aerogel is below 0.025 W/mK, a tenth of the respective value for traditional plasters, while mortars with more than 30%vol aerogel show a thermal conductivity as low as 0.23 W/mK [46].

The use of innovative coatings with different thermal and optical properties is also proposed. Cool materials are a well-established option known for the wide variety and versatility [49–52]. Cool materials have higher solar reflectance and infrared emission compared to conventional materials and therefore can better dissipate the absorbed heat. Cool materials can maintain lower temperatures as opposed to conventional building and paving materials [52–54]. Cool materials can help to maintain the proper indoor temperature, with a reduced demand for air conditioning by almost 20% [55].

Among them, daytime radiative cooling materials directly dissipate the terrestrial heat (~300 K heat source) into the outer space (~3 K heat sink) by strongly emitting in the mid-infrared wavelength windows, where the atmosphere is more transparent, and by strongly reflecting in the solar range. Daytime radiative coolers can get colder than the ambient even at peak hours. Ultra-large cooling in the order of 6–7 °C below the ambient air in the daytime has been experimentally demonstrated [56, 57]. In order to unlock the extraordinary cooling potential, fervent research is currently dedicated to optically optimal and scalable demonstrators and to overcome two major limitations: overcooling during cold periods [58] and cooling power impairment under humid, polluted, and crowded conditions [59].

10.1.2.1.2 Smart HVAC Systems and Services

Smart HVAC systems and services for indoor environment quality control have advanced considerably in the past decades with the evolution of information and communication technology (ICT) for the building sector [60]. Smart HVAC systems and services are an essential part of smart buildings. Advanced building energy management systems (BEMSs) can have a major impact on the design, operation, optimization, and control of energy-influencing building elements (e.g. HVAC, solar, fuel cells, combined heat and power [CHP], shading, and natural ventilation).

BEMSs have evolved considerably in the past decades, in parallel with the growing concern about energy efficiency requirements and the demand for environmentally friendly buildings. Modern customized building energy management solutions can be exploited to enable better visual, thermal comfort, and air quality control. Research efforts in this direction focus on advanced BEMSs, which can implement sophisticated algorithms capable of predicting and evaluating a range of alternatives in the way buildings exchange energy with the ambient environment and the grid. Nowadays, state-of-the-art BEMS techniques offer the potential for applying predictive control, which may contribute to 20–30% in the reduction of energy consumption [59, 61, 62] and equivalent operational cost savings. Prediction of energy demand is becoming increasingly effective as part of an overall energy management optimization process that could be deployed in the near future [61, 62]. Simultaneously, researchers are providing new scientific evidence on how the prediction of renewable energy production can increase its usability in building integrated applications and deal with the volatility of distributed energy resources (DER) and future microgrids.

DR offers the capability to apply changes in the energy usage by the consumers to alter their normal consumption patterns in response to changes in the energy pricing over time [63–65].

It tandems with the integration of smartness in buildings' services and systems as discussed in Section 10.1.2.3 [66, 67].

The smart building user is informed of the building's energy flows and provided with tools for the dynamic management of systems installed, e.g. to adjust indoor environment conditions according to his/her preferences or specific needs and control devices remotely. Furthermore, tools assisting users to optimize the energy performance of a smart building and at the same time minimize the cost of the energy bills are envisioned.

In this context, the goal of the efficient exchange of energy and information between the building and the grid in a way that is mutually beneficial must be facilitated. At the distribution level, the energy demand in buildings forms an important asset in terms of the collective power flexibility potential.

10.1.2.2 Energy Generation Strategies

The integration of renewables in buildings is one of the most important aspects of zero-energy and decarbonized building stock. BiPV systems have a great potential for architectural use thanks to their versatility and customizability in terms of colors, transparency, and design [68]. BiPV systems support the increase of renewable energy share and improve the building envelope's performance. Integrated solar inverter and storage systems offer capabilities such as maximum power tracking and storage control [69–71]. Significant research is performed in the area of storage control and storage systems either using batteries or other technologies [47].

Various efforts have dealt with optimizing the design and operation of building integrated renewables, thermal or electrical storage, and holistic energy management using a broad range of techniques. Combined solar- and wind-driven energy system is a breakthrough technology that entered the market of building-integrated RESs [72]. Wind solar renewable energy is integrated into the building envelope coupled with rainwater harvesters as depicted in Figure 10.2. The power generated by the renewables covers partly the energy requirements of the high-rise building depicted in Figure 10.2.

The integration of renewables in the built environment has some critical barriers. Lack of space is a very critical point for renewables at the urban level [25] because of the fact that most renewable investments in the urban environment are guided by incentives on installing renewable energy technologies at the *individual* level [74, 75], while stronger motivation is expected for investors to participate in renewable energy communities' (RECs) installations related to group projects. The initial investment cost is another considerable burden [25]. People living in low-income

Figure 10.2 Integration of renewables in the building envelope. Source: Chong et al. [73]/with permission of Elsevier.

households cannot even consider the possibility of installing extensive renewables. Therefore, matching energy production and consumption is absolutely necessary while it should be supported by increased awareness and users' integration via DR capabilities [76].

Small-level CHP plants at the neighborhood scale can change the way people perceive HVAC systems at the residential level [77] and may provide the necessary technological instruments to fight increased urban energy. Although small-scale CHP plants are not widely deployed, there are some very interesting case studies especially in Denmark where 52% of the electricity demand is met by CHP plants connected to district heating and cooling systems [78].

CHP plants can also be combined with different renewable cutting edge technologies. An example of such technologies is the high-concentration photovoltaics (HCPV) [79], which harness solar radiation with a limited number of components (no need for cells, optics, and heat exchangers). CHP plants combined with concentrated solar power can be a very promising solution because their multi-junction cells have efficiency larger than 40% and the thermal energy they produce is in the form of a hot fluid [80]. This potentiality stimulates the effort to develop hybrid PV/T technology for district-level, combining high-performance and reliability technologies.

Moreover, the combination of geothermal and solar thermal energy systems may support the achievement of the highest efficiency possible in heating and cooling renewable-based plants [81]. The main energy source is the thermal solar field, which turns solar energy into heat and the ground source heat pumps, which extract energy from boreholes in the ground, to be used for heating, cooling, and storage purposes. The excess energy can be stored under the ground to preserve it until needed.

The solar source typically generates much more energy than the geothermal source, but its main drawback is the inconstant supply associated with diurnal cycles and low-power irradiation during overcast days. The adoption of the sole geothermal energy has some inconveniences as well. For example, the efficiency of the system will decrease with time, as the ground temperature cools down because of its energy use.

The two sources can be associated to solve the drawbacks they feature individually. During most of the year, solar energy warms the flowing water at a higher temperature than that needed for heating the building. Not to waste the extra heat, the solar system pipes are redirected to the borehole. In this way, the warm water heats up the ground, thus increasing the geothermal system capacity to extract more energy, as well as keeping the ground temperature from cooling down. When the water circulating through the solar system has a similar temperature to that required to cover the building thermal load, this energy is used directly to heat up the heating system. That means, the solar circuitry will be split into multiple subsystems, one toward the borehole, another to the heat pump, and a third one directly connected with the indoor heating system.

10.1.2.2.1 Integration of Renewables in Buildings and Districts

The integration of renewable technologies in single buildings or districts of different scales implies specific challenges and benefits. Looking at the district scale, there is a huge difference in terms of the difficulties in meeting the energy target depending on whether the settlement is peri-urban, suburban, urban, or even central business district (CBD). This is due, for example, to the ratio of roofspace for on-site energy generation vs. conditioned floor area. When selecting the type of renewable energy generation technology, it is extremely important to consider the local climatic context and boundary conditions as these may limit the type and size of the RES that can be integrated into the building or within the district. Section 10.2.2 recalls this topic when discussing challenges, barriers, and benefits for the transition from individual NZEBs to positive-energy districts (PEDs).

10.1.2.3 Smart Readiness

High-performance buildings work in strong partnership with ICT and Internet of Things (IoT). As the global population grows and urban metabolism gets enriched with novel, high-tech nuances, the nearly zero- and positive-energy target becomes ever more connected to the question: how "programmable" are today's buildings and districts? In the following sections, the concepts of flexible and smart buildings are revised to bring this new energy-efficient and resilient dimension into the limelight.

10.1.2.3.1 Energy Flexibility Toward New Sustainability Paradigms

With the increasing penetration of air conditioning and intermittent renewables, achieving energy flexibility has become a priority to meet the goal of transforming the existing building stocks into NZEBs by 2050 [18]. Fortunately, the ICT sector has been moving fast toward innovation, thus providing a wide spectrum of tools for enhanced resilience, versatility, and connectivity [82].

Building energy flexibility (BEF) is a measure of the ability to shift energy use over time and shave peaks via DSM and load control [83]. The room and technical specificities of BEF can be identified following standardized ontologies by defining sources and limitations both at building and building cluster scale [84, 85]. BEF identification generally requires in situ measurements, especially when a variety of flexible home appliances (e.g. electric vehicles, solar panels, washing machines and dryers, dishwashers, and domestic hot water tanks) is concerned. Although limited in number, some pilot projects exemplify the process, as those performed in Belgium [86] and in the Netherlands [87]. Several indicators, especially for heat/thermal flexibility, have been developed, and many methodological approaches and modeling methods have been proposed, as described in Ref. [88] and further expanded in Ref. [89]. Some general conclusions can be drawn based on the current level of the acquired knowledge:

- Building's thermal insulation is a key player in the magnitude of BEF potential, especially in the heating season [90–93]. Low insulation levels come with high heat flexibility for limited periods of time, whereas high levels reduce the magnitude of heat flexibility and potential services to the power grid while reducing the time of activation of both heating and electric systems. Overall, insulation increases the inertia of a building to any adjustment in heat delivery (time constant) and decreases the amount of heat energy storable in the unit time (degree hour value) [90].
- Thermal energy storage and building thermal mass are also major BEF enablers, yet the capacity is extremely case specific (e.g. type of storage, HVAC, and climate) [94, 95]. Typical storage options include batteries, water, PCMs and thermochemical material tanks, and fuel switch [96, 97]. As for thermal mass, a general rule dictates that in the presence of a high window-to-wall ratio, as in modern design, the external wall thermal mass is not influential on the load shifting potential in comparison with the internal walls [98].
- Occupancy patterns and lifestyle are critical to unlock the BEF potential [94]. Beyond direct impacts on energy consumption and on the provision of smart utilities, people's comfort is a key BEF constraint and strongly depends on users' interaction with the technology itself [99, 100]. Indeed, static-occupancy assessment methods typically lead to a significant mismatch between predictions and actual energy consumption. Few papers attempt to track people's dynamic behavior, typically by co-simulation [101]. On the other side, potentially flexible users are a minority (~10%) compared to those who are unfamiliar with the BEF concepts. Those already adapted to passive house paradigms readily embrace BEF logics [102], while for most stakeholders, awareness-raising campaigns, policy and labeling improvement, novel financial incentives, and economic and regulatory frameworks are urgently needed [99, 103–106].

Buildings with a non-negligible BEF potential are called prosumers. The magnitude of flexibility is typically low for single prosumers. Therefore, once individual BEFs are characterized, aggregation is needed at cluster and city scale to pool and scale up the benefits and break markets' barriers. A variety of techniques has been investigated [107]: models that account for the diversity among prosumers are to be preferred over lumped approaches [108]. Buildings are grouped in archetypes through white box models and then aggregated at the desired scale by multiplying the archetype-specific energy demand by the number of buildings that belong to each archetype [109].

From a practical perspective, this aggregation is realized through pathways of interaction between individual buildings and (i) the physical environment, (ii) the civil infrastructure, (iii) other buildings, and (iv) vehicles [110, 111]. Such an intricate network is necessary for the design of ZEBs and districts but provides non-trivial challenges in planning and control [112].

The identification of BEF enabling actions is still underway and would help disclose feasible pathways toward practical and large-scale implementation of nZEB designs. In the next section, the role of monitoring and automation is targeted.

10.1.2.3.2 Enabling nZEB Paradigms and Flexibility Through Building Monitoring and Automation

Buildings, even nZEBs, frequently show under-optimal performance compared to that planned at the design stage (i.e. performance gap) [113]. Building intelligence (achieved by monitoring and automation) is a key factor in remedying this underperformance as well as in enabling BEF and interoperability frameworks. A variety of definitions [66] and assessment metrics has been developed for its assessment, such as the Building Intelligence Quotient [114] and the smart readiness indicator (SRI) [115] proposed by the European Commission through the EPBD recast [18] and widely incorporated in the UN 2030 Agenda for Sustainable Development [116]. SRI aims at condensing all data on the building's readiness to adapt to occupants' needs and grid's fluctuations into a single value. The SRI calculation encompasses a suite of 52 building services belonging to eight domains (heating, cooling, domestic hot water, controlled ventilation, lighting, dynamic building envelope, energy generation, DSM, electric vehicle charging, and monitoring and control), each weighted according to eight impact criteria (energy, flexibility for the grid, self-generation, comfort, convenience, well-being and health, maintenance and fault prediction, and information to occupants) with functionality levels ranging from zero to four, that describe the degrees of intelligence per service [115]. SRI is a cost-efficient method, yet relying on manual assessments and trained personnel. Automatic algorithms are needed to enable fast, accurate, comparable evaluation of building intelligence. Precondition is the development of digital representations of the building metadata both in terms of hardware and software. Hardware components are easily digitalized, while specific metadata schemas such as the service abstraction layer (SAL) by Hviid and Kjærgaard in Ref. [117] are needed to describe software processes. Generic queries can then be implemented for every pair of building service and functionality level in SRI calculation.

The SRI calculation is currently under test to gather evidence and recommendations toward a consolidated method. Studies on the comparability of smart readiness of nZEB buildings point to the need for benchmarks and integrated measurables, especially for energy flexibility [118, 119]. Further concerns erode the applicability of SRI as a fair rating system across the EU. These relate to the subjective nature of the proposed process for selecting SRI-relevant building services and to the inability to recognize the specific features of cold climate buildings, notably those employing advanced district heating systems [120].

On the other hand, with the clustering of buildings into larger entities, there is growing interest in extending the SRI concept beyond the building level to include synergies related to the integration of renewable energy and load shifting at the district level [121, 122]. Indeed, smart buildings have the potential to become smarter by utilizing city-scale capabilities [123]. This is especially relevant for those located on the outskirts [124].

Regardless of the scale, a prerequisite to smart nZEBs with high BEF potential and SRI is the applicability and customizability of real-time Sense-Act systems (meters, detectors, actuators) [125, 126]. Within three years of its commencement, the Horizon 2020 Program supported more than 40 different actions (mostly Innovation Actions) related to smart buildings, with a total budget exceeding €350 million [127]. The goal of each technique is to preserve the balance between user comfort and energy requirements, such that the user can maintain the desired comfort level with minimum energy consumption and maximum positive interference with the grid. A series of review papers targets different aspects of smart building operation, with a focus on instrumentation and control.

Cheng and Lee [128] examine thermo-fluidic sensors and occupancy detectors ranging from passive infrared (PIR) sensors to wearable sensors and smartphones. Their comparative analysis concludes that AC energy savings of up to 30% can be achieved with thermo-fluidic sensors and occupancy detectors. The savings can exceed 46% when human motions and metabolic rates are incorporated. Another tendency is to couple wearable sensing with personal conditioning systems (PCS) so as to enable increased thermal comfort and acceptability conditions in a wider temperature range, resulting in additional energy savings. A review of recent publications is presented in Ref. [129]. Shah et al. focus on smart controllers, notably those based on genetic algorithms, proportional integral derivative (PID), and fuzzy techniques [130]. Fuzzy controllers emerge as a promising comfort-oriented and intuitive logic. However, being real-time expert systems, they require human experience and knowledge to be properly set [131–134]. Further energy optimization comes by applying fog and edge computing along with IoT networks, which provide computing power to avoid latency and delay in the grid–prosumer interaction [135, 136]. A quite innovative approach is that of combined bat-fuzzy logic. For instance, in Ref. [137], thermal, visual, and air quality comfort are optimized via the bat algorithm, and the error difference with the measured environmental parameters is used to inform the control action of the fuzzy controller and change the status of concerned actuators. Artificial neural network, model-predictive, and adaptive neural fuzzy inference systems are also widely used [138–140]. Moreover, to compensate for the growing vulnerability of classical building control systems to increasing complexities in contemporary-built environments and energy systems, reinforcement learning (RL) methods are becoming distinctive features of control networks, especially for nZEB buildings [141]. Finally, predictive controls play a major role in the presence of energy storage systems, as reviewed in Ref. [142].

Hardware-wise, finding the best combination of sensing technology helps to achieve an energy-efficient and healthy built environment. Dong et al. discuss different types of sensors and how they influence the indoor-built environment and occupant productivity, focusing on energy saving, thermal comfort, visual comfort, and indoor air quality [143]. Beyond accurate microclimatic mapping, fine-grained occupancy information is critical to perfect predictions and control actions.

Indeed, although a one-fits-all recipe to achieve adequate intelligence and flexibility in current and future nZEBs is still undisclosed, a rule of thumb emerges: the "human dimension" governs most energy-related behaviors and enables most BEF and smart strategies. Key stakeholders, who directly or indirectly influence the acts of designing, constructing, living, operating, managing, and regulating the built environments, ultimately dictate their success (Figure 10.3) [144, 145].

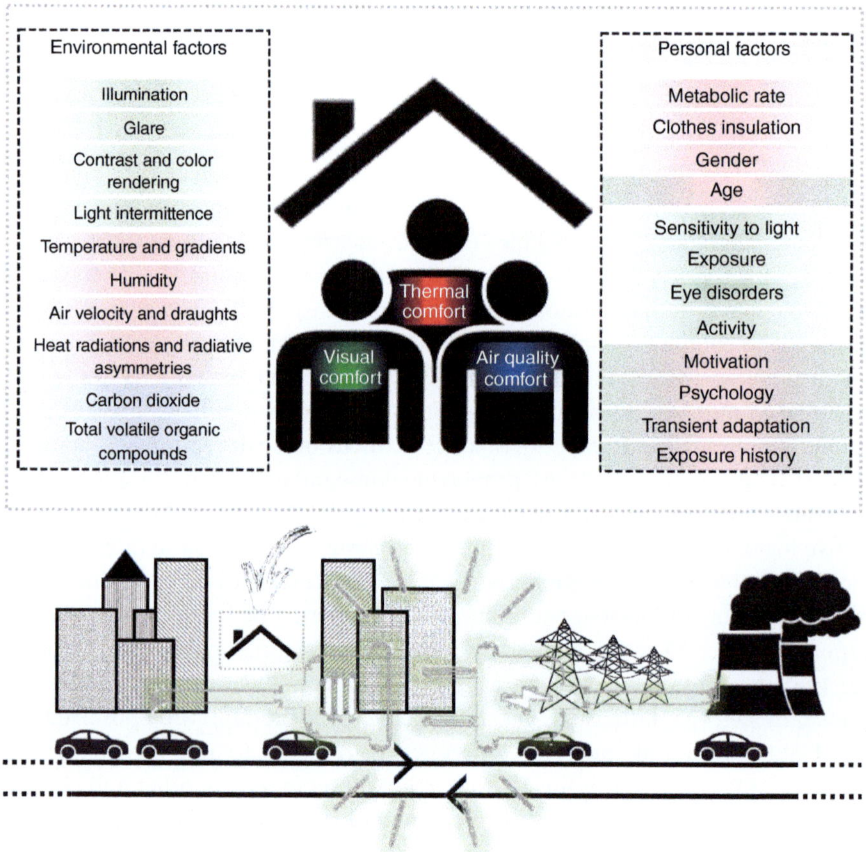

Figure 10.3 The human-centric approach toward future-proof nZEB paradigms: comfort dimensions, personal and environmental players, and interconnected, flexible frameworks between buildings, vehicles, grid, and power production. Source: Based on D'Oca et al. [144], Park et al. [145].

10.2 Status and Research Directions on High-Performance Buildings for the Coming Decade

The present section aims to build on top of what has been already discussed in the previous sections by providing new references around case studies, research projects, and initiatives around the topic of high-performance buildings. Challenges and barriers that reside behind their implementation are presented here as important opportunities and drivers for the transition to PEDs, and the development of emerging trends aimed at achieving a more affordable, energy-efficient, and sustainable built environment.

10.2.1 Overview of Case Studies and Research Projects

In recent years, several initiatives have taken place to investigate, build, and monitor high-performance buildings implementing advanced energy-efficient strategies and renewable and smart energy management systems under different targets and regulations.

In his book [36], Attia presented the case studies of nZEBs across the world designed in view of different climatic conditions and social contexts. Similarly, Voss et al. [19] analyzed more than 200 nZEB projects across the world (mainly Europe) including new and renovated buildings.

In Ref. [20], a case study building has been presented by Magrini et al. comparing the monitored energy consumption with the energy produced by the PV system between 2014 and 2020. The results of the study demonstrated that the building involuntary complied with the European Directives and presented a PEB potential (i.e. the monitored generated renewable energy resulted to be higher than the building energy needs). In the same study, some of the completed and ongoing research projects dealing with new buildings and retrofits that have received funding from the European Union's Horizon 2020 research and innovation program are discussed. Three recent and relevant projects covering different scales of applications (in order from top to bottom: PEBs, net zero-energy settlements [nZESs], and PEDs), targets, and metrics are introduced as follows:

- EXESS (acronym of "FleXible user-CEntric Energy poSitive houseS") (2019–2023) [146]. This project aims to showcase how NZEBs can be transformed into PEBs, by focusing on four demonstration projects in Europe, under different climatic conditions (Nordic, Continental, Oceanic, and Mediterranean climate zones). A user-centric approach is used to test, validate, and facilitate the integration of innovative building technologies in a way that they become replicable PEB solutions. Figure 10.4 shows the distribution map of existing PEBs in Europe that has been elaborated within the project.
- ZERO-PLUS (acronym of "Achieving near Zero and Positive-Energy Settlements in Europe using Advanced Energy Technology") (2015–2020) [147]. Four net zero-energy demonstration settlements (in Italy, France, Cyprus, and the UK) have been designed and completed during the course of the project and have

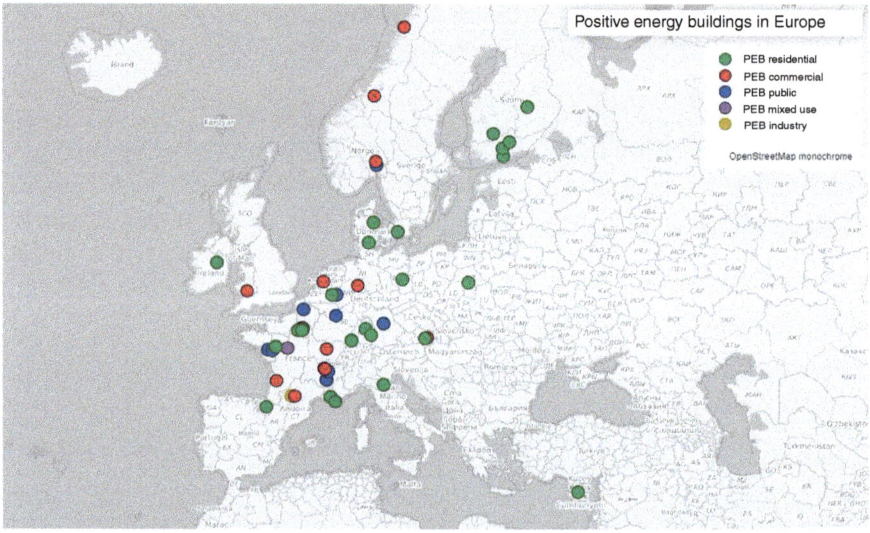

Figure 10.4 Existing positive-energy buildings across Europe identified by the EXCESS EU Horizon 2020 project team (last updated on 26 July 2020) and grouped for different building typologies (residential, commercial, public, mixed-use, and industrial) [146].

successfully achieved the energy, environmental, and cost reduction targets set in the ZERO-PLUS framework (e.g. net regulated energy consumption lower than 20 kWh/m^2/yr, renewable energy generation produced at the settlement scale equal to or higher than 50 kWh/m^2/yr, total CO_2 emission reduction for the four nZESs up to 200 tons, and initial cost reduction of 16% with respect to a nZEB implementing the same energy performance achieved with conventional technologies). Figure 10.5 shows the financial, energy, and environmental gaps between cost-optimal levels and nZEB levels in Europe and the ambition of the ZERO-PLUS project.

- ATELIER (acronym of "AmsTErdam BiLbao citizen drivEn smart cities") (2019–2024) [148]. This project focuses on PED development and establishes long-term roadmaps for the spread of positive-energy solutions. The goals set in this project consist in the achievement of 1340 MWh of primary energy surplus, 1.7 kton of CO_2 emissions, and 23 tons of total NO_x emission.

Seven existing net zero-energy districts (nZEDs) in Europe have been identified and analyzed in Ref. [149], where lessons learned are summarized and compared.

Recently, numerous partners from around the world have joined collaborative initiatives such as the International Energy Agency (IEA) EBC Annex 83 for the "PEDs" [150] and the European Cooperation on Science and Technology (COST) Action Positive Energy Districts European Network (PED-EU-NET) (2020–2024) [151] to discuss, collaborate, and support the establishment of key professionals, researchers, and practitioners and share new and ongoing development in the context of emerging PEDs [152] and PED Lab concepts. PED Labs, in particular, are "pilot actions that provide opportunities to experiment with planning and development of PEDs

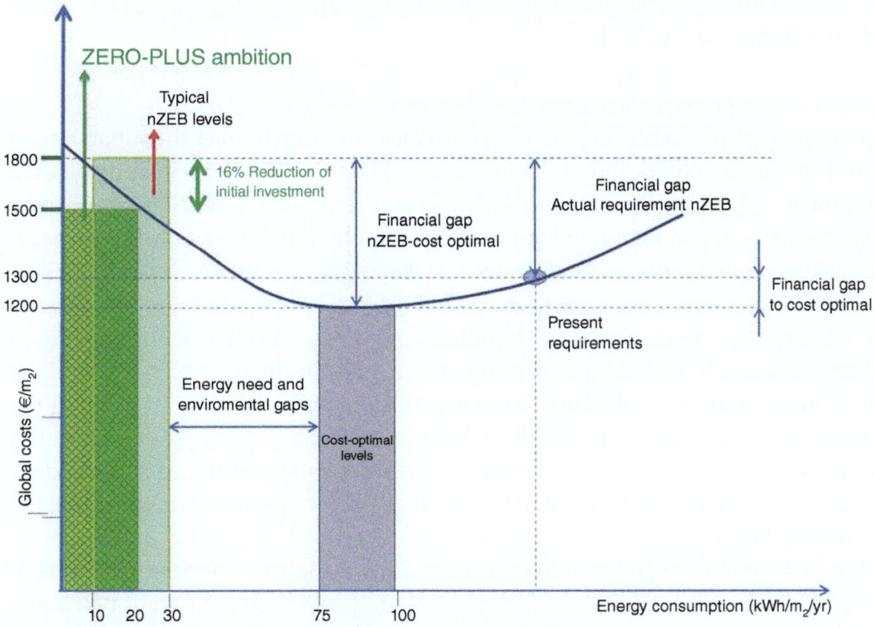

Figure 10.5 Financial, energy, and environmental gaps between cost-optimal levels and nZEB levels in Europe and the ambition of ZERO-PLUS project [147].

as well as provide seeding ground for new ideas, solutions, and services to develop" and are considered as living laboratories that "follow an integrative approach including technology, spatial, regulatory, financial, legal, social, and economic perspectives" [153].

These pioneering concepts are discussed and refined with the aim to provide a framework that can help to achieve the goals set by the recent energy policies oriented toward long-term climate-neutral energy systems, sustainable urban development, and energy transition. One example is the EU's Strategic Energy Technology (SET) Plan Action 3.2 "Smart Cities and Communities" that aims to obtain 100 PEDs and neighborhoods around Europe by 2025 [153]. A dedicated PED booklet elaborated by the JPI Urban Europe [154] presents the following framework definition for PEDs [155]: "Positive Energy Districts are energy-efficient and energy-flexible urban areas which produce net zero greenhouse gas emissions and actively manage an annual local or regional surplus production of renewable energy. They require integration of different systems and infrastructures and interaction between buildings, the users and the regional energy, mobility and ICT systems, while optimising the livability of the urban environment in line with social, economic and environmental sustainability." Similar to PEBs, energy efficiency, energy flexibility, and energy production are identified as the main functions of PEDs. The booklet also provides data end information of projects in Europe that have a PED ambition. Bossi et al. have analyzed those projects and summarized the results in Ref. [156]. A recent review on net zero-energy communities of different scales presents an overview of the main technological advancements implemented around the globe to achieve the net zero-energy target [157].

10.2.1.1 Challenges, Drivers, and Best Practices

Generally speaking, achieving high-performance buildings implies the collaboration of multiple stakeholders and decision-makers. The transition from an independent traditional and conventional building design and construction procedure to a more advanced, green, and integrated approach inevitably calls for new project management skills, tools, platforms, and forms of stakeholder communications and collaborations to overcome the increasing level of project complexity [158–160]. This is not an easy transition, and many building designers and construction practitioners are reluctant to the change. A number of studies in the literature report about the common barrier to adapting to change and the difficulties in interrupting the business-as-usual practice in building design and constructions [161–163]. Policy and regulations must drive the change in project management and stakeholder's engagement and should be supported by adequate communications campaigns to be effective [164].

Across the globe, several studies report on nZEB challenges, barriers, drivers, benefits, lessons learned, and best practices [165–167], addressing different typologies of buildings [168, 169].

Taherahmadi et al. presented a number of barriers [25] associated with the implementation of energy efficiency measures (e.g. the high price of land in urban areas and high-efficiency technologies) and renewable energy generation solutions (e.g. the limited area of roof and/or facade to install the RESs such as PVs and wind turbines or hybrid solar–wind systems and grid stability problems associated with the distribution of the generated energy). The implementation of renewable energy technologies within the building can reduce considerably the buildings' total energy intensity, although using renewable energy sources can increase the embodied energy and the investment cost of the building. The same applies at a larger scale, e.g. at the scale of the neighborhood, settlement, or district.

At the scale of the settlement and neighborhood, additional challenges have been identified compared to nZEBs, as stressed by Mavrigiannaki et al. in Ref. [170] and by Nematchoua et al. in Ref. [171]. For example, ZERO-PLUS demonstrated some of the barriers to nZES implementation dealing with the stakeholders' interactions and regulatory constraints, limitations, and delays.

The implementation of PEDs is also associated with a number of challenges, such as regulatory framework, standardization, energy systems integration, citizen participation, education and training, and business and sustainable funding models. Nowadays, two of the most significant challenges are the integration of conventional and emerging technologies for PEDs and the complex integration of PEDs in urban planning. A preliminary result from an online survey that has been elaborated and disseminated by the PED-EU-NET collaborative team [151] has identified 13 barriers to the implementation of PEDs. These barriers have been grouped into eight categories (Information and Awareness, Administrative, Legal and Regulatory, Financial, Market, Environmental, Technical, and Social). Most of the categories received high scores for their identified barriers (average of four out of five points), highlighting that several are the barriers that need to be overcome. The highest score has been recorded for the Social inertia (almost five points) followed by the Financial category, which includes barriers such as the high cost of design, construction, and implementation and the limited access to capital and cost disincentives. This indicates that PED solutions are still out of the market and are not competitive with traditional building solutions. Inadequate regulation for new technologies (Legal and Regulatory), lack of awareness among authorities (Information and Awareness), and lack of trust in social network (Social) also received high scores. This may be due to the fact that advanced technologies are developed faster than the legal and regulatory frameworks. The least perceived barrier is the negative effect of PEDs on the environment (Environmental). Conversely, PEDs are perceived as positive with reduced environmental impact (2.5 points). However, the environmental score is not zero, indicating that there is still a debate around what is the aggregate level of sustainability of PEDs and the environmental friendliness of the implemented technologies and solutions, as demonstrated by some preliminary studies focusing on the life cycle assessment (LCA) and environmental performance of nZESs [172, 173]. As an example, the production of lithium batteries, PV panels, and some construction materials is not environmentally friendly and has an impact on the overall LCA of PEDs.

Barriers are accompanied by drivers such as the recognized need for climate change mitigation, the reduction of electricity cost, the increased indoor environmental quality, territorial and market attractiveness, the energy autonomy, and the independence allowed by the implementation of on-site energy storage, among others. Local and national policy frameworks are preconditions to the implementation of nZEBs and PEDs. Similarly, effective financial mechanisms to reduce costs and maximize benefits are important unlocking elements.

Looking at the benefits of implementing nZEBs, it is evident how building more efficient homes can guarantee cheaper and easier building management (an important factor especially for social housing companies), lower energy bills for owners or tenants (especially for low-income tenants with the consequent reduction of energy poverty [174]), and unpaid rent for companies and provide a positive example for the general public and the private sector. However, the high deployment cost of the active and passive strategies that help to achieve the status of nZEB or PEB is still a limiting factor to the rapid spread of these high-performance buildings in all countries and for most of the building typologies [175].

Lessons learned from nine Norwegian zero emission building projects have been presented by Andresen et al. in Ref. [176]. Similarly, as a result of nZES projects, such as ZERO-PLUS [177], a number of take-home messages have been identified and can be summarized in (i) need for a clear and comprehensive approach and framework to be followed and acknowledged by all key stakeholders, (ii) need for a qualified team to initiate and complete all the phases of the nZES implementation (e.g. architects, thermal engineers, construction practitioners, technology providers, etc.), (iii) need for an early-stage collaboration across the team involving all key partners from the preliminary design phase until at least one year after the occupation phase, (iv) need for a rescue team during the first operational year, (v) need for a commissioning plan, and most importantly, (vi) need for continuous monitoring to verify the buildings and settlement performance.

10.2.2 Transition from Individual Nearly Zero-Energy Buildings to Positive-Energy Districts (PEDs)

Recent trends for the future of high-performance buildings involve the transition from re-design and energy requalification of individual buildings (NZEBs) to the larger nearly zero-energy settlements [178, 179] or PEDs. This transition will be especially helpful in overcoming the barrier of the high investment and maintenance cost for the owners, among others. Indeed, the economy of scale can help to reduce the overall cost that each member of the community is called to cover. Similar to the sharing of the costs, in PEDs also, energy, resources, and services are shared among the members of the community living within the energy district.

Independent PEDs can help to solve the "trilemma" of energy, decarbonization (sustainability), and affordability thanks to their renewable energy harnessing capacity and the possibility to implement high energy efficiency and urban heat island (UHI) mitigation solutions at a larger scale, amplifying their efficacy and positive impact on the local microclimate and thermal comfort. Therefore, extending the nearly zero or positive concept to the settlement and district scales produces an acceleration toward the decarbonization of the built environment. Urban scales are small enough to allow quick advancement and innovation (regardless of the level of complexity) as well as large enough to produce a significant impact. Additionally, as stated in previous sections, implementing these concepts at the district or city-scale helps smart buildings to become smarter.

In the past years, energy regulations and the market have rapidly changed to accommodate this transition. As anticipated in Section 10.1.1.1, the future energy system aims at being efficient, green, decentralized, and democratic toward a sustainable economy and climate neutrality, and in this context, a new concept of "energy communities" has been introduced and defined as "novel legal, technical, and social entities that involve citizens' participation in the energy systems as prosumers with the main goal to provide environmental, economic, and social benefits for the community rather than profit market" [180]. The current European legislation and regulatory framework includes the Clean Energy for all European Package (CEP), the Renewable Energy Directive (RED) 2018/2001/UE [181], and the Electricity Market Directive (EMD) 2019/944/UE [182]. The CEP presents five important dimensions of the European Union (e.g. Energy Union Governance, Energy Efficiency Directives and EPBD, Revised RED, New Electricity Market Design including risk preparedness, and energy prices and costs report) all oriented toward an innovative, socially fair, digital, investment-friendly, safe for all, inclusive, and interconnected framework and new forms of active consumers. At present, the prevalent form of consumption across Europe is the individual self-consumption, followed by the collective self-consumption or energy sharing. Looking at the future, energy communities seem to be the new form of active consumer.

The European legislative framework also recognizes and introduces new types of energy communities such as the citizen energy communities (CECs) [182] and RECs [181] that require new regulatory models to be applied on a large scale.

The market is responding to the changes in the regulations by starting to prioritize community benefits rather than profit-making. Ad hoc business and technical models become essentials to facilitate the transition among prosumers and energy communities as well as to distribute the right mix of incentives to all stakeholders and generate interest and trust among the communities.

This revolutionary energy transition presents again some policy, social, and technical barriers at the community level. One of the biggest legal barriers involves the implementation of community contracts. Other barriers can be seen in the lobby on data access for the establishment of energy communities and user acceptance and collaboration. Regarding the technical barriers, the need for the implementation of innovative renewable technologies in the common outdoor spaces is noteworthy as it requires specific environmental and/or other types of permissions and certificates and the need for operating and maintaining the shared systems that can cause privacy and intrusion issues.

10.3 Conclusions

Planning a sustainable future for the built environment is one of the main targets for the coming decade. High-performance buildings have an important role in decarbonizing the building sector.

In this chapter, principles, definitions, and net zero-energy frameworks have been introduced, highlighting that there is still a lack of common standardized definitions for the zero-energy and positive-energy concepts. Bridging this gap will help to build high-performance buildings according to specific combinations of design strategies (e.g. energy-efficient conservation strategies, energy generation options, and energy management solutions) and track the progress toward the real implementation and the achievement of the energy targets, thus supporting regulations to accomplish political energy targets.

Design strategies and technological advances to reach NZEB and PEB under different climates and contexts are presented along with a brief discussion on practical barriers, such as the high investment cost. Lifetime costs of PEBs can be reduced via technological integration, making them affordable to larger portions of society. It is extremely important to identify practical solutions to overcome market and cost barriers for PEBs (residential and non-residential) that can transform the encountered challenges into opportunities.

According to recent trends, investing in independent PEDs can be seen as a potential solution to the "trilemma" of energy, decarbonization (sustainability), and affordability. PEDs are not the sum of PEBs but rather the result of the economy of scale where there is a share of resources, cost, services, and energy that allows compensating for supply and demand of individual components of the districts within a functioning system. Again, at present, the integration of conventional and emerging technologies and RESs for PEDs remains one of the most significant challenges to be addressed together with the reduction of their environmental impact. Solving this challenge will make PEDs affordable to many, allowing a faster spread across the global market. Other potential barriers for the implementation of PEDs consist in the limited space availability, potential opposition from local authorities and citizens, as well as legal, economic, and financial obstacles.

To overcome the barriers presented in this chapter and take advantage of the drivers, cross-cutting and interdisciplinary research and effective collaboration among key stakeholders are essential. A joint and coordinated effort will help to solve issues that can occur during the different phases of the project (e.g. urban planning, energy system design, user behavior and involvement, etc.) and initiatives such as PED Labs are a good opportunity to experiment new integrated design, planning, and deployment approaches, leading to the promotion of standardized process and new policy frameworks.

To conclude, although a unique strategy to decarbonize the built environment and design high-performance buildings does not exist, a comprehensive classification of all potential alternative strategy options in relation to their context of applicability (climate, scale of application, building typology, etc.) and a human-centric approach toward nZEB and PEB can guide informed decisions and achieve the desired energy and environmental targets (among others) efficiently and in the long haul.

List of Abbreviations

AC	air conditioning
BEF	building energy flexibility
BEMS	building energy management system
BiPV	building integrated photovoltaic
CBD	central business district
CEC	citizen energy community
CHP	combined heat and power
DER	distributed energy resources
DOE	Department of Energy
DR	demand response
DSM	demand side management
EER	energy efficiency ratio
EPBD	Energy Performance of Buildings Directive
EPC	energy performance certificate
GHG	greenhouse gas
HCPV	high-concentration photovoltaics
HVAC	heating, ventilation, and air conditioning
IEA	International Energy Agency
ICT	information and communications technology
IOT	internet of things
JRC	Joint Research Centre

KPI	key performance indicator
LC-ZEB	life cycle zero-energy building
LED	light-emitting diode
NZEB	nearly zero-energy building
nZEB	net zero-energy building
nZED	net zero-energy district
nZES	net zero-energy settlement
PCS	personal conditioning systems
PEB	positive-energy building
PED	positive-energy district
PID	proportional integral derivative
PIR	passive infrared
PV	photovoltaic
PV/T	hybrid photovoltaic/thermal collectors
PCM	phase change material
REC	renewable energy community
RES	renewable energy system
RL	reinforcement learning
SAL	service abstraction layer
SRI	smart readiness indicator
SHEMS	smart home energy management system
UHI	urban heat island
ZEB	zero-energy building

References

1 International Energy Agency (IEA) (2021). *Global Energy Review 2021*. Paris: IEA www.iea.org/reports/global-energy-review-2021.

2 U.S. Energy Information Administration (EIA) (2021). *Annual Energy Outlook 2021* www.eia.gov/outlooks/aeo/ (accessed 19 October 2021).

3 International Energy Agency (IEA) (2020). *Tracking Buildings 2020*. Paris: IEA www.iea.org/reports/tracking-buildings-2020.

4 Sadineni, S.B., Madala, S., and Boehm, R.F. (2011). Passive building energy savings: a review of building envelope components. *Renew. Sustain. Energy Rev.* 15 (8): 3617–3631.

5 Chel, A. and Kaushik, G. (2018). Renewable energy technologies for sustainable development of energy efficient building. *Alexandria Eng. J.* 57 (2): 655–669.

6 Turner, W.J.N., Walker, I.S., and Roux, J. (2015). Peak load reductions: electric load shifting with mechanical pre-cooling of residential buildings with low thermal mass. *Energy* 82: 1057–1067.

7 Curtius, H.C. (2018). The adoption of building-integrated photovoltaics: barriers and facilitators. *Renew. Energy* 126: 783–790.

8 Biyik, E., Araz, M., Hepbasli, A. et al. (2017). A key review of building integrated photovoltaic (BIPV) systems. *Eng. Sci. Technol.* 20 (3): 833–858.
9 Lake, A., Rezaie, B., and Beyerlein, S. (2017). Review of district heating and cooling systems for a sustainable future. *Renew. Sustain. Energy Rev.* 67: 417–425.
10 Mahmoud, M., Ramadan, M., Naher, S. et al. (2020). Recent advances in district energy systems: a review. *Therm. Sci. Eng. Progr.* 20: 100678.
11 Sanguinetti, A., Karlin, B., Ford, R. et al. (2018). What's energy management got to do with it? Exploring the role of energy management in the smart home adoption process. *Energy Effic.* 11 (7): 1897–1911.
12 McIlvennie, C., Sanguinetti, A., and Pritoni, M. (2020). Of impacts, agents, and functions: an interdisciplinary meta-review of smart home energy management systems research. *Energy Res. Social Sci.* 68: 101555.
13 Stavrakas, V. and Flamos, A. (2020). A modular high-resolution demand-side management model to quantify benefits of demand-flexibility in the residential sector. *Energy Convers. Manage.* 205: 112339.
14 Mariano-Hernández, D., Hernández-Callejo, L., Zorita-Lamadrid, A. et al. (2020). A review of strategies for building energy management system: model predictive control, demand side management, optimization, and fault detect & diagnosis. *J. Build. Eng.* 33: 101692.
15 Chakraborty, N., Mondal, A., and Mondal, S. (2020). Efficient load control based demand side management schemes towards a smart energy grid system. *Sustain. Cities Soc.* 59: 102175.
16 Miller, A. and Edelson, J. (2016). Zero net energy buildings and the grid: the future of low energy building-grid interactions. *Proceedings of 2016 ASHRAE Winter Conference.*
17 Torcellini, P., Pless, S., Deru, M., and Crawley, D. (2006). Zero Energy Buildings: A Critical Look at the Definition *(No. NREL/CP-550-39833)*. Golden, CO (United States): National Renewable Energy Laboratory (NREL).
18 European Union Directive 2010/31/EU of the European Parliament and of Council of 19 May 2010 on the energy performance of building (EPBD recast) *Off. J. Eur. Union* (2010) L 153/13.
19 Voss, K., Musall, E., and Lichtmeß, M. (2011). From low-energy to net zero-energy buildings: status and perspectives. *J. Green Build.* 6 (1): 46–57.
20 Magrini, A., Lentini, G., Cuman, S. et al. (2020). From nearly zero energy buildings (NZEB) to positive energy buildings (PEB): the next challenge-The most recent European trends with some notes on the energy analysis of a forerunner PEB example. *Dev. Built Environ.* 3: 100019.
21 Laustsen, J. (2008). Energy efficiency requirements in building codes, energy efficiency policies for new buildings. *Buildings* 1–85.
22 Marszal, A.J., Heiselberg, P., Bourrelle, J.S. et al. (2011). Zero energy building – a review of definitions and calculation methodologies. *Energy Build.* 43 (4): 971–979.

23 Pless, S. and Torcellini, P. (2010). Net-Zero Energy Buildings: A Classification System Based on Renewable Energy Supply Options *(No. NREL/TP-550-44586)*. Golden, CO (United States): National Renewable Energy Laboratory (NREL).

24 Hernandez, P. and Kenny, P. (2010). From net energy to zero energy buildings: defining life cycle zero energy buildings (LC-ZEB). *Energy Build.* 42 (6): 815–821.

25 Taherahmadi, J., Noorollahi, Y., and Panahi, M. (2021). Toward comprehensive zero energy building definitions: a literature review and recommendations. *Int. J. Sustain. Energy* 40 (2): 120–148.

26 Sartori, I., Napolitano, A., and Voss, K. (2012). Net zero energy buildings: a consistent definition framework. *Energy Build.* 48: 220–232.

27 DOE (2015). A Common Definition for Zero Energy Buildings. *DOE/EE-1247*. http://energy.gov/eere/buildings/downloads/common-definition-zero-energy-buildings (accessed 19 October 2021).

28 Fufa, S.M., Schlanbusch, R.D., Sørnes, K. et al. (2016). *A Norwegian ZEB Definition Guideline*. SINTEF Academic Press.

29 UNFCCC (2016). "Decision 1/CP.21 Adoption of the Paris Agreement" FCCC/CP/2015/10/Add.1, Annex (Paris Agreement) art 2(1). www.un.org/en/development/desa/population/migration/generalassembly/docs/globalcompact/FCCC_CP_2015_10_Add.1.pdf (accessed 19 October 2021).

30 European Commission (2019). The European Green Deal. COM(2019) 640.

31 Kulovesi, K. and Oberthür, S. (2020). Assessing the EU's 2030 Climate and Energy Policy Framework: incremental change toward radical transformation? *Rev. Eur., Comp. Int. Environ. Law* 29 (2): 151–166.

32 European Commission (2020). Proposal for a Regulation of the European Parliament and of the Council Establishing a Framework for Achieving Climate Neutrality and Amending Regulation (EU) 2018/1999 (European Climate Law). COM(2020) 80.

33 European Union Directive 2018/844/UE of the European Parliament and of the Council of 30 May 2018 amending Directive 2010/31/EU on the energy performance of buildings and Directive 2012/27/EU on energy efficiency.

34 DOE (2012). Building Technology Program. www1.eere.energy.gov/buildings/publications/pdfs/corporate/myp08complete.pdf.

35 Zhang, S., Xu, W., Wang, K. et al. (2020). Scenarios of energy reduction potential of zero energy building promotion in the Asia-Pacific region to year 2050. *Energy* 213: 118792.

36 Attia, S. (2018). *Net Zero Energy Buildings (NZEB): Concepts, Frameworks and Roadmap for Project Analysis and Implementation*. Butterworth-Heinemann.

37 Santamouris, M. and Kolokotsa, D. (2015). On the impact of urban overheating and extreme climatic conditions on housing, energy, comfort and environmental quality of vulnerable population in Europe. *Energy Build.* 98: 125–133.

38 Folberth, G.A., Butler, T.M., Collins, W.J., and Rumbold, S.T. (2015). Megacities and climate change - a brief overview. *Environ. Pollut.* 203: 235–242.

39 Cartalis, C., Synodinou, A., Proedrou, M. et al. (2001). Modifications in energy demand in urban areas as a result of climate changes: an assessment for the southeast Mediterranean region. *Energy Convers. Manag.* 42 (14): 1647–1656.

40 IEA (2013). Transition to Sustainable Buildings - Strategies and Opportunities to 2050.

41 Op't Veld, P. (2015). MORE-CONNECT: Development and advanced prefabrication of innovative, multifunctional building envelope elements for modular retrofitting and smart connections. *Energy Procedia* 78: 1057–1062.

42 He, Y., Zhang, Y., Zhang, C., and Zhou, H. (2020). Energy-saving potential of 3D printed concrete building with integrated living wall. *Energy Build.* 222: 110110.

43 Tsoka, S., Theodosiou, T., Papadopoulou, K., and Tsikaloudaki, K. (2020). Assessing the energy performance of prefabricated buildings considering different wall configurations and the use of PCMs in Greece. *Energies* 13 (19): 5026.

44 Tavares, V., Lacerda, N., and Freire, F. (2019). Embodied energy and greenhouse gas emissions analysis of a prefabricated modular house: the "Moby" case study. *J. Cleaner Prod.* 212: 1044–1053.

45 Bonamente, E., Merico, M.C., Rinaldi, S. et al. (2014). Environmental impact of industrial prefabricated buildings: carbon and energy footprint analysis based on an LCA approach. *Energy Procedia* 61: 2841–2844.

46 Frick, J., Stipetić, M., Mielich, O., and Garrecht, H. (2020). *Studies on Thermal Performance of Advanced Aerogel-Based Materials. Sustainability in Energy and Buildings*, 641–649. Singapore: Springer.

47 Berardi, U. (2018). Aerogel-enhanced systems for building energy retrofits: insights from a case study. *Energy Build.* 159: 370–381.

48 Lucchi, E., Becherini, F., Di Tuccio, M.C. et al. (2017). Thermal performance evaluation and comfort assessment of advanced aerogel as blown-in insulation for historic buildings. *Build. Environ.* 122: 258–268.

49 Gobakis, K., Kolokotsa, D., Maravelaki-Kalaitzaki, N. et al. (2015). Development and analysis of advanced inorganic coatings for buildings and urban structures. *Energy Build.* 89: 196–205.

50 Ma, Y., Zhu, B., and Wu, K. (2001). Preparation and solar reflectance spectra of chameleon-type building coatings. *Sol. Energy* 70: 417–422.

51 Pisello, A.L., Castaldo, V.L., Piselli, C. et al. (2015). Combined thermal effect of cool roof and cool façade on a prototype building. *Energy Procedia* 78: 1556–1561.

52 Karlessi, T., Santamouris, M., Synnefa, A. et al. (2011). Development and testing of PCM doped cool colored coatings to mitigate urban heat island and cool buildings. *Build. Environ.* 46 (3): 570–576.

53 Pisello, A.L., Pignatta, G., Castaldo, V.L., and Cotana, F. (2014). Experimental analysis of natural gravel covering as cool roofing and cool pavement. *Sustainability* 6 (8): 4706–4722.

54 Pisello, A.L., Castaldo, V.L., Pignatta, G. et al. (2016). Experimental in-lab and in-field analysis of waterproof membranes for cool roof application and urban heat island mitigation. *Energy Build.* 114: 180–190.

55 Kolokotsa, D., Diakaki, C., Papantoniou, S., and Vlissidis, A. (2012). Numerical and experimental analysis of cool roofs application on a laboratory building in Iraklion, Crete, Greece. *Energy Build.* 55: 85–93.

56 Raman, A.P., Anoma, M.A., Zhu, L. et al. (2014). Passive radiative cooling below ambient air temperature under direct sunlight. *Nature* 515 (7528): 540–544.

57 Feng, J. and Santamouris, M. (2019). Numerical techniques for electromagnetic simulation of daytime radiative cooling: a review. *AIMS Mater. Sci.* 6 (6).

58 Santamouris, M. and Feng, J. (2018). Recent progress in daytime radiative cooling: is it the air conditioner of the future? *Buildings* 8 (12): 168.

59 Kolokotsa, D., Pouliezos, A., Stavrakakis, G., and Lazos, C. (2009). Predictive control techniques for energy and indoor environmental quality management in buildings. *Build. Environ.* 44 (9): 1850–1863.

60 Papantoniou, S., Kolokotsa, D., Kalaitzakis, K. et al. (2016). Adaptive lighting controllers using smart sensors. *Int. J. Sustain. Energy* 35 (6): 537–553.

61 Papantoniou, S., Kolokotsa, D., and Kalaitzakis, K. (2014). Building optimization and control algorithms implemented in existing BEMS using a web based energy management and control system. *Energy Build.* 98: 45–55.

62 Kolokotsa, D., Diakaki, C., Grigoroudis, E. et al. (2009). Decision support methodologies on the energy efficiency and energy management in buildings. *Adv. Build. Energy Res.* 3 (1): 121–146.

63 Moslehi, K. and Kumar, R. (2010). A reliability perspective of the smart grid. *IEEE Trans. Smart Grid* 1 (1): 57–64.

64 Kampelis, N., Ferrante, A., Kolokotsa, D. et al. (2017). Thermal comfort evaluation in HVAC demand response control. *Energy Procedia* 134: 675–682.

65 Dupont, B., De Jonghe, C., Olmos, L., and Belmans, R. (2014). Demand response with locational dynamic pricing to support the integration of renewables. *Energy Policy* 67: 344–354.

66 Al Dakheel, J., Del Pero, C., Aste, N., and Leonforte, F. (2020). Smart buildings features and key performance indicators: a review. *Sustain. Cities Soc.* 61: 102328.

67 Kolokotsa, D. (2016). The role of smart grids in the building sector. *Energy Build.* 116: 703–708.

68 Petter Jelle, B., Breivik, C., Røkenes, H.D. et al. (2012). Building integrated photovoltaic products: a state-of-the-art review and future research opportunities. *Sol. Energy Mater. Sol. Cells* 100 (7465): 69–96.

69 Ram, J.P., Rajasekar, N., and Miyatake, M. (2017). Design and overview of maximum power point tracking techniques in wind and solar photovoltaic systems: a review. *Renew. Sustain. Energy Rev.* 73: 1138–1159.

70 Koutroulis, E., Kolokotsa, D., Potirakis, A., and Kalaitzakis, K. (2006). Methodology for optimal sizing of stand-alone photovoltaic/wind-generator systems using genetic algorithms. *Sol. Energy* 80 (9): 1072–1088.

71 Koutroulis, E. and Kalaitzakis, K. (2006). Design of a maximum power tracking system for wind-energy-conversion applications. *IEEE Trans. Ind. Electron.* 53 (2): 486–494.

72 Li, Q.-Y., Chen, Q., and Zhang, X. (2013). Performance analysis of a rooftop wind solar hybrid heat pump system for buildings. *Energy Build.* 65: 75–83.

73 Chong, W.T., Naghavi, M.S., Poh, S.C. et al. (2011). Techno-economic analysis of a wind-solar hybrid renewable energy system with rainwater collection feature for urban high-rise application. *Appl. Energy* 88 (11): 4067–4077.

74 Marques, A.C., Fuinhas, J.A., and Manso, J.R.P. (2010). Motivations driving renewable energy in European countries: a panel data approach. *Energy Policy* 38 (11): 6877–6885.

75 Zahedi, A. (2011). A review of drivers, benefits, and challenges in integrating renewable energy sources into electricity grid. *Renew. Sustain. Energy Rev.* 15 (9): 4775–4779.

76 Bartusch, C. and Alvehag, K. (Jul. 2014). Further exploring the potential of residential demand response programs in electricity distribution. *Appl. Energy* 125: 39–59.

77 Ondeck, A.D., Edgar, T.F., and Baldea, M. (2015). Optimal operation of a residential district-level combined photovoltaic/natural gas power and cooling system. *Appl. Energy* 156: 1–14.

78 Andrews, D. et al. (2012). Background Report on EU-27 District Heating and Cooling Potentials, Barriers, Best Practice and Measures of Promotion. *European Commission*, p. 215.

79 Bonsignore, G. et al. (2016). CHP efficiency of a 2000 × CPV system with reflective optics. *AIP Conference Proceedings*, pp. 1–5.

80 Paredes, F. et al. (2015). Combined heat and power generation with a HCPV system at 2000 suns. *AIP Conference Proceedings*, pp. 2–7.

81 Hori, K., Matsui, T., Hasuike, T. et al. (2016). Development and application of the renewable energy regional optimization utility tool for environmental sustainability: REROUTES. *Renew. Energy* 93: 548–561.

82 Huang, A.Q., Crow, M.L., Heydt, G.T. et al. (2010). The future renewable electric energy delivery and management (FREEDM) system: the energy internet. *Proc. IEEE* 99: 133–148.

83 Jensen, S.Ø., Marszal-Pomianowska, A., Lollini, R. et al. (2017). IEA EBC Annex 67 energy flexible buildings. *Energy Build.* 155: 25–34. https://doi.org/10.1016/j.enbuild.2017.08.044.

84 Drysdale, B., Wu, J., and Jenkins, N. (2015). Flexible demand in the GB domestic electricity sector in 2030. *Appl. Energy* 139: 281–290.

85 You, S., Jin, L., Hu, J. et al. (2015). The Danish perspective of energy internet: from service-oriented flexibility trading to integrated design, planning and operation of multiple cross-sectoral energy systems. *Zhongguo Dianji Gongcheng Xuebao* 35: 3470–3481.

86 D'hulst, R., Labeeuw, W., Beusen, B. et al. (2015). Demand response flexibility and flexibility potential of residential smart appliances: experiences from large pilot test in Belgium. *Appl. Energy* 155: 79–90.

87 Kobus, C.B.A., Klaassen, E.A.M., Mugge, R., and Schoormans, J.P.L. (2015). A real-life assessment on the effect of smart appliances for shifting households' electricity demand. *Appl. Energy* 147: 335–343.

88 Li, R. and You, S. (2018). Exploring potential of energy flexibility in buildings for energy system services. *CSEE J. Power Energy Syst.* 4: 434–443.

89 Zhou, Y. and Cao, S. (2020). Quantification of energy flexibility of residential net-zero-energy buildings involved with dynamic operations of hybrid energy storages and diversified energy conversion strategies. *Sustain. Energy Grids Networks* 21: 100304.

90 Kensby, J., Trüschel, A., and Dalenbäck, J.-O. (2015). Potential of residential buildings as thermal energy storage in district heating systems--results from a pilot test. *Appl. Energy* 137: 773–781.

91 Masy, G., Georges, E., Verhelst, C. et al. (2015). Smart grid energy flexible buildings through the use of heat pumps and building thermal mass as energy storage in the Belgian context. *Sci. Technol. Built Environ.* 21: 800–811.

92 Le Dréau, J. and Heiselberg, P. (2016). Energy flexibility of residential buildings using short term heat storage in the thermal mass. *Energy.* 111: 991–1002.

93 Vivian, J., Chiodarelli, U., Emmi, G., and Zarrella, A. (2020). A sensitivity analysis on the heating and cooling energy flexibility of residential buildings. *Sustain. Cities Soc.* 52: 101815.

94 Lund, P.D., Lindgren, J., Mikkola, J., and Salpakari, J. (2015). Review of energy system flexibility measures to enable high levels of variable renewable electricity. *Renew. Sustain. Energy Rev.* 45: 785–807.

95 Arteconi, A., Hewitt, N.J., and Polonara, F. (2012). State of the art of thermal storage for demand-side management. *Appl. Energy* 93: 371–389.

96 Klein, K., Herkel, S., Henning, H.-M., and Felsmann, C. (2017). Load shifting using the heating and cooling system of an office building: quantitative potential evaluation for different flexibility and storage options. *Appl. Energy* 203: 917–937.

97 Finck, C., Li, R., Kramer, R., and Zeiler, W. (2018). Quantifying demand flexibility of power-to-heat and thermal energy storage in the control of building heating systems. *Appl. Energy* 209: 409–425.

98 Foteinaki, K., Li, R., Heller, A., and Rode, C. (2018). Heating system energy flexibility of low-energy residential buildings. *Energy Build.* 180: 95–108.

99 Li, R., Dane, G., Finck, C., and Zeiler, W. (2017). Are building users prepared for energy flexible buildings?—A large-scale survey in the Netherlands. *Appl. Energy* 203: 623–634.

100 D'Ettorre, F., De Rosa, M., Conti, P. et al. (2019). Mapping the energy flexibility potential of single buildings equipped with optimally-controlled heat pump, gas boilers and thermal storage. *Sustain. Cities Soc.* 50: 101689.

101 Li, R., Wei, F., Zhao, Y., and Zeiler, W. (2017). Implementing occupant behaviour in the simulation of building energy performance and energy flexibility: development of co-simulation framework and case study. *The 15th International Conference International Building Performance Simulation Association*, California, USA, pp. 1339–1346.

102 Mlecnik, E. (2016). Goodbye passive house, hello energy flexible building? *32nd International Conference on Passive and Low Energy Architecture*. https://repository.tudelft.nl/islandora/object/uuid%3A0e009e81-a339-42f7-afea-e7c3b27a2a9f (accessed 19 October 2021).

103 Mlecnik, E., Parker, J., Ma, Z. et al. (2020). Policy challenges for the development of energy flexibility services. *Energy Policy* 137: 111147.

104 Oskouei, M.Z., Mohammadi-Ivatloo, B., Abapour, M. et al. (2020). A novel economic structure to improve the energy label in smart residential buildings under energy efficiency programs. *J. Cleaner Prod.* 260: 121059.

105 Ma, Z. and Jørgensen, B.N. (2018). A discussion of building automation and stakeholder engagement for the readiness of energy flexible buildings. *Energy Inf.* 1: 1–15.

106 Kohlhepp, P., Harb, H., Wolisz, H. et al. (2019). Large-scale grid integration of residential thermal energy storages as demand-side flexibility resource: a review of international field studies. *Renew. Sustain. Energy Rev.* 101: 527–547.

107 You, S., Hu, J., and Ziras, C. (2016). An overview of modeling approaches applied to aggregation-based fleet management and integration of plug-in electric vehicles. *Energies* 9: 968.

108 Goy, S. and Finn, D. (2015). Estimating demand response potential in building clusters. *Energy Procedia* 78: 3391–3396.

109 Buttitta, G., Turner, W., and Finn, D. (2017). Clustering of household occupancy profiles for archetype building models. *Energy Procedia* 111: 161–170.

110 Zhou, Y., Cao, S., Hensen, J.L.M., and Lund, P.D. (2019). Energy integration and interaction between buildings and vehicles: a state-of-the-art review. *Renew. Sustain. Energy Rev.* 114: 109337.

111 Zhou, Y., Cao, S., Kosonen, R., and Hamdy, M. (2020). Multi-objective optimisation of an interactive buildings-vehicles energy sharing network with high energy flexibility using the Pareto archive NSGA-II algorithm. *Energy Convers. Manage.* 218: 113017.

112 Fallahi, Z. and Henze, G.P. (2019). Interactive buildings: a review. *Sustainability* 11: 1–26.

113 Frei, M., Deb, C., Nagy, Z. et al. (2020). Building energy performance assessment using an easily deployable sensor kit: process, risks and lessons learned. *Front. Built Environ.* 6: 222.

114 Volkov, A. (2013). Building intelligence quotient: mathematical description. *Appl. Mech. Mater.* 409–410: 392–395.

115 Verbeke, Y.M.S., Bogaert, S., Van Tichelen, P., and Uslar, M. (2017). Support for setting up a smart readiness indicator for buildings and related impact assessment – catalogue of smart ready services technical working document for stakeholder feedback.

116 United Nations (2015). Transforming Our World: The 2030 Agenda for Sustainable Development.

117 Hviid, J. and Kjærgaard, M.B. (2018). Service abstraction layer for building operating systems: enabling portable applications and improving system resilience. *2018 IEEE International Conference on Communications, Control, and Computing Technologies for Smart Grids*, pp. 1–6.

118 Vigna, I., Pernetti, R., Pernigotto, G., and Gasparella, A. (2020). Analysis of the building smart readiness indicator calculation: a comparative case-study with two panels of experts. *Energies* 13: 2796.

119 Horák, O. and Kabele, K. (2019). Testing of pilot buildings by the SRI method. *Heating Vent. Sanitation* 28: 331–334.

120 Janhunen, E., Pulkka, L., Säynäjoki, A., and Junnila, S. (2019). Applicability of the smart readiness indicator for cold climate countries. *Buildings* 9: 102.

121 Märzinger, T. and Österreicher, D. (2020). Extending the application of the smart readiness indicator a methodology for the quantitative assessment of the load shifting potential of smart districts. *Energies* 13: https://doi.org/10.3390/en13133507.

122 Vigna, I., Pernetti, R., Pasut, W., and Lollini, R. (2018). New domain for promoting energy efficiency: energy flexible building cluster. *Sustain. Cities Soc.* 38: 526–533.

123 Apanaviciene, R., Vanagas, A., and Fokaides, P.A. (2020). Smart building integration into a smart city (SBISC): development of a new evaluation framework. *Energies* 13: 2190.

124 de Falco, S., Angelidou, M., and Addie, J.P.D. (2019). From the "smart city" to the "smart metropolis"? Building resilience in the urban periphery. *Eur. Urban Reg. Stud.* 26: 205–223.

125 Markoska, E., Jakica, N., Lazarova-Molnar, S., and Kragh, M.K. (2019). Assessment of building intelligence requirements for real time performance testing in smart buildings. *2019 4th International Conference on Smart and Sustainable Technologies (SpliTech).*

126 Bode, G., Behrendt, S., Fütterer, J., and Müller, D. (2017). Identification and utilization of flexibility in non-residential buildings. *Energy Procedia* 122: 997–1002.

127 Moseley, P. (2017). EU support for innovation and market uptake in smart buildings under the Horizon 2020 framework programme. *Buildings* 7: 105.

128 Cheng, C.C. and Lee, D. (2016). Enabling smart air conditioning by sensor development: a review. *Sensors (Switzerland)* 16: 2028.

129 André, M., De Vecchi, R., and Lamberts, R. (2020). User-centered environmental control: a review of current findings on personal conditioning systems and personal comfort models. *Energy Build.* 222: 110011.

130 Shah, A.S., Nasir, H., Fayaz, M. et al. (2019). A review on energy consumption optimization techniques in IoT based smart building environments. *Information* 10: 108.

131 Ulpiani, G., Borgognoni, M., Romagnoli, A., and Di Perna, C. (2016). Comparing the performance of on/off, PID and fuzzy controllers applied to the heating system of an energy-efficient building. *Energy Build.* 116: 1–17.

132 Ulpiani, G. (2017). Overheating phenomena induced by fully-glazed facades: investigation of a sick building in Italy and assessment of the benefits achieved via fuzzy control of the AC system. *Sol. Energy* 158: 572–594.

133 Esmaeilzadeh, A., Zakerzadeh, M.R., and Koma, A.Y. (2018). The comparison of some advanced control methods for energy optimization and comfort management in buildings. *Sustain. Cities Soc.* 43: 601–623.

134 Liu, J., Zhang, W., Chu, X., and Liu, Y. (2016). Fuzzy logic controller for energy savings in a smart LED lighting system considering lighting comfort and daylight. *Energy Build.* 127: 95–104.

135 Hang, L. and Kim, D.H. (2018). Enhanced model-based predictive control system based on fuzzy logic for maintaining thermal comfort in IoT smart space. *Appl. Sci.* 8: 1031.

136 Park, H. and Rhee, S.B. (2018). IoT-based smart building environment service for occupants' thermal comfort. *J. Sensors* 2018, 10 pages.

137 Fayaz, M. and Kim, D.H. (2018). Energy consumption optimization and user comfort management in residential buildings using a bat algorithm and fuzzy logic. *Energies* 11: 1–22.

138 Shareef, H., Ahmed, M.S., Mohamed, A., and Al Hassan, E. (2018). Review on home energy management system considering demand responses, smart technologies, and intelligent controllers. *IEEE Access.* 6: 24498–24509.

139 Merabti, S., Draoui, B., and Bounaama, F. (2017). A review of control systems for energy and comfort management in buildings. *Proceedings of 2016 8th International Conference on Modelling, Identification and Control (ICMIC)*, pp. 478–486.

140 Ngarambe, J., Yun, G.Y., and Santamouris, M. (2020). The use of artificial intelligence (AI) methods in the prediction of thermal comfort in buildings: energy implications of AI-based thermal comfort controls. *Energy Build.* 211: 109807.

141 Han, M., May, R., Zhang, X. et al. (2019). A review of reinforcement learning methodologies for controlling occupant comfort in buildings. *Sustain. Cities Soc.* 51: 101748.

142 Thieblemont, H., Haghighat, F., Ooka, R., and Moreau, A. (2017). Predictive control strategies based on weather forecast in buildings with energy storage system: a review of the state-of-the art. *Energy Build.* 153: 485–500.

143 Dong, B., Prakash, V., Feng, F., and O'Neill, Z. (2019). A review of smart building sensing system for better indoor environment control. *Energy Build.* 199: 29–46.

144 D'Oca, S., Hong, T., and Langevin, J. (2018). The human dimensions of energy use in buildings: a review. *Renew. Sustain. Energy Rev.* 81: 731–742.

145 Park, J.Y., Ouf, M.M., Gunay, B. et al. (2019). A critical review of field implementations of occupant-centric building controls. *Build. Environ.* 165: 106351.

146 EXESS EU H2020 project. https://positive-energy-buildings.eu/ (accessed 19 October 2021).

147 ZERO-PLUS EU H2020 project. http://www.zeroplus.org/ (accessed 19 October 2021).

148 ATELIER EU H2020 project. https://smartcity-atelier.eu/ (accessed 19 October 2021).

149 Saheb, Y., Shnapp, S., and Paci, D. (2019). From nearly-zero energy buildings to net-zero energy districts. EUR 29734 EN. Luxembourg: Publications Office of the European Union. ISBN 978-92-76-02914-4. https://doi.org/10.2760/323828, JRC115188.

150 IEA (International Energy Agency) EBC (Energy in Buildings and Communities). Programme - Annex 83 "Positive Energy Districts". https://annex83.iea-ebc.org/ (accessed 19 October 2021).

151 EU COST Action – PED-EU-NET. www.cost.eu/actions/CA19126/#tabs|Name: overview (accessed 19 October 2021).

152 Lindholm, O. and Reda, F. (2021). Positioning positive energy districts in European cities. *Buildings* 11 (1): 19.

153 SET Plan Action 3.2 on "Smart Cities and Communities". https://setis.ec.europa.eu/system/files/setplan_smartcities_implementationplan.pdf (accessed 19 October 2021).

154 JPI Urban Europe, Positive Energy Districts (PED). https://jpi-urbaneurope.eu/ped/ (accessed 19 October 2021).

155 Framework Definition for Positive Energy Districts and Neighbourhoods (2020). https://jpi-urbaneurope.eu/wp-content/uploads/2020/04/White-Paper-PED-Framework-Definition-2020323-final.pdf.

156 Bossi, S., Gollner, C., and Theierling, S. (2020). Towards 100 positive energy districts in Europe: preliminary data analysis of 61 European cases. *Energies* 13 (22): 6083.

157 Ullah, K.R., Prodanovic, V., Pignatta, G. et al. (2021). Technological advancements towards the net-zero energy communities around the globe: a review. *Sol. Energy* 224: 1107–1126.

158 Löhnert, G., Dalkowski, A., and Sutter, W. (2003). Integrated Design Process - a guideline for sustainable and solar-optimized building design, IEA Task 23 Optimization of Solar Energy Use in Large Buildings Subtask B Design Process Guidelines, Berlin / Zug.

159 CEC (2015). *Improving Green Building Construction in North America: Guide to Integrated Design and Delivery*. Montreal, Canada: Commission for Environmental Cooperation.

160 Bomba, M.B. and Parrott, B. (2010). Integrated project delivery and building information modeling: a new breed of contract. *PCI J.* 2: 146–153.

161 Hoonakker, P., Carayon, P., and Loushine, T. (2010). Barriers and benefits of quality management in the construction industry: an empirical study. *Total Qual. Manag. Bus. Excell.* 21 (9): 953–969.

162 Hwang, B.G. and Ng, W.J. (2013). Project management knowledge and skills for green construction: overcoming challenges. *Int. J. Project Manage.* 31 (2): 272–284.

163 Aghimien, D.O., Oke, A.E., and Aigbavboa, C.O. (2018). Barriers to the adoption of value management in developing countries. *Eng. Constr. Archit. Manag.* 25 (7): 818–834.

164 Winston, N. (2010). Regeneration for sustainable communities? Barriers to implementing sustainable housing in urban areas. *Sustain. Dev.* 18 (6): 319–330.

165 Wells, L., Rismanchi, B., and Aye, L. (2018). A review of net zero energy buildings with reflections on the Australian context. *Energy Build.* 158: 616–628.

166 Feng, W., Zhang, Q., Ji, H. et al. (2019). A review of net zero energy buildings in hot and humid climates: experience learned from 34 case study buildings. *Renew. Sustain. Energy Rev.* 114: 109303.

167 Lin, Y., Zhong, S., Yang, W. et al. (2020). Towards zero-energy buildings in China: a systematic literature review. *J. Cleaner Prod.* 276: 123297.

168 Bandeiras, F., Gomes, M., Coelho, P., and Fernandes, J. (2020). Towards net zero energy in industrial and commercial buildings in Portugal. *Renew. Sustain. Energy Rev.* 119: 109580.

169 Wei, W. and Skye, H.M. (2021). Residential net-zero energy buildings: review and perspective. *Renew. Sustain. Energy Rev.* 142: 110859.

170 Mavrigiannaki, A., Pignatta, G., Assimakopoulos, M. et al. (2021). Examining the benefits and barriers for the implementation of net zero energy settlements. *Energy Build.* 230: 110564.

171 Nematchoua, M.K., Nishimwe, A.M.R., and Reiter, S. (2021). Towards nearly zero-energy residential neighbourhoods in the European Union: a case study. *Renew. Sustain. Energy Rev.* 135: 110198.

172 Pignatta, G. (2020). A simplified approach for designing sustainable Near Zero Energy Settlements. *Planning Post Carbon Cities: 35th PLEA Conference on Passive and Low Energy Architecture, A Coruña, 1st-3rd September 2020: Proceedings, A Coruña, Spain, presented at PLEA2020, A Coruña*, Spain, 01 September 2020 - 03 September 2020.

173 Cardinali, M., Pisello, A.L., Piselli, C. et al. (2020). Microclimate mitigation for enhancing energy and environmental performance of Near Zero Energy Settlements in Italy. *Sustain. Cities and Society* 53: 101964.

174 Pignatta, G., Chatzinikola, C., Artopoulos, G. et al. (2017). Analysis of the indoor thermal quality in low income Cypriot households during winter. *Energy Build.* 152: 766–775.

175 Huang, P., Huang, G., and Sun, Y. (2018). Uncertainty-based life-cycle analysis of near-zero energy buildings for performance improvements. *Appl. Energy* 213: 486–498.

176 Andresen, I., Wiik, M.K., Fufa, S.M., and Gustavsen, A. (2019). The Norwegian ZEB definition and lessons learnt from nine pilot zero emission building projects. *IOP Conference Series: Earth and Environmental Science* (Vol. 352, No. 1, p. 012026). IOP Publishing.

177 Mavrigiannaki, A., Gobakis, K., Kolokotsa, D. et al. (2020). Measurement and verification of zero energy settlements: lessons learned from four pilot cases in Europe. *Sustainability* 12 (22): 9783.

178 Artopoulos, G., Pignatta, G., and Santamouris, M. (2018). *From the Sum of Near-Zero Energy Buildings to the Whole of a Near-Zero Energy Housing Settlement: The Role of Communal Spaces in Performance-Driven Design*. Architecture_MPS.

179 Meir, I.A., Isaac, S., Kolokotsa, D. et al. (2020). Towards zero energy settlements–a brief note on commissioning and POE within the EU Zero-Plus Settlements. *IOP Conference Series: Earth and Environmental Science* (Vol. 410, No. 1, p. 012038). IOP Publishing.

180 Caramizaru, A. and Uihlein, A. (2020). *Energy Communities: An Overview of Energy and Social Innovation*. Publications Office of the European Union.

181 European Parliament & Council of the European Union (2018). Renewable Energy Directive (EU) 2018/2001.

182 European Parliament & Council of the European Union (2019). Internal Electricity Market Directive (EU) 2019/944.

11

Transition Potential of Local Energy Communities

Gabriele Comodi[1], *Gianluca Spinaci**[2], *Marialaura Di Somma*[3], *and Giorgio Graditi*[3]

[1] *Università Politecnica delle Marche, Dipartimento di Ingegneria Industriale e Scienze Matematiche (DIISM), via Brecce Bianche, 60100, Ancona, Italy*
[2] *European Committee of the Regions, Rue Belliard 101, 1040, Bruxelles, Belgium*
[3] *Italian National Agency for New Technologies, Energy and Sustainable Economic Development, ENEA, Department of Energy Technologies and Renewable Sources, Rome, Italy*

11.1 Introduction

Energy transition is a pattern toward a 100% carbon-free energy system. Energy transition entails a change of paradigm in all the supply chain of the energy sector from production to transmission, distribution, and to final consumption. Indeed, in the past decades, national energy systems were featured by a centralized power generation paradigm in which electricity was produced in large power plants and then transmitted and distributed to final users who were considered as passive actors. In this context, the concept of energy communities was limited to the main paradigm of supplying electricity to geographically constrained communities such as remote villages or islands. Thanks to distributed generation (DG) and to the mini-grid concept, it was possible to electrify rural areas of less developed or developing countries and bring services to enable economic development in zones not served by the national electric power system [1]. The usual configuration was the so-called "second-generation" mini-grids [2, 3], a standalone electric power generation and distribution system composed by a diesel generator for supplying programmable and reliable electricity, coupled with renewable energy sources (RESs), mostly photovoltaic (PV) but also wind and mini-hydro, as well as battery energy storages [4–6]. Thermal energy production for heating and cooking was mainly supplied by solid biomass and charcoal [5]. Often, cross-sectoral collaboration led to develop synergies between agriculture and energy sectors by using wastes from agriculture processes to produce biogas to feed an internal combustion

* The contents of this publication express the views of its author and do not necessarily represent the opinion of the European Committee of the Regions. The European Committee of the Regions cannot be held responsible in any way whatsoever for any use that may be made of the information contained within.

Technologies for Integrated Energy Systems and Networks, First Edition.
Edited by Giorgio Graditi and Marialaura Di Somma.

engine (ICE) generator [1, 5]. In the context of "islanded mini-grids" [6], namely off-grid mini-grid, "geographical" islands need a particular mention as they anticipate features and motivation of energy communities both in developing and in developed countries. First of all, geographical islands necessitate to maximize renewable energy production driven by the need of reducing the costs for fossil fuel supply [7], which are much higher in islands than in mainland [8]. Secondly, most of the islands worldwide present sector integration of at least two networks (water network and electricity network), which are strictly related as water is usually produced by desalination plant fed by electricity (reverse osmosis [RO] plants) or by waste heat recovery, usually from energy production plants (thermal desalination) [9]. Finally, geographical islands are particularly sensitive to resiliency aspects of their energy systems, e.g. to face destructive weather extreme [10]. In both the case of electrification of rural areas and energy supply to geographical energy islands, the use of integrated energy systems consisting of fossil fuel generators, renewable sources, and energy storage was mainly driven by economic objectives rather than by environmental motivations. Indeed, the mini-grid configuration allows electrification of sparsely populated communities avoiding huge investments in transmission and distribution networks.

In recent years, the concept of local energy communities (LECs) has been gaining a new interest as a key enabler of the energy transition in both developing and developed countries [11, 12]. In this context, it is worth to mention two important international initiatives: the United Nation (UN) 2030 Agenda on Sustainable Development Goals (SDGs) and the European Green New Deal.

11.1.1 "2030 Agenda for Sustainable Development" of United Nations

The "2030 Agenda for sustainable development" of United Nations is a global partnership among all countries in order to deploy urgent actions by 2030 in order to make the world more sustainable by facing poverty and climate change [13]. The main actions are collected in 17 SDGs spanning from environment, health, poverty, education, gender quality, water, and energy. Among these 17 goals, SDG 7 aims at "ensuring access to affordable, reliable, sustainable, and modern energy for all" [14]. In developed countries, this goal translates mainly into increasing the share of global energy mix (Target 7.2) and into improve energy efficiency (Target 7.3), while, in least developed and developing countries, it translates into expanding infrastructure and upgrading technology for supplying modern and sustainable energy services for all (https://sdgs.un.org/goals/goal7). The main idea underpinning SDG 7 is that reducing energy poverty by ensuring energy access will enable energy services that will boost other key actions/services such as healthcare facilities, education, access to clean water [15], and better conservation of food and drugs [15, 16]. In 2018, 789 million people faced lack of electricity, and in some developing countries, one out of four health facilities was not electrified [14]; in 2017, 2.90 billion people did not have access to clean cooking [14]. In this context, mini-grids are well-recognized tools to ensure energy access. Today, 47 million people are connected to 19 000 mini-grids. In order to reach universal energy access by 2030, the goal of the World

Bank, through its Energy Sector Management Assistance Program (ESMAP), is to serve 490 million people with 210 000 "third-generation" mini-grids [2]. With respect to second-generation mini-grids, the third one is featured by solar hybrid technologies, energy-efficient technologies in final uses, smart meters, and energy management systems (EMS), as well as community engagement. Several initiatives have been developed in the past years. SDG 7 is strictly related to SDG 11: "Make cities inclusive, safe, resilient, and sustainable" and in particular to the topic of "sustainable cities and communities." Indeed, the electrification of rural areas and the consequent boost of local economies is a way to reduce rural depopulation and mitigate the problem of rapid urbanization of cities and metropolitan areas in which energy consumption and local pollution are one of the major challenges. Decarbonization and sustainable planning of urban districts and cities are pivotal in the path toward energy transition.

11.1.2 Clean Energy for All European Package: Renewable and Citizen "Energy Communities"

In 2019, EU Commission published the Clean Energy for all European package [17], a set of laws aiming at accelerating the European clean energy transition, thus putting Europe as a forefront leader in energy and climate actions. As discussed in Chapter 1 of this book, the Clean Energy package recognizes the pivotal role of consumers in the energy transition; empowering citizens and making them active in the energy supply chain is one of the pillars of the EU strategy toward energy transition. This strategy envisages a user-centered smart energy system where final consumers will have the equivalent relevance of centralized power plants, transmission, and distribution networks. This idea is based on the fact that the energy transition toward a zero-carbon energy system will be achieved both by increasing renewable energy production and by acting on final users energy demand. The latter point can be achieved in two ways: on the one hand, by increasing the energy efficiency in final uses [18], thus moderating the energy demand doing the same things while consuming less; on the other hand, by empowering final users (industries, tertiary sector, and households), in order to provide flexibility to the energy system by adapting their patterns of consumption to the pattern of renewable energy production from non-programmable RES. This is even more true in the case of prosumers that are final "users," which are also "producers" after having installed distributed energy resources (DERs) such as PV, micro-wind, and combined heat and power (CHP) plant. In any case, as previously mentioned, the road leading to active final users/prosumers is far from being achieved, the main barriers being related to the difficulties to engage final users, especially the less skilled, and to their reluctance toward new technologies.

The main tool identified by the European Union to engage and empower citizens in the energy sector is the energy community. Energy communities are defined in two EU Directives: the revised Renewable Energy Directive (RED II) [19] and the Internal Electricity Market Directive (EMD II) [20].

The RED II establishes a common framework for the promotion of energy from renewable sources. In this context, it promotes the opportunity of final users to become self-consumers of renewable energy or to participate to renewable energy communities (RECs). In both cases, the overall goal is to promote the self-consumption of distributed renewable energy production through the participation of final consumers/prosumers.

The EMD II establishes common rules for the generation, transmission, distribution, energy storage, and supply of electricity, together with consumer protection provisions, with a view to creating truly integrated competitive, consumer-centered, flexible, fair, and transparent electricity markets in the Union. In this context, the Directive defines the citizen energy communities (CECs), namely an enabling framework for cooperation of citizens in energy initiatives. With respect to the renewable ones, limited to self-consumption of local renewable energy, CECs have a larger scope.

In both directives, the European Union defines energy communities as legal entities providing "environmental, economic, or social community benefits for its shareholders or members or for the local areas where it operates, rather than financial profits." The last sentence is particularly meaningful about the role of energy communities in the approach of European Union to the energy transition.

First of all, the environment. The main goal is to foster energy transition toward a zero-carbon energy system in order to reverse the course of climate change. The added value of energy communities is to increase awareness and engagement of citizens through their direct participation in energy initiatives, thus increasing their acceptance of enabling technologies and renewable energy.

Secondly, citizens should be encouraged to participate to energy communities thanks to economic benefits that are different from financial profits, especially in citizens' perception.

Finally, social community benefits. In the EU purpose, energy communities will play a role in hindering energy poverty, in increasing energy democracy, and in building human capital and enhancing citizen skills. Indeed, energy communities allow democratic participation to energy initiatives of final users [21], also those vulnerable, otherwise excluded, which can take advantage of economic benefits such as reduced energy tariffs. According to the European Committee of the Regions (CoR), the energy community paradigm can "have a positive impact on sustainable local economic and social development and can thus also contribute to tackle energy poverty and promote job creation within the community" [22] thanks to the spread of DERs that will be installed and will require maintenance services.

11.1.3 Human Capital for Local Energy Communities

The concept of LEC is based on two main pillars: (i) the installation of low/zero-carbon distributed technologies and (ii) active and empowered final users/prosumers. Both pillars imply an improvement or even building of human capital in several sectors and at different levels.

First of all, the spread of DG technologies (PV, cogeneration and trigeneration, mini-hydro, biogas, etc.) requires technical knowledge by local technicians, engineers, and designers who must design, build, operate, and maintain the DG plants and networks within the communities. Moreover, in order to foster energy communities in the energy transition, skills related to medium–long-term energy planning under both a technical and a policy point of view will also play a pivotal role to support local policy makers. This entails that, while designing the future LEC, energy planners should consider not only the state of the art of technologies but also the new ones entering the market such as energy storage, electric vehicles (EVs), and in future hydrogen and power-to-X technologies. It goes without sayings that these new technologies entering in the future market will require local technicians, engineers, and designers duly skilled.

Another important aspect, strictly related to the LECs in the energy transition, is the definition of new business models that require new skills mainly related to qualifications of the management team of the LEC [23]. In particular, beyond the know-how strictly related to energy technologies and renewables, other competences are needed such as business and legal know-how, project development, and management.

The development of new business models for LECs is also strictly related to the second aforementioned pillar, which is the empowerment of final users. Indeed, most of business models proposed for LECs require an active participation of final users/prosumers who are requested to join community programs designed to maximize local self-consumption of renewable energy and to increase flexibility of their consumption patterns in order to face the uncertainty of renewable production and unburden the electric grid. In any case, final users/prosumers are (and will be even more in future) part of a market-driven change in which they will be asked to use a set of emerging technologies as building energy management systems (BEMS), smart thermostats, smart sensors, electric batteries, and EVs, which require a minimum set of skills to be properly managed.

For all these reasons, the enhancement of human capital in LECs is widely considered a key factor to overcome barriers to the energy transition deployment such as the technology/RES acceptance and the poor skills of final users hampering their opportunity of an active role in the energy system.

11.1.4 Local Energy Communities: An Organizational Bottom-Up Model to Empower Final Users

The LEC paradigm improves the traditional concept of mini-grid and DG both by including active consumers/prosumers as part of the energy systems with equal status as generation and transmission/distribution and by enabling sector integration at the local scale. In this sense, LECs are a bottom-up organizational model, which can foster the deployment of sector coupling to provide greater flexibility to the energy system so that a cost-effective decarbonization can be achieved [24, 25]. Indeed, as will be discussed later in detail, LECs play a role both in the "end-user" and in "cross-vector" sector coupling: the former consists in an electrification of final users' energy demand, the latter consists, on demand side, in an integrated use of different energy infrastructures and vectors, in particular electricity, heat, and gas [26]. Typical examples of technologies fostering electrification of final users' energy demand are heat pumps (for heating and cooling), EVs (for mobility), electric boilers (for sanitary hot water production), and electric hob (for cooking). A typical example of technology fostering cross-vector sector coupling at the demand side is cogeneration in district heating (DH) applications: in this case, CHP plant, usually fed by natural gas but also biogas or solid biomass, simultaneously produces electricity and thermal energy [26].

Hence, in this context, the LEC becomes a legal entity providing an umbrella under which final users and prosumers can progressively grow in the path of transformation from passive to active users. Usually, the first step is the simple awareness of final users about both their way to consume energy and their role within the energy system. This first step can be fostered by awareness campaigns organized by the LEC or even better by installing smart meters.

The second step consists in the acceptance of technologies and in the engagement by final users [27, 28]. In this phase, final users install energy-efficient technologies, simple monitoring systems, and DG technologies such as rooftop PV, thus becoming prosumers, and, finally, they activate simple demand side management strategies such as (i) trying to maximize self-consumption of renewable energy by consuming more while PV plant is producing electricity or (ii) trying to minimize the energy bill by shifting consumptions in hours where electricity price is lower. LECs can foster this second step by promoting collective purchase of technologies, thus reducing costs and possibly including more vulnerable people.

Finally, the final step consists in the total empowerment of final users. In this case, the LEC can be the legal entity coordinating a set of active prosumers (able to modify their consumption patterns) and community technologies (e.g. community storage, CHP-district heating [DH] power plant, etc.) or setting up demand response (DR) programs in order to provide flexibility to the grid to reduce grid congestion and to maximize local self-consumption of local renewable energy production.

As mentioned above, this final step, namely the empowerment of final users, will be stressed in future by the launch of new business models for LEC.

11.2 Local Energy Communities Making the Green Deal Going Local

11.2.1 Game Changer of the Green Deal

As illustrated beforehand, LECs have been already promoted by the European Union through its *Clean Energy for All Europeans* package. Consequent to that, however, a major political project came to birth by the end of 2019: the European Green Deal, a plan to make Europe the first climate-neutral continent by 2050. Following the European elections in May 2019, the European Parliament voted in favor of the new European Commission led by President Ursula Von Der Leyen. The European Green

Deal, a far-reaching transformative agenda, was indicated at the very top of the political program for the whole legislature (2019–2024).

This deal resulted from a combination of political drivers:

1) The need to meet the clear request expressed by the electors and by the youngest generations to urgently address the climate emergency and set a new path for sustainable development, capable of avoiding the tipping point of the climate disaster.
2) The geopolitical approach of the Von Der Leyen Commission, which identified the "climate-energy-sustainable development" nexus as one of the key areas where the European Union needed to assume a leading role at global level, following the Paris Agreement.
3) The very concrete urgency to move up a gear on the European policy, regulatory, and financial framework in order to enable European technological and economic champions to compete in the new global "green markets" with rapidly emerging competitors from other continental powerhouses.

While the European Green Deal strategy set in December 2019 was very comprehensive across policy areas (as shown in Figure 11.1), one could argue that its geopolitical approach was a substantial conceptual element at its very outset and still remains an essential condition for its success at global level.

The timeline of the European Green Deal was then, unforeseeably, marked by the emergency of the COVID-19 health and socio-economic crisis, which unfolded as from spring 2020. The political and policy answer was however up to the

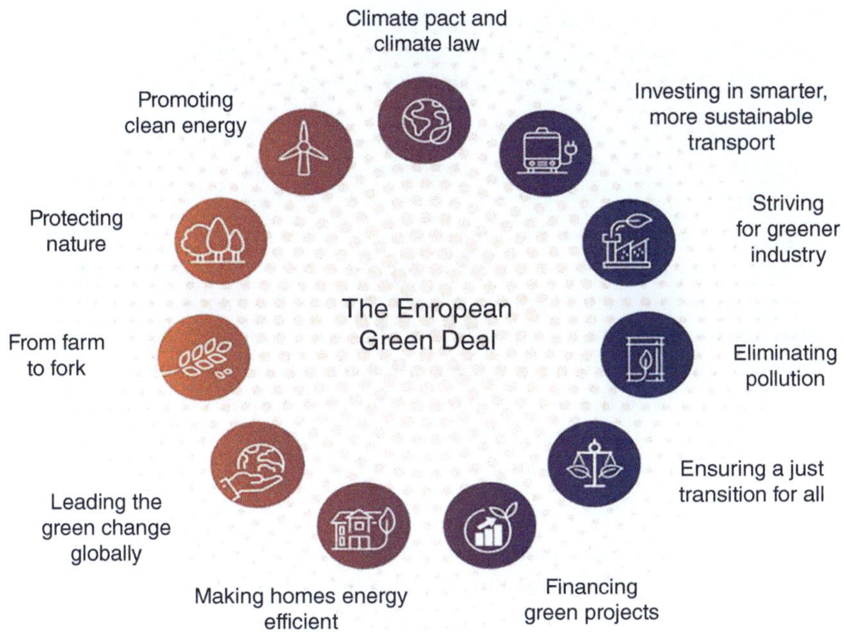

Figure 11.1 Key objectives of the European Green Deal. Source: European Commission.

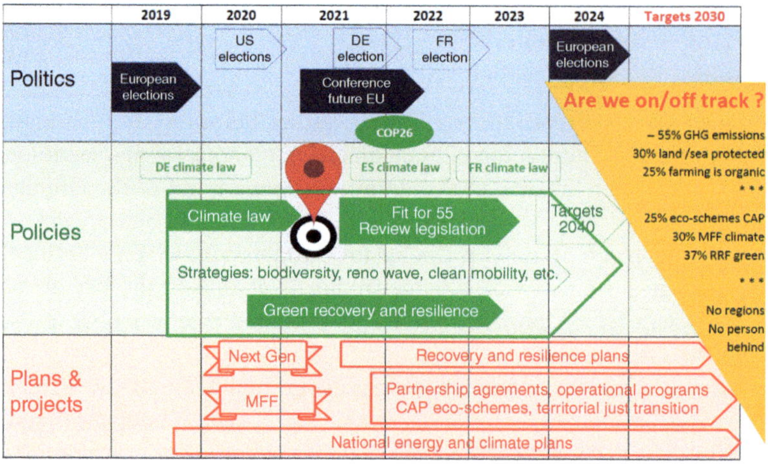

Figure 11.2 Timeline of the European Green Deal. Source: European Committee of the Regions.

expectations. Against a possible concrete risk of marginalization of the Green Deal agenda because of the new looming crisis, the European Union decided to put even more emphasis and investments into it. The green transition became a pillar of the EU Recovery and Resilience Plans with green investments being allocated up to 37% of the €750 billion from Next Generation EU and 30% of the €1080 billion from the new programs financed via the Multiannual Financial Framework (MFF) 2021–2027.

As illustrated by Figure 11.2, at the end of 2021, just two years after its launch in December 2019, the European Green Deal had therefore taken a very substantial shape, on the basis of

(a) a comprehensive set of thematic strategies and action plans (launched in 2019/2020);
(b) an unprecedented firepower in terms of financial package (agreed in 2020, Next Gen EU for the period 2021–2026 Next Gen; MFF 2021–2027); and
(c) the proposal to profoundly review the EU legislative frameworks regarding energy, transport, agriculture, manufacturing, and trading sector (*Fit for 55*), which was presented by the European Commission on 14 July 2021 to give concrete follow-up to the target of reducing greenhouse gas (GHG) emissions by 55% before 2030 (compared to 1990): via its European Climate Law, for the first time, the EU set in motion a legally binding framework and path toward climate neutrality.

While all the cards seem now on the table, there is still a long way to go in order to strike all the targets by the finishing line. In addition, this will be the real challenge for the rest of the current EU legislature, until the next European elections in May 2024, most importantly in terms of concrete implementation on the ground in the Members States and in their different territories.

11.2.2 Green Deal Going Local

It is exactly with the spirit of DELIVERING the Green Deal on the ground that the European local and regional authorities provided their very strong political support, bottom-up and from the very beginning. In particular, the European CoR adopted its resolution "The Green Deal in partnership with local and regional authorities"[1] on 5 December 2019, one week ahead of the European Commission (EC) Communication itself.

The CoR political requests were clearly directed to ensure that the European Green Deal would not lose its grass-root origin and would instead be built upon a set of tools strongly place-based, enabling action from the bottom, citizens' engagement and investment at the level of local communities. In parallel to these requests, the CoR took itself the initiative to launch an ambitious action called "Green Deal going local".[2]

The initiative consists of three main pillars:

(a) **Policy advocacy**: Ensure policy and legislative consistency across the different thematic strands of the European Green Deal, notably to facilitate their viability also for the smaller scale territories and administrations and more in general for an integrated roll-out at the local and regional level also via direct funding.
(b) **Partnership at the EU level:** Strengthen cooperation among the different branches of the EU institutions to increase the impact of the European Green Deal package across the different layers: proposal and execution by the European Commission; legislation by the European Parliament and Council; and mobilization on the ground via the CoR.
(c) **Enable and engage local actors**: Assist local and regional authorities to seize all the new opportunities of the European Green Deal by fostering awareness, capacity building, access to funding, and exchanges of best practices.

The initiative was launched in June 2020, and it has already achieved a number of accountable results:

(a) The European Commission has followed a number of CoR suggestions in several of its policy and legislative proposals (in Section 11.2.3, the Renovation Wave's example is given).
(b) A substantial number of 200 best practices of Green Deal going local has been collected and made accessible online.[3]
(c) A conceptual methodology for tailor made "Local Green Deals – A blueprint for action" has been developed, with the support of the European Commission, DG GROW.[4]
(d) The mobilization has reached out well beyond the 329 members of the CoR (mayors, presidents of regions, and councilors from all the 27 Member States).

1 https://cor.europa.eu/en/Documents/COR-2019-04351-00-00-RES-TRA-EN.pdf.
2 http://www.cor.europa.eu/GreenDealGoingLocal.go.
3 https://cor.europa.eu/EN/regions.
4 https://www.intelligentcitieschallenge.eu/sites/default/files/2021-06/Local%20Green%20Deals-8.pdf.

Active partnerships notably with the Covenant of Mayors allowed us to realize a dense calendar of activities to bring awareness, access to assistance, and knowledge on funding at the local level in every Member State.

11.2.3 Neighborhood Approach and Local Energy Communities in the Green Deal

Although the European Green Deal was on the rollercoaster of the Member States negotiations on the financing of the Next-Generation EU across summer 2020, it became soon evident that a large-scale intervention of renovation of buildings would become a key pillar of the relaunch package. In all countries, such an intervention had a strong rationale with its huge potential to both reduce GHG emission via energy efficiency and boost economy and jobs hardly hit at local level by the COVID-generated crisis. In this context it shall be highlighted the political relevance of the communication "A Renovation Wave for Europe – greening our buildings, creating jobs, improving lives"[5] which the European Commission adopted in October 2020. It is arguably the clearest signal from the Commission of embracing the logics of the *Green Deal going local* and of ensuring continuity with certain paths launched by the *Clean Energy for All Europeans* package back in 2018/2019. As a matter of fact, the Communication declared "Placing an integrated, participatory and neighbourhood based approach at the heart of the renovation wave." The Commission clearly reaffirmed the value of promoting an active participation of citizens in the energy system as prosumers via "integrated digital renovation that combines energy storage and demand-side flexibility, on-site energy generation from renewable resources, Internet of Things of the system components, appliances and recharging points for e-mobility." Other important policy orientations were provided in the same Communication, which had the destiny to become a reference point also for the Member States, when asked to prepare their national recovery and resilience and plans few months later.

The Commission included strong references to the following:

- "district and community approaches" hinting to zero-energy or even positive energy districts (e.g. advanced district heating and cooling systems with large potential for renewables and waste heat recovery);
- "energy communities" underlying their capacity to generate, consume, store, and sell energy, while offering tools for the most vulnerable citizens to lift them out of energy poverty[6];
- "affordable housing initiative" launching a pilot of 100 lighthouse renovation districts fighting energy poverty and promoting social engagement models.

  In the same Communication, the European Commission also indicated a very concrete toolbox through which similar approaches could become viable for the local actors, in particular municipal authorities:

5 COM(2020)662 of 14 October 2020 https://ec.europa.eu/energy/sites/ener/files/eu_renovation_wave_strategy.pdf.

6 A collection of good practices is provided by the European Federation of Citizen Energy Cooperatives, https://www.rescoop.eu/.

- **Assistance and capacity building** via a revamped model of the technical assistance facility European Local Energy Assistance (ELENA)[7], managed by the European Investment Bank (EIB) and to be deployed in a more decentralized way through Member States and regions.
- **Financing** via earmarking of funds both in the national recovery and resilience plans and in the national and regional operational program to be financed by the regional cohesion policy 2021–2027.
- **Mobilization and ownership** via "a close partnership with the CoR" to ensure ownership of cities, local and regional authorities, stakeholders, and national authorities.

As a matter of fact in March 2021, the European Commission (via its Directorate General for Energy) and the European Committee of the Regions signed a "joint action plan for enhanced cooperation on the renovation wave," taking stock of a number of concrete initiatives already launched as from the fall 2020: webinars, information campaign on funding, collection of best practices, etc.

11.3 Local Energy Communities as Integrated Energy Systems at Local Level

11.3.1 Local Energy Communities as Promoters for Sector Coupling

It is widely recognized that achieving integrated energy systems is one of the pillars of the medium/long-term strategy for the decarbonization of the planet. Integrating energy systems means to link different sectors in order to better exploit synergies among different energy vectors and networks and consequently be able to host an ever-increasing production of renewable energy. LECs can be considered as multi-energy environments characterized by the presence of multiple energy vectors and networks to satisfy the multi-energy demand of end-users. They can thus contribute to the integration of energy systems at local scale by promoting "end-user" and "cross-vector" sector coupling and by exploiting renewable energy production on-site.

Figure 11.3 shows a representative scheme of an integrated energy system for an LEC in terms of sectors, distribution networks, energy conversion, and generation technologies. At the far right of the figure, there is the demand (electricity, natural gas, heating and cooling, mobility, and water) of different sectors (residential, tertiary/services, industry, etc.). At the far left of the figure, there is the mix of generation technologies fed by both fossil and renewable energy that allow satisfying the user demand. The goal of the energy transition is to minimize (low carbon) or even eliminate (zero-carbon/carbon free) the fossil fuel contribution on the generation side.

In the center of Figure 11.3, between energy generation and energy demand, there is a variety of energy conversion technologies enabling the "connection" between at

7 https://www.eib.org/en/products/advising/elena/index.htm.

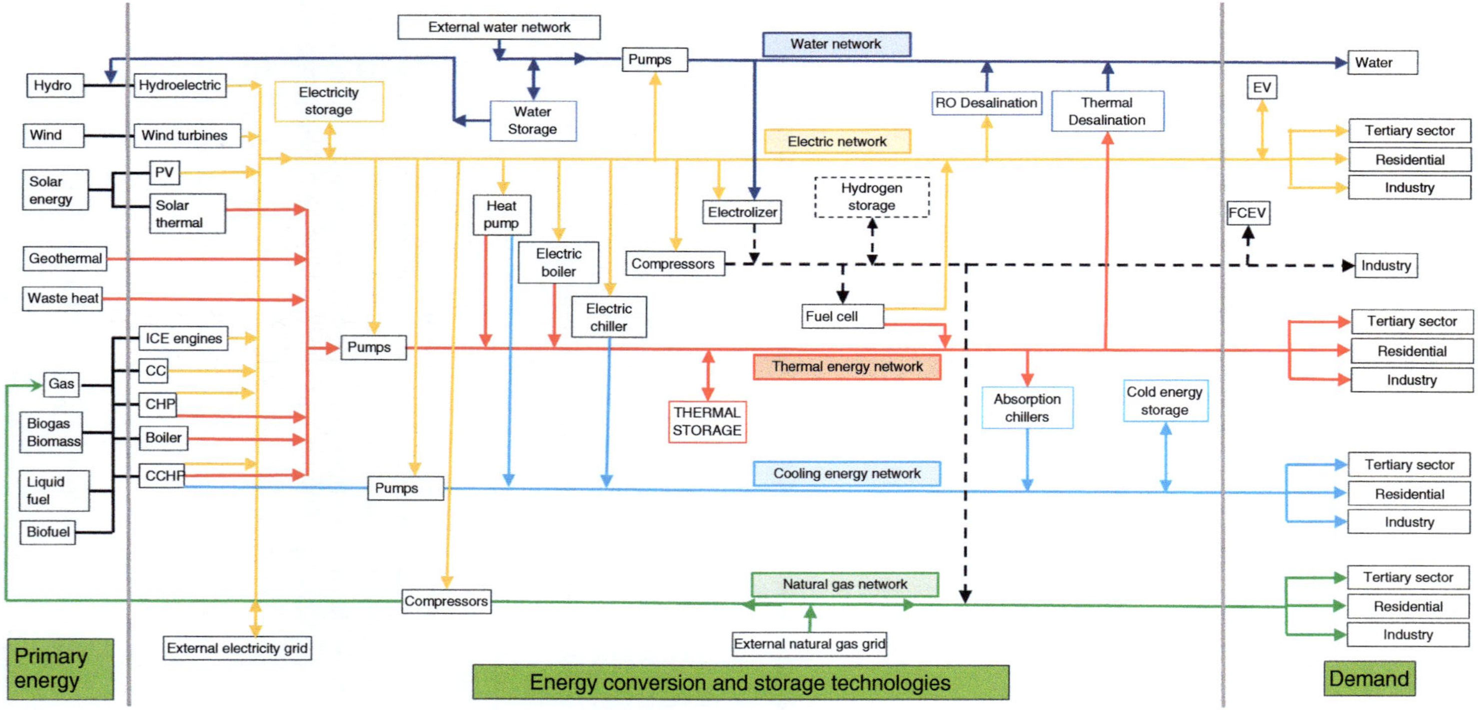

Figure 11.3 Representative scheme of an integrated energy system for a local energy community.

least two energy vectors. Most of these technologies have electricity as input energy vector and are leading the "end-user" sector coupling transformation, namely the electrification of energy end uses: heat pumps, electric chillers, electric boilers, and EVs turn electricity into a different vector, thus covering energy needs (domestic hot water [DHW], heating, cooling, and mobility), which are often met by fossil fuels (natural gas, coal, oil derivatives, etc.).

Other technologies are more suitable for "cross-vector" sector coupling, by enabling also synergies between different physical networks. A practical example is represented by cogeneration plants that link together gas, electricity, and thermal networks or trigeneration plants that also add cooling network to the previous three; reverse osmosis and thermal desalination systems that link water network with electric and heating networks, respectively; electrolyzers, that in future, will connect electric and gas networks.

In the center of Figure 11.3, there are also storage technologies (e.g. thermal and electric storage) that even if dealing with only one vector play an important role in providing flexibility to the integrated energy system as a whole by offsetting possible mismatch between not-dispatchable renewable energy production and energy demand.

Nearly zero- or zero-carbon LEC can be achieved by increasing the local energy production of renewable energy (PV, rooftop PV, mini-wind, mini-hydro, solar thermal, biomass, biogas, etc.) and by optimally managing local energy demand exploiting flexibility potential enabled by both electrification of final uses and energy storage.

Anyway, each community has its own peculiarities in terms of climate, renewable energy exploitation potential, energy demand, local economy, national prices of commodities, and so on. For this reason, the general set of technologies shown in Figure 11.3 must be tailored to local characteristics and needs.

Under the technical point of view, two steps are essentials in order to let an LEC to become reality: an optimal medium–long-term planning; digitalization of generation, distribution, storage technologies, and of final users/prosumers.

11.3.2 Optimal Medium–Long-Term Planning for Local Energy Communities

In order to achieve all the aforementioned benefits deriving LECs, the optimal medium–long-term planning in charge of local energy planners is crucial. In fact, it is necessary to design these multi-energy systems appropriately, defining, among the numerous alternatives offered by the market, both the optimal mix of technologies in terms of types, numbers, and sizes and their operation strategies, to meet the energy needs of end-users. This decision-making process is not a trivial task because of the complicated interactions occurring among different vectors and networks as also demonstrated by Figure 11.3. The complexity is also due to the typical imbalance between supply and energy demand, which is caused by the fast changing in time of users' energy demand and the limited operational flexibility of some technologies to meet users' demand. Moreover, the stochasticity of renewable sources is another key factor to take into account for the optimal planning of these energy systems.

Last but not least, another element that adds complexity to the optimal planning problem is the co-existence of multiple and often conflicting objectives to take into account. A design solution that allows minimization of investment costs, that is of interest of investors, is most probably in contrast with the long-term sustainability of the energy supply system, for instance represented by minimization of environmental impacts or maximization of primary energy saving (PES), which, as already discussed, are extremely important for energy legislations worldwide. Environmental and economic priorities are most often in contrast to each other, and the need to take both into account to ensure the short- and long-term sustainability of LECs leads to the implementation of multi-objective approaches. In fact, solutions with low environmental impacts are generally expensive, and conversely, low-cost solutions may be characterized by poor environmental performance. In this perspective, the optimal planning of LECs appears as a multi-objective problem for which there is not a single solution that is satisfactory for all the objectives. In that case, local energy planners are asked to find trade-off design solutions between the environmental and economic priorities.

11.3.3 Key Technologies in the Context of Local Energy Communities

This section briefly describes the main technologies that can be deployed in LECs. They have been classified as follows:

- Distributed generation technologies: Renewable technologies to decarbonize LEC.
- "End-user" sector coupling technologies: Energy conversion technologies for electrification of final users' energy demand, thus enabling flexibility at final users/prosumers.
- "Cross-vector" sector coupling technologies: Technologies enabling integration between energy carrier networks within LECs. The main technology that can be easily deployed in most of LEC is cogeneration and trigeneration, which can be installed both at prosumer (buildings, malls, industries) and city/district level (district heating and cooling). The other two cross-vector sector coupling technologies are presented: electrolyzer and desalination. The former links electricity and hydrogen, and it is considered a key technology for a "low-carbon energy future" as already discussed in Chapter 4. The latter is a technology already widely deployed in geographical energy islands but of growing interest, also considering the UN-SDG 6 "Ensure availability and sustainable management of water and sanitation for all."
- Storage technologies: They were already described in Chapter 5 of this book. Here, their role in the context of LECs is presented.

Distributed generation technologies

Technology	Short description	Local energy communities context
PV	**Technology** Photovoltaic technology converts solar radiation (both direct and diffuse) into electricity. Being related to solar radiation, electricity production of PV plants is uncertain and not predictable. **Energy carrier involved** Electricity **Energy demand satisfied** Electricity **Energy networks involved** Electricity	PV technology is one of the main distributed generation technologies for decarbonizing LEC and converting final users in prosumers. Indeed, this technology can be easily installed at user level in rooftop or directly integrated in buildings (e.g. facades). Small-size PV plants are usually installed in residential sector, whereas the medium-size ones are installed in large buildings (e.g. factories). If enough area is available, utility-scale PV (>1 MW_e) can also be installed in LECs. Anyway, land use of PV systems must be considered.
Solar thermal panels	**Technology** Solar thermal technology converts solar radiation in thermal energy. Solar radiation is used to heat up hot water as thermal energy vector. The technology is heavily affected by external weather conditions not only because of the uncertainty of solar radiation but also because of thermal energy losses due to external temperatures. Because of the difference between production and consumption patterns, solar thermal panels are usually coupled with thermal energy storage. Three main technologies are available in the market: Unglazed collectors: economic panels used for seasonal sanitary hot water production Glazed collectors: technology used for production of sanitary hot water all over the year. Vacuum tube collectors: technology used both for sanitary hot water production and for heating purposes (usually coupled with a boiler or a heat pump).	Solar thermal panels contribute to decarbonization of LECs by reducing the consumption of fossil fuels when hot water is produced by natural gas or oil boilers or by reducing the consumption of electricity when hot water is produced by heat pumps. In any case, the contribution to the overall decarbonization of LECs is limited. Some examples exist of large-scale solar district heating plants in smart thermal grids [29].

Technology	Short description	Local energy communities context
	Energy carrier involved Hot water **Energy demand satisfied** Heating, hot water production **Energy networks involved** –	
Mini- and micro-hydro in water supply systems (WSS)	**Technology** Mini-hydro technologies convert potential energy of water in electricity using hydraulic turbines. In recent years, mini-hydro technologies have also been used to recover energy in municipal water supply systems where the pressure of water should be reduced before being distributed to final users. **Energy carrier involved** Electricity **Energy demand satisfied** Electricity **Energy networks involved** Electricity	Water distribution network is a non-energy network always present in LECs. Micro- and mini-hydro generation technologies bring a small contribution to the decarbonization of LEC. When possible, they are usually utilized to decrease energy consumption of pumping stations in water supply systems.
Biogas/ biomethane production plant	**Technology** Biogas is a mixture of methane and CO_2 produced by anaerobic digestion of organic material. Anaerobic digestion is a process needing thermal energy. It is considered a renewable energy source. Biogas is usually utilized to produce electricity in a power plant consisting of a generation set (usually an internal combustion engines) whose waste heat provides the thermal energy required for the biogas production itself. Recently, biogas is also purified from CO_2 in order to produce biomethane, which can be directly injected in natural gas network.	Biogas allows us to decarbonize municipal scale LEC. Indeed, in order to achieve a proper size of the plants (from hundreds of kW up to some MWs), biogas power plants need a continuous input of organic material. Biogas can activate and promote job creation related to circular economy connected to organic material such as municipal/agricultural waste. Biogas can be used to decarbonize electricity production thanks to biogas power plants. If after biogas production process, there still is excess of thermal energy, this can be furtherly recovered in district heating systems. Biomethane is used to decarbonize natural gas network.

Technology	Short description	Local energy communities context
	Energy carrier involved Electricity **Energy demand satisfied** Electricity (thermal if coupled with waste heat recovery) **Energy networks involved** Electricity, district heating, and natural gas	
Biomass	**Technology** Biomass boilers burn solid biomass in order to produce sanitary how water and thermal energy for heating purposes. In residential sector, they are the technological evolution of traditional fireplaces: they are fed by wooden pellet of different sizes, and they can be remotely controlled and can interact with smart thermostats, such as a natural gas boiler. Large biomass boilers can be equipped with district heating networks also in combination with cogeneration systems. **Energy carrier involved** Hot water **Energy demand satisfied** Heating, hot water production **Energy networks involved** District heating	Biomass boilers can contribute to decarbonization of local energy communities by replacing boilers fed by fossil fuels (natural gas, oil, and liquefied petroleum gas [LPG]). Anyway, there are some issues related to the local emissions at chimney as small biomass boilers are not equipped with advanced systems to abate particulate. Biomass can activate and promote job creation, especially in rural areas, related to circular economy connected to wood industry and forest maintenance.

End-user sector coupling technologies

Technology	Short description	Local energy communities context
Electric vehicles	**Technology** Mechanical energy to move the electric vehicle is produced by an electric motor coupled with electric batteries instead of using an internal combustion engine fed by fossil fuels. **Energy carriers involved** Electricity	EV is a key technology to electrify final uses of transport sector in a local energy community, as discussed in detail in Chapter 7. The general idea is that decarbonization of the electric sector is much easier than decarbonization of oil sector through bioenergies, as electricity to charge EV can be produced by a renewable power plant.

Technology	Short description	Local energy communities context
	Energy demand satisfied Mobility **Energy networks involved** Power grid, mobility	Moreover, EVs also represent a way to make the local energy community's electric consumption more flexible by using EV as "distributed electric energy" storage in buildings (vehicle to building [V2B] and building to vehicle [B2V]) [30] and in districts (vehicle to grid [V2G] and grid to vehicle [G2V]) [31].
Heat pumps	**Technology** Heat pump technology uses electricity to produce thermal energy for heating and for domestic hot water production. Heat pumps can also be used as chillers for cooling purposes. As heat pump performances are heavily affected by external temperature, they can be coupled with heating boilers (hybrid heat pumps) in location in which the external temperature falls below 5 °C. **Energy carriers involved** Electricity, hot water, and chilled water **Energy demand satisfied** Heating, cooling, and DHW **Energy networks involved** Electricity	Heat pump is a pivotal technology in the electrification of energy demand of a local energy community. Coupled with local renewable distributed technology, it contributes to the decarbonization of a major share of heat demand in buildings. Heat pumps can also be used to provide flexibility to the LEC electric distribution network thanks to DR program. In order to increase flexibility in operation, heat pumps can be coupled with thermal energy storage or they can take advantage from the thermal inertia of the buildings.
Electric boilers	**Technology** Electric boilers are appliances producing domestic hot water from electricity. Electric boilers are thermally insulated water storages in which water is kept at a defined temperature, usually set at 70 °C. They can be used for DHW production or for heating purposes. **Energy carriers involved** Electricity, hot water	Electric boilers can contribute to the flexibility of electricity demand of a local energy community. In particular, they use artificial intelligence to predict consumption patterns of DHW and to keep hot water at low temperature when consumption is not foreseen. Hence, they can be used as programmable loads in order to participate to DR program. Indeed, smart boilers can act as both controllable loads and as thermal storages [32].

Technology	Short description	Local energy communities context
	Energy demand satisfied Domestic hot water production Heating **Energy networks involved** Electricity	

Cross-vector sector coupling technologies

Technology	Short description	Local energy communities context
Cogeneration – combined heat and power (CHP) Trigeneration – combined cooling heat and power (CCHP)	**Technology** CHP is a distributed generation technology. CHP stands for the simultaneous production of electricity and thermal energy in a unique integrated energy system. CHP can be coupled with absorption chillers, in order to produce one more energy carrier, which is chilled water for cooling applications. Absorption chillers are chilling technologies based on the properties of a pair of substances with a different volatility where heat is provided to separate the most volatile, used as a refrigerant in the cycle. For instance, in the LiBr-H_2O absorption chiller, water is the refrigerant (cooling temperature > 0 °C), while in the NH_3-H_2O absorption chiller, ammonia is the refrigerant (cooling temperature < 0 °C). When CHP is coupled with absorption chillers, the system takes the name of trigeneration plant (CCHP). The most common cogeneration technologies are combined cycle (CC), gas turbine (GT), internal combustion engine (ICE), micro gas turbine (MGT), and fuel cells (FC)	Medium and large CHP/CCHP plants are usually installed also to decarbonize industries. CHP/CCHP is a widely used energy efficiency technology that contributes to the decarbonization of the energy systems by using less primary energy in an integrated system with respect to the one spent to produce the same amount of electricity and thermal energy in separate generation plants. The energy index to assess the decarbonization potential of CHP/CCHP systems is the primary energy saving (PES) index: when PES is higher than zero, then the system contributes to decarbonize the overall energy system [33]. CHP/CCHP is one of the key technologies for sector coupling in local energy communities by linking together natural gas, electricity, and thermal networks. At city/district-level medium and large CHP/CCHP plants are usually coupled with district heating/cooling networks.

Technology	Short description	Local energy communities context
	Energy carriers involved Electricity, fuel (e.g. natural gas, gasoline, diesel, syngas, H_2, and biofuels), hot water/steam, chilled water **Energy demand satisfied** Electricity, heating, heat for process, cooling **Energy networks involved** Electricity, local/district heating /cooling network, natural gas network	At prosumer level, small and medium CHP/CCHP plants are usually installed in office buildings, shopping malls, and large condominium. CHP/CCHP also contributes to the flexibility of the LEC as the technologies involved are programmable and can be easily scheduled and operated also to offset sudden variations of non-dispatchable renewable energy.
Electrolyzers	**Technology** Electrolysis is an electrochemical conversion process that allows to produce hydrogen and oxygen starting from water and electricity. Several technologies can be used for electrolysis, but two technologies have a market technology readiness level: proton exchange membrane (PEM) and alkaline electrolyzers. **Energy carriers involved** Electricity, hydrogen, and natural gas (blended with hydrogen) **Energy networks involved** Electricity, natural gas (natural gas blended with hydrogen), hydrogen, and water (non-energy network)	Electrolyzers are the key technology enabling power-to-hydrogen (P2H), that is the use of electricity to produce hydrogen from water. Electrolyzers can be used as controllable loads to provide flexibility to the electric network. In local energy communities, P2H can be difficult to deploy, especially in the smaller ones. Anyway, electrolyzers can be used, also coupled with a dedicated renewable plant in order to produce (green) hydrogen, which can be used to feed, blended with natural gas, utility scale generation plants such as boilers and/or CHP systems in district heating networks.
Desalination	**Technology** The desalination process produces fresh/drinkable water starting from seawater. Two main processes are available for water production:	Desalination technology is a widely used technology in local energy communities located in geographical islands in which the demand of fresh water is not or only partially met by local natural sources.

Technology	Short description	Local energy communities context
	Membrane processes: Reverse osmosis (RO): thanks to electricity salt water is pressurized at a hydraulic pressure higher than the osmotic one related to a solution: water moves from high-solute region to a low-solute one through a semi-permeable non-porous membrane (no phase change occurs) *Thermal processes:* Thermal desalination mainly consists in a vaporization of salt water; the vapor is then collected and condensed in order to produce fresh water. Most of the times, the thermal energy for vaporization is provided by waste heat recovery from power plants (e.g. condensers). The main desalination processes are as follows: multi-effect distillation (MED), multi-stage flash (MSF), and mechanical vapor compression (MVC). **Energy carriers involved** Electricity (reverse osmosis) and thermal energy (thermal processes) **Energy demand satisfied** Desalination technologies connect electric and "thermal" networks to water supply network, which is a non-energy network. Desalination technologies satisfy demand of water (both drinkable or not) **Energy networks involved** Electricity (reverse osmosis), natural gas (thermal desalination at bottoming of a power production plant), and thermal network	Thermal desalination plants are usually connected with generation plants whose waste heat is recovered for the water production process. In this case, desalination technology links together electricity network, water network, and natural gas network. Reverse osmosis desalination plants are electric loads that link together electric and water supply system networks. They can be considered as controllable loads in order to provide some flexibility to the electric network. Water supply systems of LEC using reverse osmosis desalination plant can be decarbonized by installing renewable energy technologies such as PV.

Storage technologies

Technology	Short description	Local energy communities context
Battery energy storage system (BESS)	**Energy carriers involved** Electricity **Energy demand satisfied** Electricity **Energy networks involved** Electricity	BESS is one of the key technologies to optimally manage non-dispatchable renewable energy in local energy communities. BESS can be installed at prosumer level combined with PV plant in order to store the excess of electricity produced and consume it when production is low. In this way, renewable energy self-consumption is maximized at prosumer level and congestion problem are avoided at distribution system operator (DSO) network level. BESS can also be installed in the medium-voltage/low-voltage (MV/LV) electric network in order to improve the quality of the power grid, avoid congestion problem, and maximize the renewable energy exploitation at local level.
Sensible thermal energy storage (STES)	**Energy carriers involved** Hot water **Energy demand satisfied** Heating, domestic hot water **Energy networks involved** District heating (large size)	Small thermal energy storages are usually installed coupled with solar thermal panels at prosumer level. Large-scale thermal energy storage (TES) can be used in local energy communities to increase the flexibility of operation of CHP plants connected to district heating network.

11.3.4 Digitalization to Enable Flexibility and Empower Final Users

As discussed in detail in Chapter 6 of this book, digitalization plays a pivotal role to enable flexibility services and empower final users. Digital energy defines the possibility of using digital technologies to control energy exchange. The digitalization of energy is radically transforming the energy sector offering products and services to allow end-users to become independent active customers and to be responsible for the use of energy, thereby surpassing the concept of energy manager/distributor. In that sense, LECs aim to accelerate the energy transition toward decentralized and bidirectional management systems, by employing distributed architectures, and hardware and software systems for monitoring and operating the various energy carriers and technologies at different levels. In LECs, prosumers can promote local RES generation and self-consumption of renewable resources, while also cooperating through energy sharing.

EMS such as BEMS and community EMS ensure effective control and operation of energy communities. In detail, the BEMS installed at prosumers level is a system that allows us to monitor and control the equipment of a building, combining software (e.g. communication drivers, databases, control algorithms, etc.) and hardware components (sensors, actuation devices, etc.). In detail, the BEMS monitors and controls energy-related building services such as heating, ventilation, and air conditioning (HVAC) and lighting excluding not energy-related systems such as fire, closed-circuit television (CCTV), etc.

The main functions of a BEMS can be grouped in four categories, which are monitoring, control, optimization, and reporting. The BEMS also monitors the sensor measurements and, by using proper control algorithms, can modify the behavior of the facilities in the building in order to obtain the best performance toward a specific objective, such as

- improve energy efficiency;
- provide better comfort for occupants;
- review the performance of the building;
- generate automatically alarms for failures or anomaly conditions;
- identify planned and unplanned maintenance requirements; and
- log and archive data for energy management purpose.

As shown in Figure 11.4, the EMS at prosumer level may include Internet of Things (IoT) devices to monitor the energy production and consumption (smart

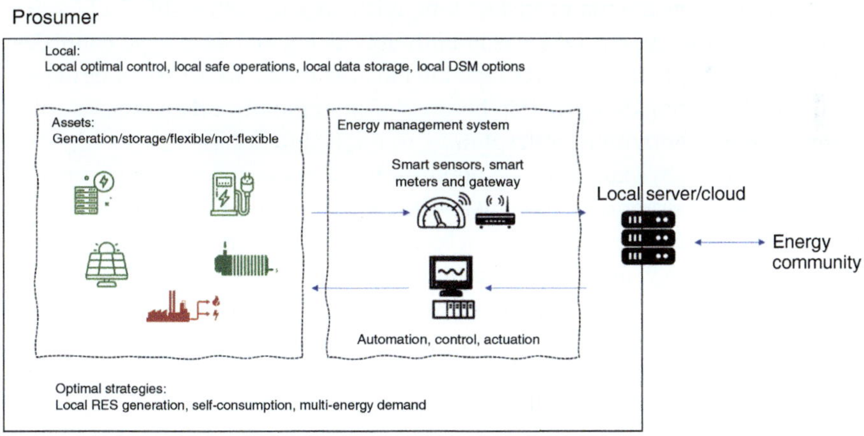

Figure 11.4 Representative scheme of the digitalization system installed at prosumer level in the LEC context.

meters and gateway), automation systems such as Supervisory Control and Data Acquisition (SCADA) and Programmable Logic Controller (PLC)/Distributed Control System (DCS)/Embedded Personal Computer, and actuators to control and monitor industrial plants. A local server can be used to aggregate data for local data storage and allow the communication and coordination with other prosumers in the LEC.

Moreover, the EMS could be provided with web applications (e.g. dashboards for analytics, reporting, and maintenance), web services (e.g. weather forecast and energy price forecast), and utility software (e.g. scripts for demand forecasting and algorithms for optimal management of energy assets also based on artificial intelligence [AI] approach).

The main goal of these systems is to manage local assets in order to optimize local objective functions (e.g. minimization of energy costs) by implementing low-level controllers that aim controlling the local assets based on the set points received from the EMS and on the actual conditions of the energy devices. As prosumers can also provide flexibility services, they can also implement demand side management strategies. Low-level controls are based mainly on pulse-width modulation, proportional integral and derivative (PID), rule-based (if-then rules), fuzzy logic, and more advanced optimization-based methodologies as model predictive control.

11.4 Local Energy Communities and Energy Transition: A Vision for the Next Future

LECs will be one of the keys to the success of the energy transition worldwide. The LEC paradigm provides a model for bottom-up approach to energy transition involving local actors and stakeholders, which can be adapted to different context. For this reason, it is recognized that LEC can contribute to achieve UN-SDG 7 "Ensure access to affordable, reliable, sustainable and modern energy for all" and SDG 11 "Make cities and human settlements inclusive, safe, resilient and sustainable." LECs are a way to provide energy in rural areas in both developing and developed countries contributing not only to their decarbonization but also to their local economic development, thus hindering depopulation because of migration in urban areas. At the same time, energy community concept must be applied to decarbonize energy districts in urban areas by exploiting the opportunity of deploying renewable energy in integrated energy systems in urban networks. Anyway, the strength of LEC in the energy transition is inherent in the idea of "community." Indeed, with respect to traditional approaches (e.g. distributed generation, mini-grid, etc.) mainly focused on technical aspects, LEC introduces the active engagement of final users leading the way toward to a consumer-centered energy system at the basis of energy transition. The concept of "community" can trigger virtuous mechanisms under technical, societal, and economic point of view. In particular, the increasing awareness of citizens and prosumers in energy questions and their active engagement in decarbonizing their community can boost the energy transition at local level.

In the discussion about a vision of the role of LEC in the energy transition, some reflections can be done looking at the European challenge undertaken via the European Green Deal.

11.4.1 Some Reflections

We cannot underestimate the scale of the challenge undertaken via the European Green Deal.

Europe has reduced its GHG emissions of around 20% over the 30 years between 1990 and 2020.

The goal is to reduce further 35% (almost twice) in just one-third of the time: 10 years from 2021 to 2030. This is the pace set by the final target of carbon neutrality by 2050.

The policy mix of the Green Deal is far reaching, and it has the ambition to activate a pervasive and transformative change in all sectors of our economy and society, getting very close to a true system resetting.

Therefore, in a race of this scale and pace, it is unavoidable that the focus is on the quantitative target to achieve. Major technological breakthroughs, rapid changes in standard setting, in the market mechanisms, or in the fiscal and incentive policy mix can make a multi-digit difference.

However, it would be very risky to pursue a strategy of "few" silver-bullet solutions. We cannot shift away from genuine locally based energy systems capable of tackling the complexity of an energy transition, which needs to be adapted to the local needs and opportunities.

While hunting the target, we need to keep real sight of the path we are on.

In particular, we need to ensure we are positively mobilizing our citizens, generating an active ownership of this green transition from the bottom-up. A new techno-economic and financial equilibrium needs to be socially and environmentally sustainable and leave room to prosumer spaces. A fresh impetus to locally tailored "social development factories" can emanate from energy becoming a factor of cohesion among neighborhoods of different revenue ranks within our urban conglomerates, between cities and rural areas, between developed economies and developing or underdeveloped economies.

In this regard, it makes great sense to provide public and private support to LECs.

The window of opportunity opened by the European Green Deal is very unique in that regard. It offers a substantial policy blueprint, a revamped legal framework, and an unprecedented amount of investment resources, which the future generations will have to pay back.

In the coming years, it will therefore be extremely interesting to see the roll-out of the National Energy and Climate Plans and of the Recovery and Resilience Plans. We might see considerable differences among countries and regions as for the boosting of LECs: where your author believes they deserve ranking up to the level of pillar priority, there is a fair chance that they become a complementary intervention; or the unfortunate scenario of a "nice to have" project to report back.

11.5 Conclusions

Energy transition is a complex pattern toward a low- or even carbon-neutral future. This final chapter focuses on the potential of LECs in the energy transition. LECs provide an opportunity to integrate energy systems at local level by deploying the existing and emerging technologies presented throughout in this book. Bringing energy transition at local level has several advantages and opens new opportunities even if some potential barriers have been identified. The core and the main key to success of LEC paradigm is the concept of "community" itself, which has various technical, social, and economic implications. The main advantage of LEC is that it provides a bottom-up approach to energy transition. This means that the integrated energy system at local level can be tailored to the needs and peculiarities of the local community; indeed, the optimal design of an LEC must take into account several factors such as local climate, energy demand of the community, community needs, national and local economy context, cost of technology, human capital, and the existing infrastructure. For this reason, the bottom-up approach of LEC entails the involvement of different actors and stakeholders such as local policy makers, energy planners, DSOs, utilities, local technology providers, citizens, and entrepreneurs. The strength of this approach is that it can be easily adapted to different contexts also very different each from the others; indeed, LEC paradigm can be deployed to decarbonize urban and rural areas in developed and developing countries. Even if each integrated energy system is tailored to the peculiarity of a particular LEC, the good practice of an energy community can represent a similar reference use case for others and can spread the best practice in a wide geographical area. A positive side effect of the LEC bottom-up approach is that it can improve the general perception and attitude of the communities toward energy transition. In particular, it can increase both the community acceptance of emerging technologies and renewable energy and the citizen awareness toward energy question at local and worldwide level, leading to behavioral changes. Indeed, in the context of an LEC, and in general, in the future energy systems, final users are active actors becoming users or prosumers empowered enabled to participate to the decarbonization process. Moreover, the LEC paradigm is that it can boost local economies and promote job creation. Indeed, integrated local energy systems consist of distributed generation technologies, energy storage, and multi-energy networks, which require technicians to be designed, installed, and maintained all over their lifetime.

Among the barriers identified, digitalization is one of the most important key enabler of integrated smart energy systems; moreover, access and visualization of data is one of the best strategies identified to increase energy awareness of citizens. Another important barrier strictly related to the local deployment of distributed energy technologies, storages, and networks is the need of human capital able to manage the several aspects related to integrated energy systems at local level. A human capital with inadequate skills can jeopardize the attitude of local actors and stakeholder toward renewable energy and energy transition.

Also, the lack of business model is a potential barrier to LEC deployment: digitalization and integration of energy systems require initial capital costs and a stable economic framework for allowing public and private actors to invest with the fair opportunity to recover the investment.

Finally, this chapter also warns about the risk of making the energy transition a technical top-down approach excluding local communities. For this reason, a particular focus has been dedicated to the efforts put in act in the European Union in order to accelerate the European clean energy transition.

Thanks "Clean Energy for All Europeans" package and the "European Green Deal" EU set up a plan to make Europe the first climate neutral continent by 2050. In these ambitious programs, EU clearly entrusts a fundamental role to local (LEC), renewable (REC), and citizens (CEC) "energy communities" as their underlying capacity to generate, consume, store, and sell energy while offering tools for the most vulnerable citizens to lift them out of energy poverty.

List of Abbreviations

AI	artificial intelligence
BEMS	building energy management system
BESS	battery energy storage system
CEC	citizen energy community
CC	combined cycle
CHP	combined heat and power (cogeneration)
CCHP	combined cooling, heat and power (trigeneration)
CCTV	closed-circuit television
COR	Committee of the Regions
DCS	distributed control system
DER	distributed energy resources
DSO	distribution system operator
DH	district heating
DHW	domestic hot water
DG	distributed generation
DR	demand response
EMS	energy management system
EU	European Union
EMD	electricity market directive
FC	fuel cell
FCEV	fuel cell electric vehicle
GT	gas turbine
ICE	internal combustion engine
IoT	internet of things
LEC	local energy community
MGT	micro gas turbine
PID	proportional integral and derivative
PLC	programmable logic control
PV	photovoltaic

REC	renewable energy community
RED	renewable energy directive
RES	renewable energy source
SCADA	supervisory control and data acquisition
SDG	sustainable development goals
TES	thermal energy storage
UN	United Nations
WSS	water supply system

References

1 Zebra, E.I.C., van der Windt, H.J., Nhumaio, G., and Faaij, A.P.C. (2021). A review of hybrid renewable energy systems in mini-grids for off-grid electrification in developing countries. *Renew. Sustain. Energy Rev.* 144: 111036. https://doi.org/10.1016/j.rser.2021.111036.

2 ESMAP (2019). Mini Grids for Half a Billion People: Market Outlook and Handbook for Decision Makers. Executive Summary. Energy Sector Management Assistance Program (ESMAP) *Technical Report 014/19*. Washington, DC: World Bank.

3 ESMAP (2017). A reprospective analysis of the role of isolated and mini grids in power system development.

4 Domenech, B., Ferrer-Martí, L., Lillo, P. et al. (2014). A community electrification project: combination of microgrids and household systems fed by wind, PV or micro-hydro energies according to micro-scale resource evaluation and social constraints. *Energy Sustain. Dev.* 23: 275–285.

5 Kankam, S. and Boon, E.K. (2009). Energy delivery and utilization for rural development: lessons from Northern Ghana. *Energy Sustain. Dev.* 13 (3): 212–218.

6 Raya-Armenta, J.M., Bazmohammadi, N., Avina-Cervantes, J.G. et al. (2021). Energy management system optimization in islanded microgrids: an overview and future trends. *Renew. Sustain. Energy Rev.* 149: 111327. https://doi.org/10.1016/j.rser.2021.111327.

7 Weir, T. (2018). Renewable energy in the Pacific Islands: its role and status. *Renew. Sustain. Energy Rev.* 94: 762–771.

8 Kuang, Y., Zhang, Y., Zhou, B. et al. (2016). A review of renewable energy utilization in islands. *Renew. Sustain. Energy Rev.* 59: 504–513.

9 Cabrera, P., Carta, J.A., Lund, H., and Thellufsen, J.Z. (2021). Large-scale optimal integration of wind and solar photovoltaic power in water-energy systems on islands. *Energy Convers. Manage.* 235: 113982.

10 Delina, L.L., Ocon, J., and Esparcia, E. (2020). What makes energy systems in climate-vulnerable islands resilient? Insights from the Philippines and Thailand. *Energy Res. Social Sci.* 69: 101703.

11 Kallis, G., Stephanides, P., Bailey, E. et al. (2021). The challenges of engaging island communities: lessons on renewable energy from a review of 17 case studies. *Energy Res. Social Sci.* 81: 102257.

12 Bandeiras, F., Pinheiro, E., Gomes, M. et al. (2020). Review of the cooperation and operation of microgrid clusters. *Renew. Sustain. Energy Rev.* 133: 110311.

13 UN General Assembly (2015). Transforming our world: the 2030 Agenda for Sustainable Development. Resolution adopted by the General Assembly on 25 September 2015. https://www.un.org/ga/search/view_doc.asp?symbol=A/RES/70/1&Lang=E (accessed 20 October 2021).

14 Tracking SDG7: The Energy Progress Report 2019. https://sdgs.un.org/sites/default/files/2021-05/Report%20-%202019%20Tracking%20SDG7%20Report.pdf (accessed 20 October 2021).

15 van Gevelt, T., Holzeis, C.C., Fennell, S. et al. (2018). Achieving universal energy access and rural development through smart villages. *Energy Sustain. Dev.* 43: 139–142.

16 Broto, V.C. and Kirshner, J. (2020). Energy access is needed to maintain health during pandemics. *Nat. Energy* 5: 419–421. https://doi.org/10.1038/s41560-020-0625-6.

17 EU Commission (2019). Clean energy for all Europeans.

18 Directive (EU) 2018/2002 of the European Parliament and of the Council of 11 December 2018 amending Directive 2012/27/EU on energy efficiency (Text with EEA relevance.).

19 European Commission. Directive (EU) 2018/2001 of the European Parliament and of the Council of 11 December 2018 on the promotion of the use of energy from renewable sources.

20 Directive (EU) 2019/944 of the European Parliament and of the Council of 5 June 2019 on common rules for the internal market for electricity and amending Directive 2012/27/EU.

21 Caramizaru, E. and Uihlein, A. (2020). Energy Communities: An Overview of Energy and Social Innovation. *EUR 30083 EN*. Luxembourg: Publications Office of the European Union.

22 The European Committee of the Regions - 131st plenary session, 5–6 December 2018 OPINION Models of local energy ownership and the role of local energy communities in energy transition in Europe.

23 Herbes, C., Rilling, B., and Holstenkamp, L. (2021). Ready for new business models? Human and social capital in the management of renewable energy cooperatives in Germany. *Energy Policy* 156: 112417. https://doi.org/10.1016/j.enpol.2021.112417.

24 Italian G20 Presidency 2021 in collaboration with ENEA and RSE. Energy Efficiency and Circularity in a Post Pandemic Economy.

25 ETIP-SNET (2021). POSITION PAPER Smart Sector Integration, towards EU System of Systems Building blocks, enablers, architectures, regulatory barriers, economic assessment.

26 EU Policy Department for Economic (2018). Scientific and Quality of Life Policies Directorate-General for Internal Policies. Sector coupling: how can it be enhanced in the EU to foster grid stability and decarbonise?

27 Broska, L.H. (2021). It's all about community: On the interplay of social capital, social needs, and environmental concern in sustainable community action. *Energy Res. Social Sci.* 79: 102165.

28 Warbroek, B., Hoppe, T., Bressers, H., and Coenen, F. (2019). Testing the social, organizational, and governance factors for success in local low carbon energy initiatives. *Energy Res. Social Sci.* 58: 101269.

29 Tian, Z., Zhang, S., Deng, J. et al. (2019). Large-scale solar district heating plants in Danish smart thermal grid: developments and recent trends. *Energy Convers. Manage.* 189: 67–80.

30 Borge-Diez, D., Icaza, D., Açıkkalp, E., and Amaris, H. (2021). Combined vehicle to building (V2B) and vehicle to home (V2H) strategy to increase electric vehicle market share. *Energy* 237: 121608.

31 Bibak, B. and Tekiner-Moğulkoç, H. (2021). A comprehensive analysis of Vehicle to Grid (V2G) systems and scholarly literature on the application of such systems. *Renew. Energy Focus* 36: 1–20.

32 Ciabattoni, L., Comodi, G., Ferracuti, F., and Foresi, G. (2020). A methodology to enable electric boiler as a storage for residential energy management. *2020 IEEE International Conference on Consumer Electronics (ICCE)*, pp. 1–2.

33 Comodi, G. and Rossi, M. (2016). Energy versus economic effectiveness in CHP (combined heat and power) applications: investigation on the critical role of commodities price, taxation and power grid mix efficiency. *Energy* 109: 124–136.

Index

Technologies for Integrated Energy Systems and Networks, First Edition.
Edited by Giorgio Graditi and Marialaura Di Somma.

k

l

m

t

u

v

w